# A TEXTBOOK OF PHYSICAL CHEMISTRY

## CHEMISTRY

### Volume I

Mandeep Dalal

**DALAL INSTITUTE**

*Publisher*:

Dalal Institute, Main Market, Sector 14, Rohtak, Haryana 124001, India

(info@dalalinstitute.com, +91-9802825820)

www.dalalinstitute.com



*Disclaimer:*

Although every precaution has been taken to verify the accuracy of the information contained herein, the author and publisher assume no responsibility for any errors or omissions. No liability is assumed for damages that may result from the use of the information contained within.

*Credits:*

The dedication image is a derivative work of the famous painting "Innocence" by William-Adolphe Bouguereau, whereas the image on the front cover is a derivative work of Iain Goodyear's "Atom Tunnel".

Dedicated to my mother "Darshana Devi"

This Page is Intentionally Left Blank

# PREFACE

The preface writing has always been a wonderful feeling that cannot be expressed in words as it relates you to your audience through your work. I conceived the idea of writing a new advanced-level textbook in physical chemistry during my Ph.D. pursuit when I saw post-graduate chemistry students who were tired in search of the syllabus topics because of their ill-resourced university or college library. I also decided to write the textbooks of inorganic and organic chemistry because I think that someone who wants to teach or text one stream must have the core conceptual understanding of all the three streams of chemical science otherwise one would not be able to connect and explain the interdisciplinary topics in a comprehensive manner.

Out of the series of three textbooks, the present book, entitled "A Textbook of Physical Chemistry – Volume 1", is the first installment of "A Textbook of Physical Chemistry", which is a four-volume set in all. All the students and teachers are advised to read and consult all the four volumes in a subsequent pattern for a more efficient and thorough understanding of the subject of physical chemistry.

I also celebrate this opportunity for expressing the bottom-hearted gratitude towards the people who supported me at all stages of my work. First of all, I would like to express my sincere gratitude to my doctoral supervisors, Prof. S.P. Khatkar and Prof. V.B. Taxak for their continuous support and guidance from day one. Then I would like to record appreciation to my lovely sister, Jyoti Dalal, for her unconditional love, support and for being the guiding light when the life threw me in the darkest of corners. I am very much thankful to my beautiful wife, Anita Sangwan, who always stands shoulder to shoulder with me in my good and bad times. I especially want to thank my brother Sandeep Dalal for his positive criticism, encouragement, motivation and truly selfless support. A special thanks to my dearest sister Garima Sheoran for her love, care, and all-time encouragement. I also wish to thank my entire family, friends, and teachers for providing a loving environment for me.

Lastly, and most importantly, I wish to thank my mother, Darshana Devi, who bore me, raised me, supported me, taught me, and loved me.

**Mandeep Dalal**

# Table of Contents

# CHAPTER 1

## Quantum Mechanics – I

### ❖ Postulates of Quantum Mechanics

In modern quantum theory, the postulates of quantum mechanics are simply the step-to-step procedure to solve a simple quantum mechanical problem. In other words, it is like the manual that must be followed to retrieve the information about various states of any quantum mechanical system. We will first learn about the nature and the significance of these postulates, and then we will apply them to some real problems like the particle in a one-dimensional box or the harmonic oscillator.

> #### *The First Postulate*

*All time-independent states of any quantum mechanical system can be described mathematically as long as the function used is single-valued, continuous and finite.*

**Explanation:** The systems around us can be broadly classified into two categories; the first is classical and the other one as quantum mechanical. The classical systems simply refer to the systems which are governed by the classical or the Newtonian mechanics. Now because all the macroscopic objects follow Newton's laws of motion, they fall in the category of classical systems; for example, a rotating gym dumbbell, the vibrating spring of steel, or an athlete running in the playground. Every classical system can possess many states which belong to a continuous domain, and each state can be described mathematically.

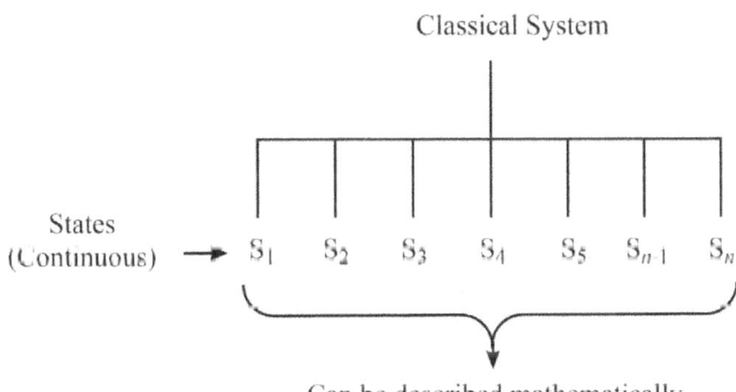


However, if the rotating gym dumbbell is replaced by the rotating diatomic molecule, the system would not remain classical anymore and would start violating classical laws. The states of such microscopic systems (here it just means the extremely small) belong to a discontinuous domain and can also be described mathematically. These mathematical descriptions are labeled as $\psi_1$, $\psi_2$, $\psi_3$ ….. $\psi_n$ and generally called as the "wave functions". The term "wave function" is used because as we go from the macroscopic to the microscopic world i.e. from classical to the quantum mechanical world, things start behaving like waves rather particle. All of the states are wave-like; and because every wave we see around us is continuous, single-valued and finite;

only continuous, single-valued and finite expressions can represent those states. For instance, when you drop a stone in a standstill pond, the waves are generated which travel from the center to the boundary of the pond; and you don't see any discontinuity in it.

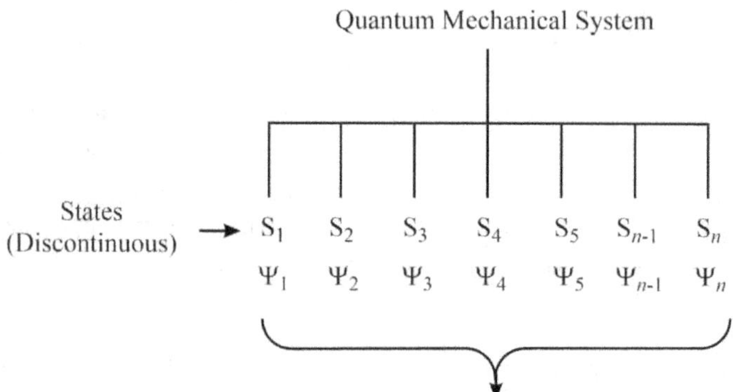


$\Psi_1, \Psi_2, \Psi_3 \dots \Psi_n$ are the mathematical description for $S_n$ states

Hence, if a function is not single-valued, continuous and finite; it will not be able to represent any wave-like behavior at all. That is why every function that correlates a quantum mechanical state must be single-valued, continuous and finite; and this function describes the corresponding state completely.

> ### The Second Postulate

*For every physical property like linear momentum or the kinetic energy, a particular operator exists in quantum mechanics, the nature of which depends upon the classical expression of the same property.*

**Explanation:** In classical mechanics, there are simply straight forward formulas for all physical properties; like linear momentum can simply be calculated by multiplying the mass with velocity. However, in case of quantum mechanical systems, the value of a certain physical property for a particular state cannot be calculated simply by using its classical formula but from an operator. It does sound silly but the classical formulas which are so well-tested on the scale of time fail in quantum world. For instance, you can use the $mv^2/2$ to calculate the kinetic energy of a moving particle in classical world by just putting its mass and velocity; but if the mass of the moving particle is extremely less, you will not get any rational results.

It is also worthy to note it again that though the classical formulas fail to give the value of physical property, they are still important as they form the basis of the derivations for corresponding quantum mechanical operators. For instance, the operator for kinetic energy (T) along *x*-axis can be derived as:

$$K.E.(T) = \frac{1}{2}mv^2 = \frac{(mv)^2}{2m} = \frac{p^2}{2m} \tag{1}$$

Where *m* and *v* are mass and the velocity, respectively; and *p* represents the angular momentum whose squared operator is:

$$\hat{p}_x^2 = \frac{-h^2}{4\pi^2}\frac{\partial^2}{\partial x^2} \tag{2}$$

Now putting the value of momentum squared from equation (2) into equation (1), we get:

$$\hat{T}_x = \frac{-h^2}{8\pi^2 m}\frac{\partial^2}{\partial x^2} \tag{3}$$

The expressions of various quantum mechanical operators are given below.

Table 1. Various important physical properties and their corresponding quantum mechanical operators.

| Physical property | | Operator | |
|---|---|---|---|
| *Name* | *Symbol* | *Symbol* | *Operation* |
| Position | $x$ | $\hat{x}$ | Multiplication by $x$ |
| Position squared | $x^2$ | $\hat{x}^2$ | Multiplication by $x^2$ |
| Momentum | $p_x$ | $\hat{p}_x$ | $\dfrac{h}{2\pi i}\dfrac{\partial}{\partial x}$ |
| Momentum squared | $p_x{}^2$ | $\hat{p}_x^2$ | $\dfrac{-h^2}{4\pi^2}\dfrac{\partial^2}{\partial x^2}$ |
| Kinetic energy | $T = \dfrac{P^2}{2m}$ | $\hat{T}_x$ | $\dfrac{-h^2}{8\pi^2 m}\dfrac{\partial^2}{\partial x^2}$ |
| Potential energy | $V(x)$ | $\hat{V}(x)$ | Multiplication by $V(x)$ |
| Total energy | $E = T + V(x)$ | $\hat{H}$ | $\dfrac{-h^2}{8\pi^2 m}\dfrac{\partial^2}{\partial x^2} + V(x)$ |

For three dimensional systems, the total operator can be obtained by summing the individual operators along three different axes. For instance, some important three-dimensional operators are:

$$\hat{T} = \frac{-h^2}{8\pi^2 m}\left(\frac{\partial^2}{\partial x^2} + \frac{\partial^2}{\partial y^2} + \frac{\partial^2}{\partial z^2}\right) \tag{4}$$

$$\hat{p} = \frac{h}{2\pi i}\left(\frac{\partial}{\partial x} + \frac{\partial}{\partial y} + \frac{\partial}{\partial z}\right) \tag{5}$$

$$\hat{H} = \frac{-h^2}{8\pi^2 m}\left(\frac{\partial^2}{\partial x^2} + \frac{\partial^2}{\partial y^2} + \frac{\partial^2}{\partial z^2}\right) + V(x, y, z) \tag{6}$$

> ### *The Third Postulate*

*If ψ is a well-behaved function for the given state of system and $\hat{A}$ is a suitable operator for a particular physical property, then the operation on ψ by the operator $\hat{A}$ gives the function ψ multiplied by the value of the physical property which can be constant or variable but always real (R). Mathematically, it can be shown as:*

$$\hat{A}\psi = R\psi \tag{7}$$

**Explanation:** The third postulate of quantum mechanics actually connects the first and second postulate of quantum mechanics. The first postulate talks about the possibility of describing a quantum mechanical state mathematically, while the second postulate says that the values of all physical properties in the quantum world are obtained by the operator rather than the simple classical formula. Now the third postulate says that if we operate the operator (from second postulate) over the wave function (from first postulate), we will get the value of the corresponding physical property.

However, at this point, a new problem arises as we do not know the exact mathematical description i.e. the wave function of any quantum mechanical state; and the operators need the absolute mathematical description of the quantum mechanical state to yield any actual result. Now though we know the expressions of different operators proposed by the second postulate; the first postulate speaks only about the presence of a single-valued, continuous and finite mathematical function but does not give actual function itself; and without the knowledge of actual "wave functions", the operators are pretty much useless.  Therefore, one would think that there must be some route by which the wave functions are obtained first, which would be used as operand afterward. However, the procedure to find the exact mathematical descriptions of various quantum mechanical states is somewhat more synergistic. The "magic mystery" is that all the operators need absolute expression of the wave function that defines the quantum mechanical state except one, the most famous "Hamiltonian operator". The special thing about the Hamiltonian operator is that it does not necessarily need the absolute form but the symbolic form only to yield the value of its physical property i.e. energy. Nevertheless, in the process of applying the Hamiltonian operator over the symbolic form of the wave function, the absolute expression is also obtained. Mathematically,

$$\hat{H}\psi = E\psi \tag{8}$$

After putting the expression of the Hamiltonian operator in equation (8) and then rearranging, we get:

$$\frac{\partial^2\psi}{\partial x^2} + \frac{\partial^2\psi}{\partial y^2} + \frac{\partial^2\psi}{\partial z^2} + \frac{8\pi^2 m(E-V)\psi}{h^2} = 0 \tag{9}$$

The second-order differential equation i.e. equation (9) is the famous Schrodinger wave equation, the solution of which gives not only the energy but the wave function as well. Now, once the exact expression of the wave function representing a particular state is known, other operators can be operated over it to find their values.

> ### *The Fourth Postulate*

*If the value of the physical property obtained after multiplying the wave function by the corresponding operator is constant (postulate 3), the value is called as the eigen-value and is directly reportable; and the wave function will be labeled as the eigen-function of the operator used.*

**Explanation:** The third postulate said that when the wave function of a particular quantum mechanical state is multiplied by the operator of an observable quantity, we get a real value multiplied by the wave function itself; however, the value obtained so can be constant or variable. Mathematically,

$$\hat{O}\,\Psi = \text{Value }\Psi$$

Eigen Function / Eigen Value  ⟵ Constant   Variable ⟶  Non-eigen Function / Non-eigen Value

The constant value of the observable quantity can be reported directly, and the function is called the eigenfunction of the operator under consideration.

> ### *The Fifth Postulate*

*If the value of the physical property obtained after multiplying the wave function by the corresponding operator is variable i.e. non-eigen, the value can be reported only after averaging it over the whole configurational space.*

$$< a > or\ \bar{a} = \frac{\oint \psi^* \hat{O}\psi\ d\tau}{\oint \psi^*\psi\ d\tau} \tag{10}$$

**Explanation:** As we have seen in the fourth postulate that the value obtained by multiplying the Hermitian operator with any quantum mechanical state can also be variable in nature. For instance, if we multiply a wave function simply by position operator, we will get

$$\hat{x}\psi = x\psi \tag{11}$$

or

$$\hat{x} = \frac{x\psi}{\psi} \tag{12}$$

Now because "$x$" is a variable number, it must have reported as an average value before any further rational argument is made.

Therefore, we can say that the fifth postulate is simply an extension of the fourth postulate; i.e. the fourth postulate is used to obtain the value of a particular physical property if it is an eigenvalue, however, the fifth postulate is employed to calculate all non-eigenvalues.

## ❖ Derivation of Schrodinger Wave Equation

The Schrodinger wave equation can be derived from the classical wave equation as well as from the third postulate of quantum mechanics. Now though the two routes may appear completely different, the final result is just the same indicating the objectivity of the quantum mechanical system.

### ➤ *The Derivation of Schrodinger Wave Equation from Classical Wave Equation*

After the failure of the Bohr atomic model to comply with the Heisenberg's uncertainty principle and dual character proposed by Louis de Broglie in 1924, an Austrian physicist Erwin Schrodinger developed his legendary equation by making the use of wave-particle duality and classical wave equation. In order to understand the concept involved, consider a wave traveling in a string along the $x$-axis with velocity $v$.

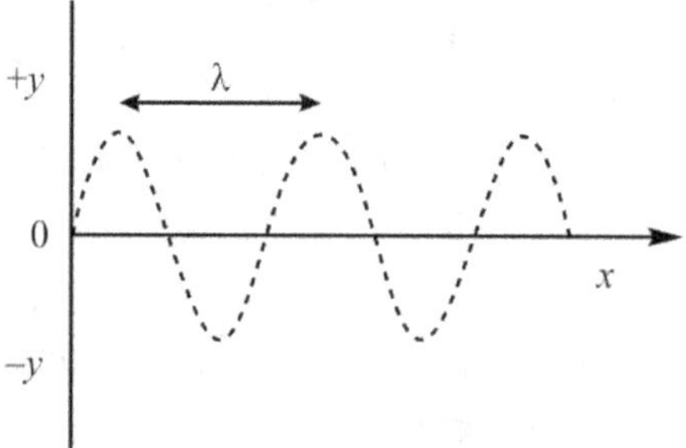

Figure 1. The wave motion in a string.

It can be clearly seen that the amplitude of the wave at any time $t$ is the function of displacement $x$, and the equation for wave motion can be formulated as given below.

$$\frac{\partial^2 y}{\partial x^2} = \frac{1}{v^2}\frac{\partial^2 y}{\partial t^2} \tag{13}$$

Therefore, we can say that y is a function of $x$ well at $t$.

$$y = f(x)f'(t) \tag{14}$$

Where $f(x)$ and $f'(t)$ are the functions of coordinate $x$ and time, respectively. The nature of the function $f(x)$ can be understood by taking the example of stationary or the standing wave.

A standing wave is created in a string fixed between two points with a wave traveling in one direction, and when it strikes the other end, it gets reflected with the same velocity but in negative amplitude. This would create vibrations in that string with or without nodes depending upon the frequency incorporated. We can create fundamental mode (0 node), first overtone (1 node) or second overtone (2 nodes) just by changing the vibrational frequency. The nature of these standing or stationary waves can be understood more clearly by the diagram given below.

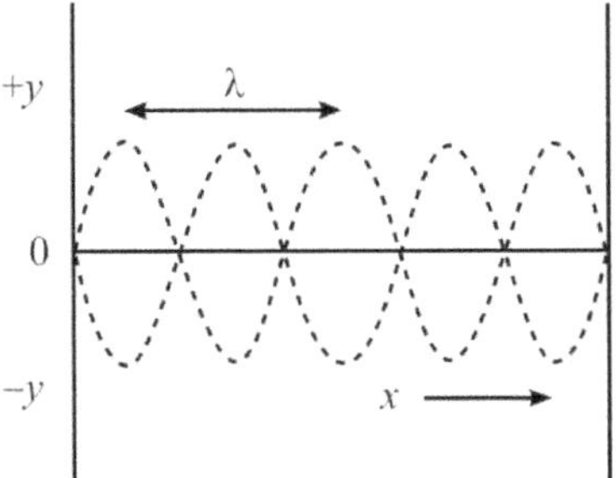


Figure 2. Standing waves in a string.

The mathematical description for such a wave motion is

$$f'(t) = A \sin 2\pi vt. \tag{15}$$

Where $A$ is a constant representing maximum amplitude and $v$ is the frequency of the vibration. Now putting the value of $f'(t)$ from equation (15) in equation (14), we get

$$y = f(x)\, A \sin 2\pi vt \tag{16}$$

Differentiating the above equation w.r.t. $t$, we are left with

$$\frac{\partial y}{\partial t} = f(x)\, A 2\pi v \cos 2\pi vt \tag{17}$$

Differentiating again

$$\frac{\partial^2 y}{\partial t^2} = -f(x)\, 4\pi^2 v^2\, A \sin 2\pi vt \tag{18}$$

$$\frac{\partial^2 y}{\partial t^2} = -4\pi^2 v^2 f(x)\, f'(t) \tag{19}$$

Now differentiating equation (14) w.r.t. $x$ only, we get

$$\frac{\partial y}{\partial x} = f'(t)\, \frac{\partial f(x)}{\partial x} \tag{20}$$

Differentiating again

$$\frac{\partial^2 y}{\partial x^2} = f'(t)\, \frac{\partial^2 f(x)}{\partial x^2} \tag{21}$$

Now put the value of equation (19) and (21) in equation (13), we get

$$f'(t)\frac{\partial^2 f(x)}{\partial x^2} = \left(\frac{1}{v^2}\right)\left[-4\pi^2 v^2 f(x)\, f'(t)\right] \tag{22}$$

$$\frac{\partial^2 f(x)}{\partial x^2} = \frac{-4\pi^2 v^2}{v^2} f(x) \tag{23}$$

The equation (23) is now time-independent; and therefore, shows the amplitude dependence only upon the coordinate $x$. Since $c = v\lambda$ ($v = c/\lambda$), the velocity of the wave can also be replaced by the multiplication of frequency and wavelength i.e. $v = v\lambda$.

$$\frac{\partial^2 f(x)}{\partial x^2} = \frac{-4\pi^2 v^2}{v^2 \lambda^2} f(x) \tag{24}$$

$$\frac{\partial^2 f(x)}{\partial x^2} = \frac{-4\pi^2}{\lambda^2} f(x) \tag{25}$$

The symbol of the function $f(x)$ is replaced by popular $\psi(x)$ or simply the $\psi$.

$$\frac{\partial^2 \psi}{\partial x^2} = \frac{-4\pi^2}{\lambda^2} \psi \tag{26}$$

Also, as we know that $\lambda = h/mv$, the equation (26) becomes

$$\frac{\partial^2 \psi}{\partial x^2} = \frac{-4\pi^2 m^2 v^2}{h^2} \psi \tag{27}$$

$$\frac{\partial^2 \psi}{\partial x^2} + \frac{4\pi^2 m^2 v^2}{h^2} \psi = 0 \tag{28}$$

Furthermore, as the total energy ($E$) is simply the sum of the potential ($V$) and kinetic energy, we can say that

$$E = \frac{mv^2}{2} + V \tag{29}$$

$$mv^2 = 2(E - V) \tag{30}$$

After putting the value of $mv^2$ from equation (30) in equation (28), we get

$$\frac{\partial^2 \psi}{\partial x^2} - \frac{8\pi^2 m}{h^2}(E - V)\psi = 0 \tag{31}$$

For three-dimension i.e. $\psi(x, y, z)$, the above equation can be extended to following

$$\frac{\partial^2 \psi}{\partial x^2} + \frac{\partial^2 \psi}{\partial y^2} + \frac{\partial^2 \psi}{\partial z^2} - \frac{8\pi^2 m}{h^2}(E - V)\psi = 0 \tag{32}$$

The above-mentioned second order differential equation i.e. equation (32) is our popular form of the Schrodinger wave equation.

> ➤ *The Derivation of Schrodinger Wave Equation from the Postulates of Quantum Mechanics*

The Schrodinger wave equation can be derived using the first three postulates of quantum mechanics. In other words, we can say that the Schrodinger wave equation is nothing but the rearranged form of the following equation:

$$\hat{H}\psi = E\psi \tag{12}$$

In order to prove the above claim, consider a single particle having "*m*" mass that moves with a velocity "*v*" in the three-dimensional region. The sum of its kinetic and potential energy can be given as:

$$E = T + V \tag{13}$$

However, we know that

$$T = \frac{1}{2}mv^2 = \frac{p^2}{2m} \tag{14}$$

Where "p" represents the total linear momentum of the particle under consideration. Furthermore, as we also know that

$$p^2 = p_x^2 + p_y^2 + p_z^2 \tag{15}$$

Where $p_x$, $p_y$ and $p_z$ are the magnitudes of total linear momentum along *x*, *y* and *z*-axis, respectively. Now putting the value of $p^2$ from equation (15) into equation (14), we get the following

$$T = \frac{p_x^2 + p_y^2 + p_z^2}{2m} \tag{16}$$

And now put the value of kinetic energy from equation (16) into equation (13). We get

$$E = \frac{p_x^2 + p_y^2 + p_z^2}{2m} + V \tag{17}$$

However, from the second postulate of quantum mechanics, we know that the expressions for linear momentum operator along three different directions are:

$$\widehat{p_x} = \frac{h}{2\pi i}\frac{\partial}{\partial x} \tag{18}$$

$$\widehat{p_y} = \frac{h}{2\pi i}\frac{\partial}{\partial y} \tag{19}$$

$$\widehat{p_z} = \frac{h}{2\pi i}\frac{\partial}{\partial z} \tag{20}$$

The operator of "V" is simply itself as it is a function of position coordinates only.

Hence, after putting values of linear momentum operators and potential energy operator in equation (17), the operator for total energy (Hamiltonian operator) becomes

$$\widehat{H} = \frac{1}{2m}\left[\left(\frac{h}{2\pi i}\frac{\partial}{\partial x}\right)^2 + \left(\frac{h}{2\pi i}\frac{\partial}{\partial y}\right)^2 + \left(\frac{h}{2\pi i}\frac{\partial}{\partial z}\right)^2\right] + V \tag{21}$$

$$\widehat{H} = -\frac{h^2}{8\pi^2 m}\left(\frac{\partial^2}{\partial x^2} + \frac{\partial^2}{\partial y^2} + \frac{\partial^2}{\partial z^2}\right) + V \tag{22}$$

$$\widehat{H} = -\frac{h^2}{8\pi^2 m}\nabla^2 + V \tag{23}$$

Where

$$\nabla^2 = \frac{\partial^2}{\partial x^2} + \frac{\partial^2}{\partial y^2} + \frac{\partial^2}{\partial z^2}$$

represents the Laplacian operator.

Now, after putting the value of Hamiltonian operator from equation (22) into equation (12) i.e. given by the third postulate of quantum mechanics, get

$$\left[-\frac{h^2}{8\pi^2 m}\left(\frac{\partial^2}{\partial x^2} + \frac{\partial^2}{\partial y^2} + \frac{\partial^2}{\partial z^2}\right) + V\right]\psi = E\psi \tag{24}$$

$$\left[-\frac{h^2}{8\pi^2 m}\left(\frac{\partial^2}{\partial x^2} + \frac{\partial^2}{\partial y^2} + \frac{\partial^2}{\partial z^2}\right) + V\right]\psi - E\psi = 0 \tag{25}$$

$$-\frac{h^2}{8\pi^2 m}\frac{\partial^2\psi}{\partial x^2} - \frac{h^2}{8\pi^2 m}\frac{\partial^2\psi}{\partial y^2} - \frac{h^2}{8\pi^2 m}\frac{\partial^2\psi}{\partial z^2} + V\psi - E\psi = 0 \tag{26}$$

Multiplying the equation (26) throughout by

$$-\frac{8\pi^2 m}{h^2}$$

we get

$$\frac{\partial^2\psi}{\partial x^2} + \frac{\partial^2\psi}{\partial y^2} + \frac{\partial^2\psi}{\partial z^2} - \frac{8\pi^2 m}{h^2}V\psi + \frac{8\pi^2 m}{h^2}E\psi = 0 \tag{27}$$

$$\frac{\partial^2\psi}{\partial x^2} + \frac{\partial^2\psi}{\partial y^2} + \frac{\partial^2\psi}{\partial z^2} + \frac{8\pi^2 m}{h^2}(E - V)\psi = 0 \tag{28}$$

Equation (28) is the most popular form of the Schrodinger wave equation for three dimensional systems. In the case of two and one dimensional systems first three terms can be reduced to two and one, respectively.

### ❖ Max-Born Interpretation of Wave Functions

In 1926, a German physicist Max Born formulated a rule which is generally called as the Born law or Born's rule of quantum mechanics, giving the probability that a measurement on a quantum system will yield a given result. In other words, it states that the probability density of finding the particle at a given point is proportional to the square of the magnitude of the particle's wavefunction at that point. The Max-Born interpretation is one of the key concepts of quantum mechanics to understand wave-particle duality.

Let us suppose that the particle under consideration is an electron whose way of existence is represented by a mathematical expression $\psi$ which is a function of the electron's coordinates i.e. $x$, $y$, and $z$. Max Born actually suggested that because this mathematical expression is single-valued, continuous and finite i.e. wave-like; one can opt the same route to find it's intensity as we use in case of light or sound waves. In the case of light or sound waves, the intensity at any point can simply be obtained by squaring the amplitude i.e. $\psi$ of the wave at the same point. Therefore, in the case of the electron, the square of the amplitude of electron wave ($\psi^2$) at a particular point also gives the intensity of the electron wave at the same point. In other words, the density of electron wave (probability density) at a point for a quantum mechanical state is simply obtained by the square of the magnitude of the corresponding wavefunction at the same point. Mathematically, we can show this as:

$$Probability\ density = |\psi|^2 = \psi\psi^* \tag{29}$$

Where $\psi^*$ designates a complex conjugate of the wave function $\psi$. The reason for using $\psi^*$ lies in the fact that the wave function representing a quantum mechanical state is not always real but be imaginary as well. However, as the probability density should always be real, $\psi\psi^*$ is more appropriate than simple $\psi^2$. In other words, if the wave function defining the quantum mechanical state is real, we can use $\psi^2$ as the probability density; nevertheless, if the wave function does contain the imaginary part (like $\psi = a + ib$), $\psi\psi^*$ must be used to yield real values. This can be explained by taking an imaginary expression $\psi$ and then multiplying it by its complex conjugate $\psi^*$ to yield real value.

$$\psi = a + ib; \qquad \psi^* = a - ib \tag{30}$$

or

$$\psi\psi^* = (a + ib) \times (a - ib) \tag{31}$$

or

$$\psi\psi^* = a^2 + b^2 \tag{32}$$

Moreover, if $\psi$ is real, $\psi = \psi^*$, $\psi\psi^*$ becomes $\psi^2$, the value we have already discussed.

Now though the probability density in space is not a constant parameter ($\psi$ is not constant), in a very small segment it can be considered constant. Now let us discuss the Max-Born interpretation for one, two and three dimensional systems.

> ### *One Dimensional Systems*

The probability of finding the particle in any one-dimensional system in the region from $x$ to $x+dx$ must be obtained by integrating $\psi^2$ from $x$ to $x+dx$ i.e. by finding the area under the curve from $x$ to $x+dx$.

Now although the $\psi^2$ or $\psi\psi^*$ (because $\psi$ is continuous) varies continuously with $x$, the decrease or increase in $\psi^2$ can be neglected and it can be assumed that it remains constant as we move from $x$ to $x+dx$.

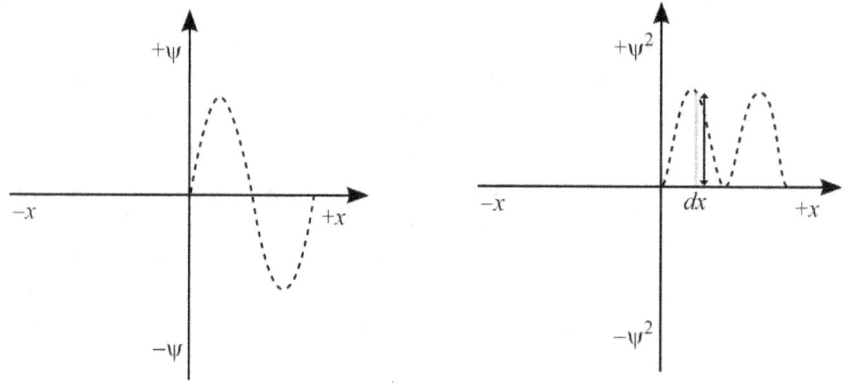


Figure 3. Born interpretation of wave function and probability density in a one-dimensional system.

Therefore, the area of the shaded region and hence the probability of finding the particle can be obtained just by multiplying the length (or height i.e. $\psi^2$) with the width ($dx$) of the narrow rectangle. Thus we can say that

$$Probability = |\psi|^2 \times (dx) = \psi\psi^* dx = \psi^2 dx \tag{33}$$

Since the chances of finding the particle over whole length (whole configurational space) must be unity, we get the following

$$\oint |\psi|^2 \times (dx) = \oint \psi\psi^* dx = \oint \psi^2 dx \tag{34}$$

> ### *Two Dimensional Systems*

The probability of finding the particle in an area element $dx\,dy$ ($dA$), situated at a distance $r$ distance from the center, would be $\psi\,(x, y) \times \psi^*\,(x, y) \times dx \times dy$; or in short can be written as $\psi\psi^* dA$. Hence, it must be obtained by integrating $\psi^2$ from $(x, y)$ to $(x+dx, y+dy)$ i.e. by finding the area under the curve $dA$. Now although the $\psi^2$ or $\psi\psi^*$ (because $\psi$ is continuous) varies continuously with coordinates $(x\,y)$, the decrease or increase in $\psi^2$ can be neglected and it can be assumed that it remains constant as we move from $(x, y)$ to $(x+dx, y+dy)$.

Therefore, the area of the shaded region and hence the probability of finding the particle can be obtained just by multiplying the magnitude of the wave function ($\psi^2$) with the area ($dA$) of the area element. Thus we can say that

$$Probability = |\psi|^2 \times (dA) = \psi\psi^* dA = \psi^2 dA \tag{35}$$

Since the chances of finding the particle over the whole area (whole configurational space) must be unity, we get the following

$$\oint |\psi|^2 \times (dA) = \oint \psi\psi^* dA = \oint \psi^2 dA \tag{36}$$

The pictorial representation of the area element is given below.

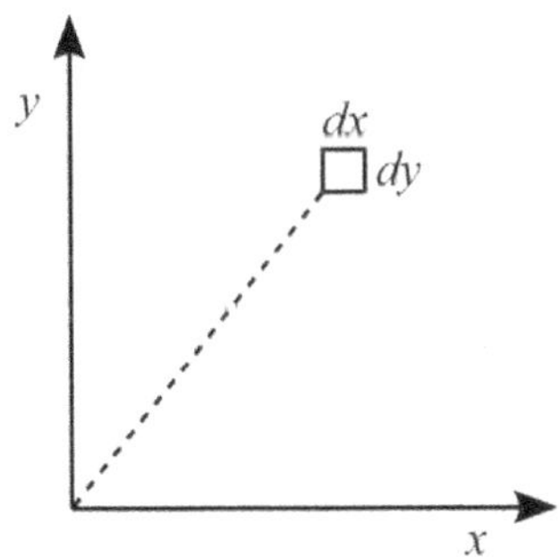


Figure 2. Born interpretation of wave function and probability density in two dimensional systems.

> ### *Three Dimensional System*

The probability of finding the particle in a volume element $dx\ dy\ dz\ (dV)$, situated at a distance $r$ distance from the center, would be $\psi\,(x, y, z) \times \psi^*\,(x, y, z) \times dx \times dy \times dz$; or in short can be written as $\psi\psi^* dV$. must be obtained by integrating $\psi^2$ from $(x, y, z)$ to $(x+dx, y+dy, z+dz)$ i.e. by finding the area under the curve from $(x, y, z)$ to $(x+dx, y+dy, z+dz)$. Now although the $\psi^2$ or $\psi\psi^*$ (because $\psi$ is continuous) varies continuously with coordinates $(x, y, z)$, the decrease or increase in $\psi^2$ can neglected and it can be assumed that it remains constant as we move from $(x, y, z)$ to $(x+dx, y+dy, z+dz)$.

Therefore, the probability of finding the particle can be obtained just by multiplying the magnitude of the wave function ($\psi^2$) with the volume ($dV$) of the area element. Thus we can say that

$$Probability = |\psi|^2 \times (dV) = \psi\psi^* dV = \psi^2 dV \tag{35}$$

Since the chances of finding the particle over the whole area (whole configurational space) must be unity, we get the following

$$\oint |\psi|^2 \times (dV) = \oint \psi\psi^* dV = \oint \psi^2 dV \tag{36}$$

The pictorial representation of the area element is given below.

**DALAL**
**INSTITUTE**

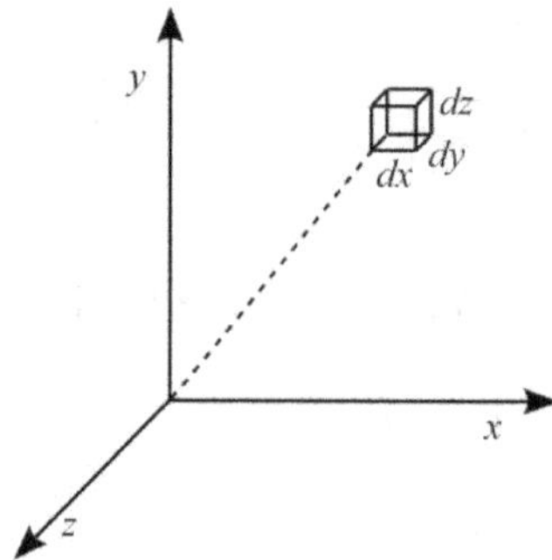

Figure 3. Born interpretation of wave function and probability density in three dimensional systems.

## ❖ The Heisenberg's Uncertainty Principle

In quantum mechanical world, the Heisenberg's uncertainty principle (or simply the uncertainty principle) is one of a variety of mathematical inequalities asserting a fundamental limit to the precision with which certain pairs of physical properties of a particle, known as complementary variables or canonically conjugate variables such as position $x$ and momentum $p$, can be known. The concept was first introduced in 1927, by a German physicist Werner Heisenberg.

*The Heisenberg's uncertainty principle states that the more precisely the position of some particle is determined, the less precisely its momentum can be known, and vice versa.*

The formal inequality relating the standard deviation of position $\Delta x$ and the standard deviation of momentum $\Delta p_x$ was derived by Earle Hesse Kennard later that year and by Hermann Weyl in 1928:

$$\Delta x . \Delta p_x \geq \frac{h}{4\pi} \tag{37}$$

or

$$\Delta x . \Delta p_x \geq \hbar/2 \tag{38}$$

Where $\hbar$ is the reduced Planck's constant, which is obviously equal to the Planck's constant divided by $2\pi$. Besides the equation (37), the is also an energy-time uncertainty relation given by W. Heisenberg which states that higher the lifetime of a quantum mechanical state, less uncertain would be the energy value. Mathematically, it can be shown as:

$$\Delta E . \Delta t \geq \frac{h}{2\pi} \tag{39}$$

Where $\Delta E$ and $\Delta t$ represent the uncertainties in the energy and time respectively.

> *Position Momentum Uncertainty*

Among various kinds of uncertainties, the position-momentum uncertainty is one of the popular kind that arises as a consequence of wave-particle duality. In order to understand the relation, we first need to study the effect of wave behavior on the simultaneous measurement of position about $x$-coordinate and the linear momentum component along the $x$-axis for a microscopic particle.

Consider a beam of particles traveling with a momentum "$p$" along the $y$-direction, and this beam finally strikes a narrow slit of width "$w$". Now, from the principles of optics, we know that the uncertainty in the position of the particle along $x$-axis must be equal to the slit width. In other words, as the width of the slit is along $x$-axis, any particle that strikes the detector must have crossed the $\Delta x$ region i.e. $w$, the slit width available. However, we exactly don't know where it does cross from. It could be along the center of the slit, or along a line slightly above or below the central trajectory. Therefore, the slit width ($w = \Delta x$) would be equal to a crossing domain that we are uncertain about. However, a diffraction pattern will be observed in the case of microscopic particles because of their wave-like character. The amplitude of the wave at a particular point on the detector represents the number of the particles reaching that point. Now because of this diffraction, the incident beam does not strike only at the central point O but also at the above and below to it. It means that some particles do reach upward and downward to O, suggesting that the part of their linear momentum is transferred along $x$-axis also.

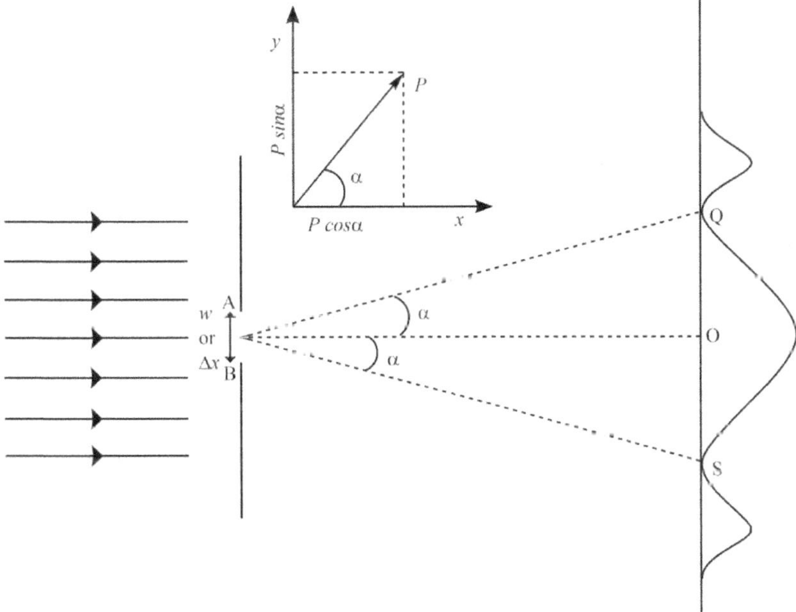

Figure 4. The diffraction of electron waves by single slit systems.

The $x$-component of linear momentum of the wave (aka particle) diffracted at an angle α can be obtained by the rectangular resolution of the linear momentum vector. The particles diffracted upward and downward at an angle α will yield the $x$-component as $P\,sin\alpha$ and $-P\,sin\alpha$, respectively. Now because a large number of particles reach the plate in between +α to −α i.e in between the first minimums, half of the momentum spread in the central diffraction peak should give the uncertainty in the momentum along $x$-axis. Mathematically, we can say that

$$\Delta p_x = P\,sin\alpha \tag{40}$$

Multiplying the above equation by the uncertainty in the position i.e. width of the slit used for the measurement purpose, we get

$$\Delta x.\,\Delta p_x = w.\,P\,sin\alpha \tag{41}$$

Here, it is very important to recall the fact that the condition which must be satisfied to obtain the first minima is that the path difference between the waves reaching the minima point should be an integral multiple of $\lambda/2$.

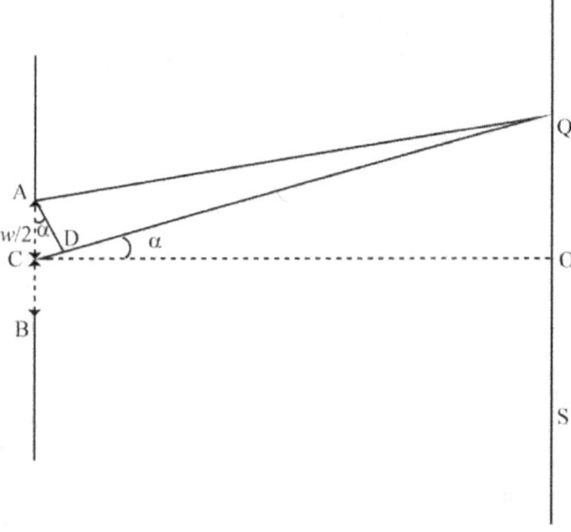


Figure 5. The calculation for 1$^{st}$ order diffraction for electron wave in single slit systems.

Hence we have the following equalities from the diagram given above.

$$AQ = DQ \tag{42}$$

$$CQ = difference\ in\ the\ path\ length \tag{43}$$

Now because the distance of the detector is very large as compared to the slit width, AQ and CQ can be considered parallel to each other i.e. AQ ∥ DQ. Hence, we can say that

$$< ADC = 90° \tag{44}$$

$$< CAD = \alpha \tag{45}$$

also

$$AC = \frac{w}{2} \tag{46}$$

$$CD = \frac{\lambda}{2} \tag{47}$$

From the trigonometric relations, we get

$$\frac{CD}{AC} = Sin\ \alpha \tag{48}$$

$$CD = AC\ Sin\ \alpha \tag{49}$$

Putting the values of AC and CD from equation (46) and (47) in equation (49), we get

$$\frac{\lambda}{2} = \frac{w}{2}\ Sin\ \alpha \tag{50}$$

$$\lambda = w\ Sin\ \alpha \tag{51}$$

Now, after putting the value of $w$ from equation (51) in equation (41), we get

$$\Delta x.\Delta p_x = \frac{\lambda}{Sin\ \alpha}.P\ sin\alpha \tag{52}$$

$$\Delta x.\Delta p_x = \lambda.P \tag{53}$$

Using the de Broglie relation ($\lambda = h/p$) in equation (53), we get

$$\Delta x.\Delta p_x = \frac{h}{P}.P \tag{54}$$

$$\Delta x.\Delta p_x = h \tag{55}$$

Now because we didn't define the uncertainty very precisely, we should not use the "equal" sign. Therefore, the above equation can be reduced to the following.

$$\Delta x.\Delta p_x \approx h \tag{56}$$

This eventually means that decreasing the uncertainty in the position of the incident particle (decreasing the slit width) would result in a higher uncertainty in the momentum along $x$-axis; while the higher slit width does give more precise momentum but small precision in the calculation of the position of the incident particle.

> ➢   ***Energy Time Uncertainty***

The uncertainty principle doesn't limit itself to position-momentum only but can also be applied to some other pairs of conjugate variables. All the variable pairs whose products have the same dimension as the Plank's constant $h$ (Js) are said to be a conjugate pair. Besides the position-momentum, another famous uncertainty is relation energy-time because the product of these two quantities (energy × time) also has the unit of $h$ (Js).

$$\Delta E . \Delta t \approx h \tag{57}$$

Where $\Delta E$ and $\Delta t$ are uncertainties in energy and time, respectively. This popular relation can be derived directly from the concept of wave-particle duality. In the quantum mechanical world, a particle is supposed to possess a wave packet. Now, let us consider that this wave packet occupies the $\Delta x$ region along the direction $x$-direction and travels with a velocity $v$. The time it needs to pass a certain point in $x$-direction has an uncertainty magnitude of $\Delta t$, and can be formulated as:

$$\Delta t = \frac{\Delta x}{v} \tag{58}$$

Now because this wave packet occupies the region $\Delta x$, the momentum uncertainty along $x$-axis can be given by the following relation.

$$\Delta p_x = \frac{h}{\Delta x} \tag{59}$$

or

$$\Delta x = \frac{h}{\Delta p_x} \tag{60}$$

Putting the value of $\Delta x$ from equation (60) in equation (58), we get

$$\Delta t = \frac{h}{v \Delta p_x} \tag{61}$$

Moreover, we also know that

$$E = \frac{p_x^2}{2m} \tag{62}$$

Differentiating the above equation w.r.t $p_x$, we get

$$\frac{dE}{dp_x} = \frac{\Delta E}{\Delta p_x} = \frac{p_x}{m} = \frac{mv}{m} = v \tag{63}$$

$$\Delta E = \frac{dE}{dp_x} \Delta p_x = v . \Delta p_x \tag{64}$$

Multiplying equation (63) and (64), we get

$$\Delta E . \Delta t = v\Delta p_x . \frac{h}{v\Delta p_x} \tag{65}$$

$$\Delta E . \Delta t \approx h \tag{66}$$

The physical interpretation of the above relation can be viewed in terms of fluctuating energy level with a total $\Delta E$ uncertainty if the system does not stay in it longer than $\Delta t$ interval of time i.e. lifetime of the state.

## ❖ Quantum Mechanical Operators and Their Commutation Relations

*An operator may be simply defined as a mathematical procedure or instruction which is carried out over a function to yield another function.*

$$\text{(Operator) . (Function) = (Another function)} \tag{67}$$

The function used on the left-hand side of the equation (67) is called as the operand i.e. the function over which the operation is actually carried out. The operator alone has no significance but when operated over a certain mathematical description, these operators can provide very detailed insights into those functions. Some of the simple illustrations of equation (67) are given below.

i) Consider the differential operator $d/dx$ whose operation has to be studied over the function $y = x^5$. The mathematical treatment is

$$\frac{dy}{dx} = \frac{d}{dx} x^5 = 5x^4 \tag{68}$$

The operation of $d/dx$ on $y$ means that the rate of change of function $y$ w.r.t. the variable $x$. The expression $x^5$ is the operand while the $5x^4$ is the final result of our differential operator.

ii) Consider the integral operator $\int (y)\, dx$ whose operation has to be studied over the function $y = x^5$. The mathematical treatment is

$$\int y(dx) = \int x^5(dx) = \frac{x^6}{6} \tag{69}$$

The operation of $\int dx$ on $y$ means that we can find the function whose derivative is $x^5$. The expression $x^5$ is the operand while the $x^6/6$ is the final result of our integral operator.

In a similar way, the multiplication of a function by a constant number, or taking the square and cube roots of any function are also the operators which give some other function after operating them over the operand. The symbol of the operator typically carries a cap over it ($\hat{A}$) which differentiates it from the function used in the whole procedure.

> ### *Algebra of Operators*

Just like the normal algebra, the resultants like addition or the multiplication of operators also follow certain rules; however, these rules are different from the typical algebra. Some of the most important rules of operator algebra are given below.

**1. Addition and subtraction of operators:** Let A and B as two different operators; $f$ as the function that has to be used as the operand. Then, the addition and subtraction of these two operators must be carried out in the manner discussed below.

$$(\hat{A} + \hat{B})f = \hat{A}f + \hat{B}f \tag{70}$$

and

$$(\hat{A} - \hat{B})f = \hat{A}f - \hat{B}f \tag{71}$$

**2. Multiplication of operators:** If A and B as two different operators; and $f$ as the function that has to be used as operand. Then, the multiplication of these two operators must be carried out in the manner discussed below.

$$\hat{A}\hat{B}f = f'' \tag{72}$$

The interpretation of the above equation is that first we need to operate B on $f$, which would give us another function $f'$, which in turn is further used as the operand for operator giving the final result $f''$. In other words, we can say that when multiplication of two or more operators is used, we should follow from left to right. Moreover, the square or cube of a particular operator must be considered as double or triple multiplication of the operator itself; mathematically, it can be shown as given below.

$$\hat{A}^2 f = \hat{A}\hat{A}f \tag{73}$$

At this point it also very important to discuss one of the most fundamental properties of operator multiplication, the commutation relation or the commutation rule. Consider two operators, A and B which can be operated over the function $f$.

$$\hat{A} = \frac{d}{dx}; \quad \hat{B} = x; \quad f = x^3 \tag{74}$$

Now

$$\hat{A}\hat{B}f = \frac{d}{dx}x(x^3) = \frac{d}{dx}x^4 = 4x^3 \tag{75}$$

And

$$\hat{B}\hat{A}f = x\frac{d}{dx}(x^3) = x(3x^2) = 3x^3 \tag{76}$$

From equation (75) and (76), it the clear that in this case

$$\hat{A}\hat{B}f \neq \hat{B}\hat{A}f \qquad (77)$$

These operators are said to be non commutating with the commutator given below.

$$\hat{A}\hat{B} - \hat{B}\hat{A} = 4x^3 - 3x^3 \qquad (78)$$

However, the two operators are said to be commute if their result is the same even after reverting their order of application. Mathematically, it can be stated as given by equation (79).

$$\hat{A}\hat{B}f = \hat{B}\hat{A}f \qquad (79)$$

This is quite different from the normal algebra in which the product of two numbers is always the same irrespective of the order of multiplication ($x.y = y.x$). Summarizing the commutation rule, it can be concluded that

$$[\hat{A},\hat{B}] = \hat{A}\hat{B} - \hat{B}\hat{A} = 0 \;\rightarrow\; Commutating \qquad (80)$$

and

$$[\hat{A},\hat{B}] = \hat{A}\hat{B} - \hat{B}\hat{A} \neq 0 \;\rightarrow\; Non\text{-}commutating \qquad (81)$$

**3. Linear Operator:** An operator $\hat{A}$ is said to be a linear operator if its application on the sum of two functions $f$ and $g$ gives the same result as the sum of its individual operations. Mathematically, it can be shown as given below.

$$\hat{A}(f + g) = \hat{A}f + \hat{A}g \qquad (82)$$

For example, consider the differential operator A; with $f$ and $g$ as the functions which have to be used as the operand.

$$\hat{A} = \frac{d}{dx}; \quad f = 2x^2; \quad g = 3x^2 \qquad (83)$$

or

$$\hat{A}(f + g) = \frac{d}{dx}(2x^2 + 3x^2) = \frac{d}{dx}(5x^2) = 10x \qquad (84)$$

or

$$\hat{A}f + \hat{A}g = \frac{d}{dx}(2x^2) + \frac{d}{dx}(3x^2) = 4x + 6x = 10x \qquad (85)$$

Hence, from equation (84) and equation (85), it is clear that the differential operator is clearly linear in nature. On the other hand, the "square root" operator is not linear as it does not give the same result when operated individually.

> ## Some Important Quantum Mechanical Operators

One of the most basic and very popular operators in quantum mechanics is the Laplacian operator, typically symbolized as $\nabla^2$, and is given by the following expression.

$$\nabla^2 = \frac{\partial^2}{\partial x^2} + \frac{\partial^2}{\partial y^2} + \frac{\partial^2}{\partial z^2} \tag{86}$$

The popular form of the Schrodinger equation can be written in terms of Laplacian operator as well.

$$\frac{\partial^2 \psi}{\partial x^2} + \frac{\partial^2 \psi}{\partial y^2} + \frac{\partial^2 \psi}{\partial z^2} + \frac{8\pi^2 m}{h^2}(E - V)\psi = 0 \tag{87}$$

or

$$\nabla^2 \psi + \frac{8\pi^2 m}{h^2}(E - V)\psi = 0 \tag{88}$$

The Laplacian operator is pronounced as "del squared". This operator is also a part of the "mighty" Hamiltonian operator which forms the basis for value evaluation for other operators, as we have already discussed in the postulates of quantum mechanics. The Hamiltonian operator is typically symbolized as $\hat{H}$ and is given by the following expression.

$$\hat{H} = -\frac{h^2}{8\pi^2 m}\left(\frac{\partial^2}{\partial x^2} + \frac{\partial^2}{\partial y^2} + \frac{\partial^2}{\partial z^2}\right) + V \tag{89}$$

or

$$\hat{H} = -\frac{h^2}{8\pi^2 m}\nabla^2 + V \tag{90}$$

The popular form of the Schrodinger equation is written in terms of the Hamiltonian operator as well.

$$\hat{H}\psi = E\psi \tag{91}$$

or

$$\left[-\frac{h^2}{8\pi^2 m}\left(\frac{\partial^2}{\partial x^2} + \frac{\partial^2}{\partial y^2} + \frac{\partial^2}{\partial z^2}\right) + V\right]\psi = E\psi \tag{92}$$

or

$$\left(-\frac{h^2}{8\pi^2 m}\nabla^2 + V\right)\psi = E\psi \tag{93}$$

Furthermore, we know from the third postulate of quantum mechanics that owing to the constant value of E (eigenvalue) the wave function $\psi$ can be labeled as eigenfunction.

Therefore, the Schrodinger equation is also called as the "eigen value equation". Simplifying this, we can say that

$$(Energy\ operator)(Wave\ function) = (Energy)(Wave\ function) \tag{94}$$

The equation (94) is applicable to observables in the quantum mechanical world.

For three dimensional systems, like the Hamiltonian, the operator can be obtained by summing the individual operators along three different axes. For instance, some important three-dimensional operators are:

$$\hat{T} = \frac{-h^2}{8\pi^2 m}\left(\frac{\partial^2}{\partial x^2} + \frac{\partial^2}{\partial y^2} + \frac{\partial^2}{\partial z^2}\right) \tag{95}$$

$$\hat{p} = \frac{h}{2\pi i}\left(\frac{\partial}{\partial x} + \frac{\partial}{\partial y} + \frac{\partial}{\partial z}\right) \tag{96}$$

The list of various important quantum mechanical operators in one dimension, along with their mode of operation is given below.

Table 2. Name and symbols of various important physical properties and their corresponding quantum mechanical operators.

| Physical property | | Operator | |
|---|---|---|---|
| *Name* | *Symbol* | *Symbol* | *Operation* |
| Position | $x$ | $\hat{x}$ | Multiplication by $x$ |
| Position squared | $x^2$ | $\hat{x}^2$ | Multiplication by $x^2$ |
| Position cubed | $x^3$ | $\hat{x}^2$ | Multiplication by $x^3$ |
| Momentum | $p_x$ | $\hat{p}_x$ | $\dfrac{h}{2\pi i}\dfrac{\partial}{\partial x}$ |
| Momentum squared | $p_x^2$ | $\hat{p}_x^2$ | $\dfrac{-h^2}{4\pi^2}\dfrac{\partial^2}{\partial x^2}$ |
| Kinetic energy | $T = \dfrac{P^2}{2m}$ | $\hat{T}_x$ | $\dfrac{-h^2}{8\pi^2 m}\dfrac{\partial^2}{\partial x^2}$ |
| Potential energy | $V(x)$ | $\hat{V}(x)$ | Multiplication by $V(x)$ |
| Total energy | $E = T + V(x)$ | $\hat{H}$ | $\dfrac{-h^2}{8\pi^2 m}\dfrac{\partial^2}{\partial x^2} + V(x)$ |

Besides the record of different operators presented in 'Table 2', there still many operators which are extremely important like angular momentum, parity, or the step-up–step-down operators. The discussion of every operator is beyond the scope of this book; however, a brief discussion of the essential operators in quantum mechanics is given below.

**1. Angular momentum operator:** In order to understand the angular momentum operator in the quantum mechanical world, we first need to understand the classical mechanics of one particle angular momentum. Let us consider a particle of mass $m$ which moves within a cartesian coordinate system with a position vector "$r$". Hence, we can say that

$$r = ix + jy + kz \tag{97}$$

The coordinates $x$. $y$ and $z$ are the functions of time, and therefore, we can define the velocity as the time derivative of the position vector as given below.

$$v = \frac{dr}{dt} = i\frac{dx}{dt} + j\frac{dy}{dt} + k\frac{dz}{dt} \tag{98}$$

or

$$v = v_x + v_y + v_z \tag{99}$$

Now, since we that $p = mv$, we can say that

$$p_x = mv_x; \quad p_y = mv_y; \quad p_z = mv_z \tag{100}$$

The angular momentum of a particle with mass $m$ and distance $r$ from the origin is given by the following relation.

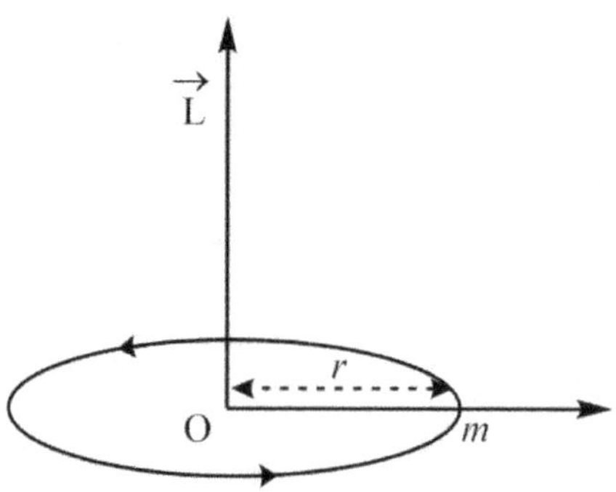


Figure 6. The angular momentum vector.

$$\vec{L} = \vec{v} \times m \times \vec{r} \tag{101}$$

$$\vec{L} = \vec{p} \times \vec{r} \tag{102}$$

Equation (102) can also be written in the form of a matrix as:

$$L = \begin{bmatrix} i & j & k \\ x & y & z \\ p_x & p_y & p_z \end{bmatrix} \tag{103}$$

$$L_x = yp_z - zp_y; \quad L_y = zp_x - xp_z; \quad L_z = xp_y - yp_x \tag{104}$$

Where $i, j, k$ are the unit vectors along $x, y, z$ axis and $L_x, L_y, L_z$ are the component of angular momentum along $x, y, z$ axis. Moreover, it is also worthy to note that the angular momentum vector is always perpendicular to the direction of the position vector of the particle i.e. the plane in which the particle is moving.

Now since the mathematical nature of any quantum mechanical operator is dependent upon the classical expression of the same observable, the angular momentum is not any exception. The quantum mechanical operator for angular momentum is given below.

$$\hat{L} = -i\frac{h}{2\pi}(r \times \nabla) = -i\hbar(r \times \nabla) \tag{105}$$

The angular momentum can be divided into two categories; one is orbital angular momentum (due to the orbital motion of the particle) and the other is spin angular momentum (due to spin motion of the particle). Moreover, being a vector quantity, the operator of angular momentum can also be resolved along different axes.

$$\hat{L} = \hat{L}_x + \hat{L}_y + \hat{L}_z \tag{106}$$

And we know that

$$\hat{L}_x = yp_z - zp_y = y\left(\frac{h}{2\pi i}\frac{\partial}{\partial z}\right) - z\left(\frac{h}{2\pi i}\frac{\partial}{\partial y}\right) = \frac{h}{2\pi i}\left(y\frac{\partial}{\partial z} - z\frac{\partial}{\partial y}\right) \tag{107}$$

or

$$\hat{L}_y = zp_x - xp_z = z\left(\frac{h}{2\pi i}\frac{\partial}{\partial x}\right) - x\left(\frac{h}{2\pi i}\frac{\partial}{\partial z}\right) = \frac{h}{2\pi i}\left(z\frac{\partial}{\partial x} - x\frac{\partial}{\partial z}\right) \tag{108}$$

or

$$\hat{L}_z = xp_y - yp_x = x\left(\frac{h}{2\pi i}\frac{\partial}{\partial y}\right) - y\left(\frac{h}{2\pi i}\frac{\partial}{\partial x}\right) = \frac{h}{2\pi i}\left(x\frac{\partial}{\partial y} - y\frac{\partial}{\partial x}\right) \tag{109}$$

$$\hat{L} = \frac{h}{2\pi i}\left[\left(y\frac{\partial}{\partial z} - z\frac{\partial}{\partial y}\right) + \left(z\frac{\partial}{\partial x} - x\frac{\partial}{\partial z}\right) + \left(x\frac{\partial}{\partial y} - y\frac{\partial}{\partial x}\right)\right] \tag{110}$$

It is also worthy to recall that equation (107) to (110) can also be reported in terms of $\hbar$; or by multiplying and dividing by $i$, or both.

**2. Ladder operator:** These operators are also called as step-up–step-down or rising-lowering operators. The reason for such terminology lies in the fact that these operators can increase or decrease the eigenvalues. Moreover, it should also be noted that this increase or decrease is always quantized in nature.

$$\hat{J}_+ = \hat{J}_x + i\hat{J}_y \tag{111}$$

and

$$\hat{J}_+ = \hat{J}_x - i\hat{J}_y \tag{112}$$

The equation (111) and (112) represent the step-up and step-down operators respectively. These operators can be used to increase or decrease the eigen values.

> #### Operator Evaluation

The operator evaluation simply means that we need to find the result by applying the operator over a given function. Some general examples are given below.

*i)* $(d/dx)\,(x^5)$: In this case $d/dx$ is the operator while the function $x^5$ is the operand.

$$\frac{d}{dx}x^5 = 5x^4 \tag{113}$$

*ii)* $\int(x^5)$: In this case, $\int$ is the operator while the function $x^5$ is the operand.

$$\int x^5 = \frac{x^6}{6} \tag{114}$$

*iii)* $(d^2/dt^2)\,(A\,Sine\ 2\pi vt)$: In this particular case, $(d^2/dt^2)$ is the operator while the function $(A\,Sin\ 2\pi vt)$ is the operand.

Let the function is symbolized by $y$. Then, we have

$$y = A\,Sin\ 2\pi vt \tag{115}$$

Differentiating with respect to $t$, we get

$$\frac{dy}{dt} = A\,2\pi v\,Cos\ 2\pi vt \tag{116}$$

Differentiating again

$$\frac{d^2y}{dt^2} = -A\,4\pi^2v^2\,Sin\ 2\pi vt \tag{117}$$

The operator evaluation is frequently used as a part of the commutator calculation and will be discussed in detail in this chapter.

> *Calculation of Resultant Operator*

Sometimes the operator is simplified to another form which is easy to apply over a function. This resultant operator is obtained by the rules of operator algebra. For instance, consider the following cases.

*i*) Find the resultant expression for the following operator

$$\left(\frac{d}{dx}x\right)^2 \tag{118}$$

In order to find the resultant operator, suppose a function $\psi(x)$ which is used as an operand, then we can say

$$\left(\frac{d}{dx}x\right)^2 \psi = \left(\frac{d}{dx}x\right)\left(\frac{d}{dx}x\right)\psi \tag{119}$$

or

$$\left(\frac{d}{dx}x\right)^2 \psi = \left(\frac{d}{dx}x\right)\left(\frac{d}{dx}x\psi\right) \tag{120}$$

or

$$\left(\frac{d}{dx}x\right)^2 \psi = \left(\frac{d}{dx}x\right)\left(x\frac{d\psi}{dx} + \psi\frac{dx}{dx}\right) \tag{121}$$

$$\left(\frac{d}{dx}x\right)^2 \psi = \frac{d}{dx}\left(x^2\frac{d\psi}{dx} + x\psi\right) \tag{122}$$

$$\left(\frac{d}{dx}x\right)^2 \psi = \left[x^2\frac{d^2\psi}{dx^2} + \frac{d\psi}{dx}(2x)\right] + \left[x\frac{d\psi}{dx} + \psi\frac{dx}{dx}\right] \tag{123}$$

$$\left(\frac{d}{dx}x\right)^2 \psi = x^2\frac{d^2\psi}{dx^2} + 2x\frac{d\psi}{dx} + x\frac{d\psi}{dx} + \psi \tag{124}$$

$$\left(\frac{d}{dx}x\right)^2 \psi = \left[x^2\frac{d^2}{dx^2} + 3x\frac{d}{dx} + 1\right]\psi \tag{125}$$

Removing $\psi$ from both sides, we get

$$\left(\frac{d}{dx}x\right)^2 = x^2\frac{d^2}{dx^2} + 3x\frac{d}{dx} + 1 \tag{126}$$

*ii*) Find the resultant expression for the following operator

$$\left(x + \frac{d}{dx}\right)\frac{d}{dx} \tag{127}$$

In order to find the resultant operator, suppose a function $\psi(x)$ which is used as operand, then we can say that



**DALAL INSTITUTE**

$$\left[\left(x + \frac{d}{dx}\right)\frac{d}{dx}\right]\psi = \left(x + \frac{d}{dx}\right)\frac{d\psi}{dx} \tag{128}$$

$$\left[\left(x + \frac{d}{dx}\right)\frac{d}{dx}\right]\psi = x\frac{d\psi}{dx} + \frac{d^2\psi}{dx^2}$$

Removing $\psi$ from both sides, we get

$$\left(x + \frac{d}{dx}\right)\frac{d}{dx} = x\frac{d}{dx} + \frac{d^2}{dx^2} \tag{129}$$

*iii*) Find the resultant expression for the following operator

$$\left(\frac{d}{dx} + x\right)^2 \tag{130}$$

In order to find the resultant operator, suppose a function $\psi(x)$ which is used as operand, then we can say that

$$\left[\left(\frac{d}{dx} + x\right)^2\right]\psi = \left[\left(\frac{d}{dx} + x\right)\left(\frac{d}{dx} + x\right)\right]\psi \tag{131}$$

$$\left[\left(\frac{d}{dx} + x\right)^2\right]\psi = \left(\frac{d}{dx} + x\right)\left(\frac{d\psi}{dx} + x\psi\right) \tag{132}$$

$$\left[\left(\frac{d}{dx} + x\right)^2\right]\psi = \frac{d^2\psi}{dx^2} + \frac{d}{dx}x\psi + x\frac{d\psi}{dx} + x^2\psi \tag{133}$$

$$\left[\left(\frac{d}{dx} + x\right)^2\right]\psi = \frac{d^2\psi}{dx^2} + x\frac{d\psi}{dx} + \psi\frac{dx}{dx} + x\frac{d\psi}{dx} + x^2\psi \tag{134}$$

$$\left[\left(\frac{d}{dx} + x\right)^2\right]\psi = \frac{d^2\psi}{dx^2} + 2x\frac{d\psi}{dx} + x^2\psi + \psi \tag{135}$$

Removing $\psi$ from both sides, we get

$$\left(\frac{d}{dx} + x\right)^2 = \frac{d^2}{dx^2} + 2x\frac{d}{dx} + x^2 + 1 \tag{136}$$

*iv*) Find the resultant expression for the following operator

$$\left(x + \frac{d}{dx}\right)\left(x - \frac{d}{dx}\right) \tag{137}$$

In order to find the resultant operator, suppose a function $\psi(x)$ which is used as operand, then we can say that

$$\left[\left(x + \frac{d}{dx}\right)\left(x - \frac{d}{dx}\right)\right]\psi = \left(x + \frac{d}{dx}\right)\left(x\psi - \frac{d\psi}{dx}\right) \tag{138}$$

$$\left[\left(x + \frac{d}{dx}\right)\left(x - \frac{d}{dx}\right)\right]\psi = xx\psi - x\frac{d\psi}{dx} + \frac{d}{dx}x\psi - \frac{d^2\psi}{dx^2} \tag{139}$$

$$\left[\left(x + \frac{d}{dx}\right)\left(x - \frac{d}{dx}\right)\right]\psi = x^2\psi - x\frac{d\psi}{dx} + x\frac{d\psi}{dx} + \psi\frac{dx}{dx} - \frac{d^2\psi}{dx^2} \tag{140}$$

$$\left[\left(x + \frac{d}{dx}\right)\left(x - \frac{d}{dx}\right)\right]\psi = x^2\psi + \psi\frac{dx}{dx} - \frac{d^2\psi}{dx^2} \tag{141}$$

$$\left[\left(x + \frac{d}{dx}\right)\left(x - \frac{d}{dx}\right)\right]\psi = \left[x^2 + \frac{dx}{dx} - \frac{d^2}{dx^2}\right]\psi \tag{142}$$

Removing $\psi$ from both sides, we get

$$\left(x + \frac{d}{dx}\right)\left(x - \frac{d}{dx}\right) = x^2 + 1 - \frac{d^2}{dx^2} \tag{143}$$

The resultant operator calculation is frequently used as a part of the commutator calculation and will be discussed in detail in this chapter.

> ➤ *Commutation Relations of Various Quantum Mechanical Operators*

As we have discussed previously that one of the most fundamental properties of operator multiplication is the commutation relation or the commutation rule. two operators, A and B, are said to be commutating or non-commutating depending upon the value of their commutator.

$$[\hat{A}, \hat{B}] = \hat{A}\hat{B} - \hat{B}\hat{A} = 0 \;\rightarrow Commutating \tag{144}$$

$$[\hat{A}, \hat{B}] = \hat{A}\hat{B} - \hat{B}\hat{A} \neq 0 \;\rightarrow Non\text{-}commutating \tag{145}$$

The physical significance of the commutation relations is that when two operators commute, it means they are having a simultaneous set of eigenfunctions; and their corresponding physical properties can be calculated simultaneously and accurately. However, if the commutator is non-zero, the respective physical properties cannot be obtained simultaneously and accurately. Some important commutation relations are given below.

**1. Commutators of some simple operators:**

*i*) Calculate the commutator of the following

$$\left[x, \frac{d}{dx}\right] \tag{146}$$

Let it be operated over a function $\psi$. We have

$$\left[x,\frac{d}{dx}\right]\psi = x\,\frac{d}{dx}\psi - \frac{d}{dx}x\,\psi \tag{147}$$

$$\left[x,\frac{d}{dx}\right]\psi = x\,\frac{d\psi}{dx} - \psi - x\,\frac{d\,\psi}{dx} \tag{148}$$

$$\left[x,\frac{d}{dx}\right]\psi = -\psi \tag{149}$$

or

$$\left[x,\frac{d}{dx}\right] = -1 \tag{150}$$

*ii*) Calculate the commutator of the following

$$\left[y,\frac{d}{dx}\right] \tag{151}$$

Let it be operated over a function ψ. We have

$$\left[y,\frac{d}{dx}\right]\psi = y\,\frac{d}{dx}\psi - \frac{d}{dx}y\,\psi \tag{152}$$

$$\left[y,\frac{d}{dx}\right]\psi = y\,\frac{d\psi}{dx} - y\,\frac{d\,\psi}{dx} - \psi\,\frac{d\,y}{dx} \tag{153}$$

$$\left[x,\frac{d}{dx}\right]\psi = 0 \tag{154}$$

*iii*) Calculate the commutator of the following

$$\left[\frac{d}{dx},\frac{d^2}{dx^2}\right] \tag{155}$$

Let it be operated over a function ψ. We have

$$\left[\frac{d}{dx},\frac{d^2}{dx^2}\right]\psi = \frac{d}{dx}\frac{d^2}{dx^2}\psi - \frac{d^2}{dx^2}\frac{d}{dx}\,\psi \tag{156}$$

or

$$\left[\frac{d}{dx},\frac{d^2}{dx^2}\right]\psi = \frac{d^3\psi}{dx^3} - \frac{d^3\psi}{dx^3} \tag{157}$$

$$\left[\frac{d}{dx},\frac{d^2}{dx^2}\right]\psi = 0 \tag{158}$$

**2. Commutators of position and linear momentum operators:**

*i*) Find the commutator of the following

$$[\hat{x}, \hat{p}_x] \tag{159}$$

Let it be operated over a function ψ. We have

$$[\hat{x}, \hat{p}_x]\psi = \hat{x}\,\hat{p}_x\,\psi - \hat{p}_x\,\hat{x}\,\psi \tag{160}$$

$$[\hat{x}, \hat{p}_x]\psi = x\frac{h}{2\pi i}\frac{\partial}{\partial x}\psi - \frac{h}{2\pi i}\frac{\partial}{\partial x}x\,\psi \tag{161}$$

$$[\hat{x}, \hat{p}_x]\psi = \frac{h}{2\pi i}x\frac{\partial\psi}{\partial x} - \frac{h}{2\pi i}x\frac{\partial\psi}{\partial x} - \frac{h}{2\pi i}\psi\frac{\partial x}{\partial x} \tag{162}$$

$$[\hat{x}, \hat{p}_x]\psi = -\frac{h}{2\pi i}\psi$$

$$[\hat{x}, \hat{p}_x] = -\frac{h}{2\pi i} = \frac{hi}{2\pi} = i\hbar \tag{163}$$

*ii*) Find the commutator of the following

$$[\hat{x}^n, \hat{p}_x] \tag{164}$$

Let it be operated over a function ψ. We have

$$[\hat{x}^n, \hat{p}_x]\psi = \hat{x}^n\,\hat{p}_x\,\psi - \hat{p}_x\,\hat{x}^n\,\psi \tag{165}$$

$$[\hat{x}^n, \hat{p}_x]\psi = x^n\frac{h}{2\pi i}\frac{\partial}{\partial x}\psi - \frac{h}{2\pi i}\frac{\partial}{\partial x}x^n\,\psi \tag{166}$$

$$[\hat{x}^n, \hat{p}_x]\psi = \frac{h}{2\pi i}x^n\frac{\partial\psi}{\partial x} - \frac{h}{2\pi i}x^n\frac{\partial\psi}{\partial x} - \frac{h}{2\pi i}nx^{n-1}\psi \tag{167}$$

$$[\hat{x}^n, \hat{p}_x]\psi = \frac{h}{2\pi i}nx^{n-1}\psi \tag{168}$$

Removing ψ from both sides, we get

$$[\hat{x}^n, \hat{p}_x] = -\frac{h}{2\pi i}nx^{n-1} \tag{169}$$

The commutation relations between position and linear momentum can mainly be divided into three categories as discussed below.

*(a) When position and momentum are along the same axis:*

$$[\hat{x}^n, \hat{p}_x] = ni\hbar x^{n-1} \tag{170}$$

$$[\hat{p}_x, \hat{x}^n] = -ni\hbar x^{n-1} \tag{171}$$

and

$$[\hat{x}, \hat{p}_x^n] = ni\hbar p_x^{n-1} \tag{172}$$

$$[\hat{p}_x^n, \hat{x}\,] = -ni\hbar p_x^{n-1} \tag{173}$$

*(b) When position and momentum are along different axis:*

$$\left[\hat{x}, \hat{p}_y\right] = 0 \tag{174}$$

$$[\hat{x}, \hat{p}_z] = 0 \tag{175}$$

$$[\hat{y}, \hat{p}_x] = 0 \tag{176}$$

$$[\hat{y}, \hat{p}_z] = 0 \tag{177}$$

$$[\hat{z}, \hat{p}_x] = 0 \tag{178}$$

$$\left[\hat{z}, \hat{p}_y\right] = 0 \tag{179}$$

*(b) When positions are along the different axis:*

$$[\hat{x}, \hat{y}] = 0 \tag{180}$$

$$[\hat{x}, \hat{z}] = 0 \tag{181}$$

$$[\hat{y}, \hat{z}] = 0 \tag{182}$$

*(b) When positions are along the different axis:*

$$\left[\hat{p}_x, \hat{p}_y\right] = 0 \tag{183}$$

$$[\hat{p}_x, \hat{p}_z] = 0 \tag{184}$$

$$\left[\hat{p}_y, \hat{p}_z\right] = 0 \tag{185}$$

### 3. Commutators of angular momentum operators:

*i*) The commutator of orbital angular momentum operators along $x$ and $y$-axis.

$$\left[\hat{L}_x, \hat{L}_y\right] = \hat{L}_x\hat{L}_y - \hat{L}_y\hat{L}_x \tag{186}$$

Finding the values of $\hat{L}_x\hat{L}_y$ , we get

$$\hat{L}_x\hat{L}_y = \left[\frac{h}{2\pi i}\left(y\frac{\partial}{\partial z} - z\frac{\partial}{\partial y}\right)\right]\left[\frac{h}{2\pi i}\left(z\frac{\partial}{\partial x} - x\frac{\partial}{\partial z}\right)\right] \tag{187}$$

$$= -\frac{h^2}{4\pi^2}\left[\left(y\frac{\partial}{\partial z} - z\frac{\partial}{\partial y}\right)\left(z\frac{\partial}{\partial x} - x\frac{\partial}{\partial z}\right)\right] \tag{188}$$

$$= -\frac{h^2}{4\pi^2}\left(y\frac{\partial}{\partial z}z\frac{\partial}{\partial x} - z\frac{\partial}{\partial y}z\frac{\partial}{\partial x} - y\frac{\partial}{\partial z}x\frac{\partial}{\partial z} + z\frac{\partial}{\partial y}x\frac{\partial}{\partial z}\right) \tag{189}$$

$$= -\frac{h^2}{4\pi^2}\left(y\frac{\partial}{\partial x}\frac{\partial z}{\partial z} + yz\frac{\partial^2}{\partial^2 zx} - z^2\frac{\partial^2}{\partial^2 yx} - yx\frac{\partial^2}{\partial z^2} + zx\frac{\partial^2}{\partial yz}\right) \tag{190}$$

$$= -\hbar^2\left(y\frac{\partial}{\partial x} + yz\frac{\partial^2}{\partial^2 zx} - z^2\frac{\partial^2}{\partial^2 yx} - yx\frac{\partial^2}{\partial z^2} + zx\frac{\partial^2}{\partial yz}\right) \tag{191}$$

Similarly obtaining the value of $\hat{L}_y\hat{L}_x$, we get

$$\hat{L}_y\hat{L}_x = \left[\frac{h}{2\pi i}\left(z\frac{\partial}{\partial x} - x\frac{\partial}{\partial z}\right)\right]\left[\frac{h}{2\pi i}\left(y\frac{\partial}{\partial z} - z\frac{\partial}{\partial y}\right)\right] \tag{192}$$

$$= -\frac{h^2}{4\pi^2}\left[\left(z\frac{\partial}{\partial x} - x\frac{\partial}{\partial z}\right)\left(y\frac{\partial}{\partial z} - z\frac{\partial}{\partial y}\right)\right] \tag{193}$$

$$= -\frac{h^2}{4\pi^2}\left(z\frac{\partial}{\partial x}y\frac{\partial}{\partial z} - z\frac{\partial}{\partial x}z\frac{\partial}{\partial y} - x\frac{\partial}{\partial z}y\frac{\partial}{\partial z} + x\frac{\partial}{\partial z}z\frac{\partial}{\partial y}\right) \tag{194}$$

$$= -\frac{h^2}{4\pi^2}\left(zy\frac{\partial^2}{\partial^2 xz} - z^2\frac{\partial^2}{\partial^2 xy} - xy\frac{\partial^2}{\partial z^2} + xz\frac{\partial^2}{\partial zy} + x\frac{\partial}{\partial y}\frac{\partial z}{\partial z}\right) \tag{195}$$

$$= -\hbar^2\left(zy\frac{\partial^2}{\partial^2 xz} - z^2\frac{\partial^2}{\partial^2 xy} - xy\frac{\partial^2}{\partial z^2} + xz\frac{\partial^2}{\partial zy} + x\frac{\partial}{\partial y}\right) \tag{196}$$

Now putting the values of $\hat{L}_x\hat{L}_y$ and $\hat{L}_y\hat{L}_x$ in equation (183), we get the following.

$$[\hat{L}_x, \hat{L}_y] = \left[-\hbar^2\left(y\frac{\partial}{\partial x} + yz\frac{\partial^2}{\partial^2 zx} - z^2\frac{\partial^2}{\partial^2 yx} - yx\frac{\partial^2}{\partial z^2} + zx\frac{\partial^2}{\partial yz}\right)\right] \tag{197}$$
$$- \left[-\hbar^2\left(zy\frac{\partial^2}{\partial^2 xz} - z^2\frac{\partial^2}{\partial^2 xy} - xy\frac{\partial^2}{\partial z^2} + xz\frac{\partial^2}{\partial zy} + x\frac{\partial}{\partial y}\right)\right]$$

$$[\hat{L}_x, \hat{L}_y] = -\hbar^2\left(y\frac{\partial}{\partial x} - x\frac{\partial}{\partial y}\right) \tag{198}$$

Taking negative sign common, we get

$$[\hat{L}_x, \hat{L}_y] = \hbar^2\left(x\frac{\partial}{\partial y} - y\frac{\partial}{\partial x}\right) \tag{199}$$

$$\left[\hat{L}_x, \hat{L}_y\right] = i\hbar\left[-i\hbar\left(x\frac{\partial}{\partial y} - y\frac{\partial}{\partial x}\right)\right] \tag{200}$$

$$\left[\hat{L}_x, \hat{L}_y\right] = i\hbar\hat{L}_z \tag{201}$$

*ii*) The commutator of orbital angular momentum operators along $y$ and $z$-axis.

$$\left[\hat{L}_y, \hat{L}_z\right] = \hat{L}_y\hat{L}_z - \hat{L}_z\hat{L}_y \tag{202}$$

Finding the values of $\hat{L}_y\hat{L}_z$ , we get

$$\hat{L}_y\hat{L}_z = \left[\frac{h}{2\pi i}\left(z\frac{\partial}{\partial x} - x\frac{\partial}{\partial z}\right)\right]\left[\frac{h}{2\pi i}\left(x\frac{\partial}{\partial y} - y\frac{\partial}{\partial x}\right)\right] \tag{203}$$

$$= -\frac{h^2}{4\pi^2}\left[\left(z\frac{\partial}{\partial x} - x\frac{\partial}{\partial z}\right)\left(x\frac{\partial}{\partial y} - y\frac{\partial}{\partial x}\right)\right] \tag{204}$$

$$= -\frac{h^2}{4\pi^2}\left(z\frac{\partial}{\partial x}x\frac{\partial}{\partial y} - x\frac{\partial}{\partial z}x\frac{\partial}{\partial y} - z\frac{\partial}{\partial x}y\frac{\partial}{\partial x} + x\frac{\partial}{\partial z}y\frac{\partial}{\partial x}\right) \tag{205}$$

$$= -\frac{h^2}{4\pi^2}\left(z\frac{\partial}{\partial y}\frac{\partial x}{\partial x} + zx\frac{\partial^2}{\partial xy} - x^2\frac{\partial^2}{\partial^2 zy} - zy\frac{\partial}{\partial x^2} + xy\frac{\partial^2}{\partial zx}\right) \tag{206}$$

$$= -\hbar^2\left(z\frac{\partial}{\partial y} + zx\frac{\partial^2}{\partial xy} - x^2\frac{\partial^2}{\partial^2 zy} - zy\frac{\partial}{\partial x^2} + xy\frac{\partial^2}{\partial zx}\right) \tag{207}$$

Similarly obtaining the value of $\hat{L}_z\hat{L}_y$, we get

$$\hat{L}_z\hat{L}_y = \left[\frac{h}{2\pi i}\left(x\frac{\partial}{\partial y} - y\frac{\partial}{\partial x}\right)\right]\left[\frac{h}{2\pi i}\left(z\frac{\partial}{\partial x} - x\frac{\partial}{\partial z}\right)\right] \tag{208}$$

$$= -\frac{h^2}{4\pi^2}\left[\left(x\frac{\partial}{\partial y} - y\frac{\partial}{\partial x}\right)\left(z\frac{\partial}{\partial x} - x\frac{\partial}{\partial z}\right)\right] \tag{209}$$

$$= -\frac{h^2}{4\pi^2}\left(x\frac{\partial}{\partial y}z\frac{\partial}{\partial x} - x\frac{\partial}{\partial y}x\frac{\partial}{\partial z} - y\frac{\partial}{\partial x}z\frac{\partial}{\partial x} + y\frac{\partial}{\partial x}x\frac{\partial}{\partial z}\right) \tag{210}$$

$$= -\frac{h^2}{4\pi^2}\left(xz\frac{\partial^2}{\partial yx} - x^2\frac{\partial^2}{\partial yz} - yz\frac{\partial^2}{\partial x^2} + yx\frac{\partial^2}{\partial xz} + y\frac{\partial}{\partial z}\frac{\partial x}{\partial x}\right) \tag{211}$$

$$= -\hbar^2\left(xz\frac{\partial^2}{\partial yx} - x^2\frac{\partial^2}{\partial yz} - yz\frac{\partial^2}{\partial x^2} + yx\frac{\partial^2}{\partial xz} + y\frac{\partial}{\partial z}\right) \tag{212}$$

Now putting the values of $\hat{L}_y\hat{L}_z$ and $\hat{L}_z\hat{L}_y$ in equation (212), we get the following.

$$[\hat{L}_y, \hat{L}_z] = \left[ -\hbar^2 \left( z\frac{\partial}{\partial y} + zx\frac{\partial^2}{\partial xy} - x^2\frac{\partial^2}{\partial^2 zy} - zy\frac{\partial}{\partial x^2} + xy\frac{\partial^2}{\partial zx} \right) \right]$$
$$- \left[ -\hbar^2 \left( xz\frac{\partial^2}{\partial yx} - x^2\frac{\partial^2}{\partial yz} - yz\frac{\partial^2}{\partial x^2} + yx\frac{\partial^2}{\partial xz} + y\frac{\partial}{\partial z} \right) \right] \tag{213}$$

$$[\hat{L}_y, \hat{L}_z] = -\hbar^2 \left( z\frac{\partial}{\partial y} - y\frac{\partial}{\partial z} \right) \tag{214}$$

Taking negative sign common, we get

$$[\hat{L}_y, \hat{L}_z] = \hbar^2 \left( y\frac{\partial}{\partial z} - z\frac{\partial}{\partial y} \right) \tag{215}$$

$$[\hat{L}_y, \hat{L}_z] = i\hbar \left[ -i\hbar \left( y\frac{\partial}{\partial z} - z\frac{\partial}{\partial y} \right) \right] \tag{216}$$

$$[\hat{L}_y, \hat{L}_z] = i\hbar\hat{L}_x \tag{217}$$

*iii*) The commutator of orbital angular momentum operators along $z$ and $x$-axis.

$$[\hat{L}_z, \hat{L}_x] = \hat{L}_z\hat{L}_x - \hat{L}_x\hat{L}_z \tag{218}$$

Finding the values of $\hat{L}_z\hat{L}_x$ , we get

$$\hat{L}_z\hat{L}_x = \left[ \frac{h}{2\pi i}\left( x\frac{\partial}{\partial y} - y\frac{\partial}{\partial x} \right) \right]\left[ \frac{h}{2\pi i}\left( y\frac{\partial}{\partial z} - z\frac{\partial}{\partial y} \right) \right] \tag{219}$$

$$= -\frac{h^2}{4\pi^2}\left[ \left( x\frac{\partial}{\partial y} - y\frac{\partial}{\partial x} \right)\left( y\frac{\partial}{\partial z} - z\frac{\partial}{\partial y} \right) \right] \tag{220}$$

$$= -\frac{h^2}{4\pi^2}\left( x\frac{\partial}{\partial y}y\frac{\partial}{\partial z} - x\frac{\partial}{\partial y}z\frac{\partial}{\partial y} - y\frac{\partial}{\partial x}y\frac{\partial}{\partial z} + y\frac{\partial}{\partial x}z\frac{\partial}{\partial y} \right) \tag{221}$$

$$= -\frac{h^2}{4\pi^2}\left( x\frac{\partial}{\partial z}\frac{\partial y}{\partial y} + xy\frac{\partial^2}{\partial yz} - xz\frac{\partial^2}{\partial y^2} - y^2\frac{\partial^2}{\partial xz} + yz\frac{\partial^2}{\partial xy} \right) \tag{222}$$

$$= -\hbar^2\left( x\frac{\partial}{\partial z} + xy\frac{\partial^2}{\partial yz} - xz\frac{\partial^2}{\partial y^2} - y^2\frac{\partial^2}{\partial xz} + yz\frac{\partial^2}{\partial xy} \right) \tag{223}$$

Similarly obtaining the value of $\hat{L}_x\hat{L}_z$, we get

$$\hat{L}_x\hat{L}_z = \left[ \frac{h}{2\pi i}\left( y\frac{\partial}{\partial z} - z\frac{\partial}{\partial y} \right) \right]\left[ \frac{h}{2\pi i}\left( x\frac{\partial}{\partial y} - y\frac{\partial}{\partial x} \right) \right] \tag{224}$$

$$= -\frac{h^2}{4\pi^2}\left[\left(y\frac{\partial}{\partial z} - z\frac{\partial}{\partial y}\right)\left(x\frac{\partial}{\partial y} - y\frac{\partial}{\partial x}\right)\right] \tag{225}$$

or

$$= -\frac{h^2}{4\pi^2}\left(y\frac{\partial}{\partial z}x\frac{\partial}{\partial y} - y\frac{\partial}{\partial z}y\frac{\partial}{\partial x} - z\frac{\partial}{\partial y}x\frac{\partial}{\partial y} + z\frac{\partial}{\partial y}y\frac{\partial}{\partial x}\right) \tag{226}$$

$$= -\frac{h^2}{4\pi^2}\left(yx\frac{\partial^2}{\partial zy} - y^2\frac{\partial^2}{\partial zx} - zx\frac{\partial^2}{\partial y^2} + z\frac{\partial}{\partial x}\frac{\partial y}{\partial y} + zy\frac{\partial^2}{\partial yx}\right) \tag{227}$$

$$= -\hbar^2\left(yx\frac{\partial^2}{\partial zy} - y^2\frac{\partial^2}{\partial zx} - zx\frac{\partial^2}{\partial y^2} + z\frac{\partial}{\partial x} + zy\frac{\partial^2}{\partial yx}\right) \tag{228}$$

Now putting the values of $\hat{L}_z\hat{L}_x$ and $\hat{L}_x\hat{L}_z$ in equation (218), we get the following.

$$[\hat{L}_z,\hat{L}_x] = \left[-\hbar^2\left(x\frac{\partial}{\partial z} + xy\frac{\partial^2}{\partial yz} - xz\frac{\partial^2}{\partial y^2} - y^2\frac{\partial^2}{\partial xz} + yz\frac{\partial^2}{\partial xy}\right)\right] \tag{229}$$
$$- \left[-\hbar^2\left(yx\frac{\partial^2}{\partial zy} - y^2\frac{\partial^2}{\partial zx} - zx\frac{\partial^2}{\partial y^2} + z\frac{\partial}{\partial x} + zy\frac{\partial^2}{\partial yx}\right)\right]$$

$$[\hat{L}_z,\hat{L}_x] = -\hbar^2\left(x\frac{\partial}{\partial z} - z\frac{\partial}{\partial x}\right) \tag{230}$$

Taking negative sign common, we get

$$[\hat{L}_z,\hat{L}_x] = \hbar^2\left(y\frac{\partial}{\partial z} - z\frac{\partial}{\partial y}\right) \tag{231}$$

$$[\hat{L}_z,\hat{L}_x] = i\hbar\left[-i\hbar\left(y\frac{\partial}{\partial z} - z\frac{\partial}{\partial y}\right)\right] \tag{232}$$

$$[\hat{L}_z,\hat{L}_x] = i\hbar\hat{L}_y \tag{233}$$

*iv*) The commutator of total orbital angular momentum squared operator and orbital angular momentum along one of the three-axis.

$$[\hat{L}^2,\hat{L}_z] = [\hat{L}_x^2 + \hat{L}_y^2 + \hat{L}_z^2, \hat{L}_z] \tag{234}$$

$$= [\hat{L}_x^2\,\hat{L}_z + \hat{L}_y^2\,\hat{L}_z + \hat{L}_z^2\,\hat{L}_z - \hat{L}_z\hat{L}_x^2 - \hat{L}_z\hat{L}_y^2 - \hat{L}_z\hat{L}_z^2] \tag{235}$$

$$= [(\hat{L}_x^2\,\hat{L}_z - \hat{L}_z\hat{L}_x^2) + (\hat{L}_y^2\,\hat{L}_z - \hat{L}_z\hat{L}_y^2) + (\hat{L}_z^2\,\hat{L}_z - \hat{L}_z\hat{L}_z^2)] \tag{236}$$

$$[\hat{L}^2,\hat{L}_z] = [\hat{L}_x^2,\hat{L}_z] + [\hat{L}_y^2\,\hat{L}_z] + [\hat{L}_z^2\,\hat{L}_z] \tag{237}$$

Now finding $[\hat{L}_x^2,\hat{L}_z]$ first, we get

$$\left[\hat{L}_x^2,\hat{L}_z\right] = \hat{L}_x^2\,\hat{L}_z - \hat{L}_z\hat{L}_x^2 \tag{238}$$

$$= \hat{L}_x\hat{L}_x\,\hat{L}_z - \hat{L}_z\hat{L}_x\hat{L}_x \tag{239}$$

$$= \left[\hat{L}_x\hat{L}_x\,\hat{L}_z - \hat{L}_x\hat{L}_z\,\hat{L}_x\right] - \left[\hat{L}_z\hat{L}_x\hat{L}_x - \hat{L}_x\hat{L}_z\,\hat{L}_x\right] \tag{240}$$

$$= \hat{L}_x\left[\hat{L}_x\hat{L}_z - \hat{L}_z\,\hat{L}_x\right] - \left[\hat{L}_z\hat{L}_x - \hat{L}_x\hat{L}_z\,\right]\hat{L}_x \tag{241}$$

$$= \hat{L}_x\left[\hat{L}_x\hat{L}_z - \hat{L}_z\,\hat{L}_x\right] + \left[\hat{L}_x\hat{L}_z - \hat{L}_z\hat{L}_x\,\right]\hat{L}_x \tag{242}$$

$$= \hat{L}_x\left[-i\hbar\hat{L}_y\,\right] + \left[-i\hbar\hat{L}_y\,\right]\hat{L}_x \tag{243}$$

$$= -i\hbar\hat{L}_x\hat{L}_y - i\hbar\hat{L}_y\hat{L}_x = -i\hbar\left[\hat{L}_x\hat{L}_y + \hat{L}_y\hat{L}_x\right] \tag{244}$$

Similarly,

$$\left[\hat{L}_y^2,\hat{L}_z\right] = \hat{L}_y^2\,\hat{L}_z - \hat{L}_z\hat{L}_y^2 \tag{245}$$

$$= \hat{L}_y\hat{L}_y\,\hat{L}_z - \hat{L}_z\hat{L}_y\hat{L}_y \tag{246}$$

$$= \left[\hat{L}_y\hat{L}_y\,\hat{L}_z - \hat{L}_y\hat{L}_z\hat{L}_y\right] - \left[\hat{L}_z\hat{L}_y\hat{L}_y - \hat{L}_y\hat{L}_z\hat{L}_y\right] \tag{247}$$

$$= \hat{L}_y\left[\hat{L}_y\hat{L}_z - \hat{L}_z\,\hat{L}_y\right] - \left[\hat{L}_z\hat{L}_y - \hat{L}_y\hat{L}_z\,\right]\hat{L}_y \tag{248}$$

$$= \hat{L}_y\left[\hat{L}_y\hat{L}_z - \hat{L}_z\,\hat{L}_y\right] + \left[\hat{L}_y\hat{L}_z - \hat{L}_z\,\hat{L}_y\,\right]\hat{L}_y \tag{249}$$

$$= \hat{L}_y\left[i\hbar\hat{L}_x\,\right] + \left[i\hbar\hat{L}_x\,\right]\hat{L}_y \tag{250}$$

$$= i\hbar\hat{L}_y\hat{L}_x + i\hbar\hat{L}_x\hat{L}_y = i\hbar\left[\hat{L}_y\hat{L}_x + \hat{L}_x\hat{L}_y\right] \tag{251}$$

Similarly,

$$\left[\hat{L}_z^2,\hat{L}_z\right] = \hat{L}_z^2\,\hat{L}_z - \hat{L}_z\hat{L}_z^2 \tag{245}$$

$$= \hat{L}_z\hat{L}_z\,\hat{L}_z - \hat{L}_z\hat{L}_z\hat{L}_z \tag{246}$$

$$\left[\hat{L}_z^2,\hat{L}_z\right] = 0 \tag{247}$$

Now putting the value of $\hat{L}_x^2\hat{L}_z$, $\hat{L}_y^2\hat{L}_z$ and $\hat{L}_z^2\hat{L}_z$ in equation (237), we get

$$\left[L^2,L_z\right] = -i\hbar\left[\hat{L}_x\hat{L}_y + \hat{L}_y\hat{L}_x\right] + i\hbar\left[\hat{L}_y\hat{L}_x + \hat{L}_x\hat{L}_y\right] + 0 \tag{248}$$

$$\left[\hat{L}^2,\hat{L}_z\right] = 0 \tag{249}$$

Also

$$\left[\hat{L}^2, \hat{L}_y\right] = 0; \, and \, \left[\hat{L}^2, \hat{L}_x\right] = 0 \tag{250}$$

Hence, the commutation relations of angular momentum operators along two different directions do not commute with each other and hence cannot give eigenvalues simultaneously and accurately. One the other hand, total angular momentum squared and angular momentum along one axis do commute with each other.

The commutation relations between angular momentum operators can be mainly divided into four categories as discussed below.

*(a) Orbital angular momentum commutation:*

$$\left[\hat{L}_x, \hat{L}_y\right] = i\hbar\hat{L}_z; \quad \left[\hat{L}_y, \hat{L}_x\right] = -i\hbar\hat{L}_z \tag{251}$$

$$\left[\hat{L}_y, \hat{L}_z\right] = i\hbar\hat{L}_x; \quad \left[\hat{L}_z, \hat{L}_y\right] = -i\hbar\hat{L}_x \tag{252}$$

$$\left[\hat{L}_z, \hat{L}_x\right] = i\hbar\hat{L}_y; \quad \left[\hat{L}_x, \hat{L}_z\right] = -i\hbar\hat{L}_y \tag{253}$$

$$\left[\hat{L}^2, \hat{L}_x\right] = 0; \quad \left[\hat{L}_x, \hat{L}^2\right] = 0 \tag{254}$$

$$\left[\hat{L}^2, \hat{L}_y\right] = 0; \quad \left[\hat{L}_y, \hat{L}^2\right] = 0 \tag{255}$$

$$\left[\hat{L}^2, \hat{L}_z\right] = 0; \quad \left[\hat{L}_z, \hat{L}^2\right] = 0 \tag{256}$$

*(b) Spin angular momentum commutation:*

$$\left[\hat{S}_x, \hat{S}_y\right] = i\hbar\hat{S}_z; \quad \left[\hat{S}_y, \hat{S}_x\right] = -i\hbar\hat{S}_z \tag{257}$$

$$\left[\hat{S}_y, \hat{S}_z\right] = i\hbar\hat{S}_x; \quad \left[\hat{S}_z, \hat{S}_y\right] = -i\hbar\hat{S}_x \tag{258}$$

$$\left[\hat{S}_z, \hat{S}_x\right] = i\hbar\hat{S}_y; \quad \left[\hat{S}_x, \hat{S}_z\right] = -i\hbar\hat{S}_y \tag{259}$$

$$\left[\hat{S}^2, \hat{S}_x\right] = 0; \quad \left[\hat{S}_x, \hat{S}^2\right] = 0 \tag{260}$$

$$\left[\hat{S}^2, S_y\right] = 0; \quad \left[\hat{S}_y, \hat{S}^2\right] = 0 \tag{261}$$

$$\left[\hat{S}^2, \hat{S}_z\right] = 0; \quad \left[\hat{S}_z, \hat{S}^2\right] = 0 \tag{262}$$

*(c) Total angular momentum commutation:*

$$\left[\hat{J}_x, \hat{J}_y\right] = i\hbar\hat{J}_z; \quad \left[\hat{J}_y, \hat{J}_x\right] = -i\hbar\hat{J}_z \tag{263}$$

$$\left[\hat{J}_y, \hat{J}_z\right] = i\hbar\hat{J}_x; \quad \left[\hat{J}_z, \hat{J}_y\right] = -i\hbar\hat{J}_x \tag{264}$$

$$\left[\hat{J}_z, \hat{J}_x\right] = i\hbar\hat{J}_y; \quad \left[\hat{J}_x, \hat{J}_z\right] = -i\hbar\hat{J}_y \tag{265}$$

$$\left[\hat{J}^2, \hat{J}_x\right] = 0; \quad \left[\hat{J}_x, \hat{J}^2\right] = 0 \tag{266}$$

$$[\hat{J}^2, J_y] = 0; \quad [\hat{J}_y, \hat{J}^2] = 0 \tag{267}$$

$$[\hat{J}^2, \hat{J}_z] = 0; \quad [\hat{J}_z, \hat{J}^2] = 0 \tag{268}$$

*(d) Total angular momentum commutation:*

$$[\hat{L}_x, \hat{S}_x] = 0; \quad [\hat{S}_x, \hat{L}_x] = 0 \tag{263}$$

$$[\hat{L}_x, \hat{S}_y] = 0; \quad [\hat{S}_y, \hat{L}_x] = 0 \tag{264}$$

$$[\hat{L}_x, \hat{S}_z] = 0; \quad [\hat{S}_z, \hat{L}_x] = 0 \tag{265}$$

$$[\hat{L}_y, \hat{S}_x] = 0; \quad [\hat{S}_x, \hat{L}_y] = 0 \tag{266}$$

$$[\hat{L}_y, \hat{S}_y] = 0; \quad [\hat{S}_y, \hat{L}_y] = 0 \tag{267}$$

$$[\hat{L}_y, \hat{S}_z] = 0; \quad [\hat{S}_z, \hat{L}_y] = 0 \tag{268}$$

$$[\hat{L}_z, \hat{S}_x] = 0; \quad [\hat{S}_x, \hat{L}_z] = 0 \tag{269}$$

$$[\hat{L}_z, \hat{S}_y] = 0; \quad [\hat{S}_y, \hat{L}_z] = 0 \tag{270}$$

$$[\hat{L}_z, \hat{S}_z] = 0; \quad [\hat{S}_z, \hat{L}_z] = 0 \tag{271}$$

## 4. Commutators of Ladder operators:

*i*) Find the commutator of the following

$$[\hat{J}^2, \hat{J}_+] \tag{272}$$

Let

$$[\hat{J}^2, \hat{J}_+] = [\hat{J}^2, \hat{J}_x + i\hat{J}_y] \tag{273}$$

$$= \hat{J}^2(\hat{J}_x + i\hat{J}_y) - (\hat{J}_x + i\hat{J}_y)\hat{J}^2 \tag{274}$$

$$= \hat{J}^2\hat{J}_x + i\hat{J}^2\hat{J}_y - \hat{J}_x\hat{J}^2 - i\hat{J}_y\hat{J}^2 \tag{275}$$

$$= [\hat{J}^2\hat{J}_x - \hat{J}_x\hat{J}^2] + i[\hat{J}^2\hat{J}_y - \hat{J}_y\hat{J}^2] \tag{276}$$

$$- [\hat{J}^2, \hat{J}_x] + i[\hat{J}^2, \hat{J}_y] \tag{277}$$

$$= 0 + i(0) = 0 \tag{278}$$

Hence

$$[\hat{J}^2, \hat{J}_+] = 0 \tag{279}$$

Similarly

$$[\hat{J}^2, \hat{J}_-] = 0 \tag{280}$$

*ii*) Find the commutator of the following

$$[\hat{J}_+, \hat{J}_z] \tag{281}$$

Let

$$[\hat{J}_+, \hat{J}_z] = [\hat{J}_x + i\hat{J}_y, \hat{J}_z] \tag{282}$$

$$= (\hat{J}_x + i\hat{J}_y)\hat{J}_z - \hat{J}_z(\hat{J}_x + i\hat{J}_y) \tag{283}$$

$$= \hat{J}_x\hat{J}_z + i\hat{J}_y\hat{J}_z - \hat{J}_z\hat{J}_x - \hat{J}_z i\hat{J}_y \tag{284}$$

$$= \hat{J}_x\hat{J}_z - \hat{J}_z\hat{J}_x + i\hat{J}_y\hat{J}_z - i\hat{J}_z\hat{J}_y \tag{285}$$

$$= [\hat{J}_x\hat{J}_z - \hat{J}_z\hat{J}_x] + i[\hat{J}_y\hat{J}_z - \hat{J}_z\hat{J}_y] \tag{286}$$

$$= [\hat{J}_x, \hat{J}_z] + i[\hat{J}_y, \hat{J}_z] \tag{287}$$

$$= -i\hbar\hat{J}_y + i(i\hbar\hat{J}_x) = -i\hbar\hat{J}_y - \hbar\hat{J}_x \tag{288}$$

$$= -\hbar(\hat{J}_x + i\hat{J}_y) = -\hbar\hat{J}_+ \tag{289}$$

$$[\hat{J}_+, \hat{J}_z] = -\hbar\hat{J}_+ \tag{290}$$

Similarly

$$[\hat{J}_-, \hat{J}_z] = \hbar\hat{J}_- \tag{291}$$

*iii*) Find the commutator of the following

$$[\hat{J}_+, \hat{J}_-] \tag{292}$$

Let

$$[\hat{J}_+, \hat{J}_-] = (\hat{J}_x + i\hat{J}_y)(\hat{J}_x - i\hat{J}_y) - (\hat{J}_x - i\hat{J}_y)(\hat{J}_x + i\hat{J}_y) \tag{293}$$

$$= \hat{J}_x\hat{J}_x - i\hat{J}_x\hat{J}_y + i\hat{J}_y\hat{J}_x + \hat{J}_y\hat{J}_y - (\hat{J}_x\hat{J}_x + i\hat{J}_x\hat{J}_y - i\hat{J}_y\hat{J}_x + \hat{J}_y\hat{J}_y) \tag{294}$$

$$= \hat{J}_x\hat{J}_x - i\hat{J}_x\hat{J}_y + i\hat{J}_y\hat{J}_x + \hat{J}_y\hat{J}_y - \hat{J}_x\hat{J}_x - i\hat{J}_x\hat{J}_y + i\hat{J}_y\hat{J}_x - \hat{J}_y\hat{J}_y \tag{295}$$

$$= -i\hat{J}_x\hat{J}_y + i\hat{J}_y\hat{J}_x - i\hat{J}_x\hat{J}_y + i\hat{J}_y\hat{J}_x \tag{296}$$

$$= -i[\hat{J}_x\hat{J}_y - \hat{J}_y\hat{J}_x] + i[\hat{J}_y\hat{J}_x - \hat{J}_x\hat{J}_y] \tag{297}$$

$$= -i\left[\hat{J}_x,\hat{J}_y\right] + i\left[\hat{J}_y,\hat{J}_x\right] \tag{298}$$

$$= -i\left[i\hbar\hat{J}_z\right] + i\left[-i\hbar\hat{J}_z\right] \tag{299}$$

$$= \hbar\hat{J}_z + \hbar\hat{J}_z = 2\hbar\hat{J}_z \tag{300}$$

The commutation relations between angular-momentum and Ladder operators can be mainly divided into three categories as discussed below.

*(a) Ladder operator and total angular momentum commutation:*

$$\left[\hat{J}^2,\hat{J}_+\right] = 0; \quad \left[\hat{J}_+,\hat{J}^2\right] = 0 \tag{301}$$

$$\left[\hat{J}^2,\hat{J}_-\right] = 0; \quad \left[\hat{J}_-,\hat{J}^2\right] = 0 \tag{302}$$

$$\left[\hat{J}_+,\hat{J}_z\right] = -\hbar\hat{J}_+; \quad \left[\hat{J}_z,\hat{J}_+\right] = \hbar\hat{J}_+ \tag{303}$$

$$\left[\hat{J}_-,\hat{J}_z\right] = \hbar\hat{J}_-; \quad \left[\hat{J}_z,\hat{J}_-\right] = -\hbar\hat{J}_- \tag{304}$$

$$\left[\hat{J}_+,\hat{J}_-\right] = 2\hbar\hat{J}_z; \quad \left[\hat{J}_-,\hat{J}_+\right] = -2\hbar\hat{J}_z \tag{305}$$

*(b) Ladder operator and orbital angular momentum commutation:*

$$\left[\hat{L}^2,\hat{L}_+\right] = 0; \quad \left[\hat{L}_+,\hat{L}^2\right] = 0 \tag{306}$$

$$\left[\hat{L}^2,\hat{L}_-\right] = 0; \quad \left[\hat{L}_-,\hat{L}^2\right] = 0 \tag{307}$$

$$\left[\hat{L}_+,\hat{L}_z\right] = -\hbar\hat{L}_+; \quad \left[\hat{L}_z,\hat{L}_+\right] = \hbar\hat{L}_+ \tag{308}$$

$$\left[\hat{L}_-,\hat{L}_z\right] = \hbar\hat{L}_-; \quad \left[\hat{L}_z,\hat{L}_-\right] = -\hbar\hat{L}_- \tag{309}$$

$$\left[\hat{L}_+,\hat{L}_-\right] = 2\hbar\hat{L}_z; \quad \left[\hat{L}_-,\hat{L}_+\right] = -2\hbar\hat{L}_z \tag{310}$$

*(b) Ladder operator and spin angular momentum commutation:*

$$\left[\hat{S}^2,\hat{S}_+\right] = 0; \quad \left[\hat{S}_+,\hat{S}^2\right] = 0 \tag{311}$$

$$\left[\hat{S}^2,\hat{S}_-\right] = 0; \quad \left[\hat{S}_-,\hat{S}^2\right] = 0 \tag{312}$$

$$\left[\hat{S}_+,\hat{S}_z\right] = -\hbar\hat{S}_+; \quad \left[\hat{S}_z,\hat{S}_+\right] = \hbar\hat{S}_+ \tag{313}$$

$$\left[\hat{S}_-,\hat{S}_z\right] = \hbar\hat{S}_-; \quad \left[\hat{S}_z,\hat{S}_-\right] = -\hbar\hat{S}_- \tag{314}$$

$$\left[\hat{S}_+,\hat{S}_-\right] = 2\hbar\hat{S}_z; \quad \left[\hat{S}_-,\hat{S}_+\right] = -2\hbar\hat{S}_z \tag{315}$$

## ❖ Hermitian Operators – Elementary Ideas, Quantum Mechanical Operator for Linear Momentum, Angular Momentum and Energy as Hermitian Operator

It is a quite well-known fact that all the physical properties are actually real quantities, and therefore are bound to have real values. It means that any operator which is used to represent a physical property must yield real values. In this section, we will discuss the elementary idea of Hermitian operators (named in honor of a great mathematician Charles Hermite), and will also prove that many important operators in quantum mechanics like linear momentum, angular momentum and Hamiltonian are Hermitian in nature.

### ➤ *Elementary Idea of Hermitian Operator*

Every physical property must have real eigen or expectation values, which therefore implies that the corresponding operators should have some special characteristics. One of the most important special characteristics includes a feature that the Hermitian conjugate of such an operator should be itself. In other words, if the Hermitian conjugate of an operator is itself, the operator is called as Hermitian; however, if the Hermitian conjugate of an operator is equal to its negative expression, the operator is called as anti-Hermitian or skew-Hermitian. Mathematically, we can say that

$$if\ A^\dagger = A; \qquad A\ is\ Hermitian \tag{316}$$

$$if\ A^\dagger = -A; \qquad A\ is\ anti-Hermitian \tag{317}$$

Where A is an operator whose Hermitian conjugate is represented by $A^\dagger$.

However, the obvious question regarding the aforementioned definition would be "what is a Hermitian conjugate and how is it obtained". The answer is "the operator $A^\dagger$ will be called as the Hermitian conjugate (or adjoint) of operator $A$ if the operation of $A^\dagger$ on the complex conjugate of function $\psi$ gives the same result as when the A is operated over $\psi$". Mathematically, we can say that

$$\langle\psi|A|\psi\rangle = \int_{-\infty}^{+\infty} \psi^*(x)A\psi(x)dx = \langle\psi|A\psi\rangle = \langle A^\dagger\psi|\psi\rangle \tag{318}$$

or

$$\langle A^\dagger\varphi|\psi\rangle = \langle\varphi|A\psi\rangle \tag{319}$$

**1. Hermitian conjugates of different operators:** The Hermitian conjugates of different operators can be studied in three different categories.

*i) Hermitian conjugates of quantum mechanical operators:*

Let $Q$ be any quantum mechanical operator, then by the definition of Hermitian conjugates operator, we have the following condition.

$$\langle \varphi | Q\psi \rangle = \langle Q^{\dagger}\varphi | \psi \rangle \tag{320}$$

If Q is the momentum operator, then we can proceed as discussed below.

$$\int \psi^{*}\hat{p}_{x}\psi\, dx = \int \psi\hat{p}_{x}\psi^{*}\, dx \tag{321}$$

$$\int \psi^{*}\left(\frac{h}{2\pi i}\frac{\partial}{\partial x}\right)\psi\, dx = \int \psi\left(\frac{h}{2\pi i}\frac{\partial}{\partial x}\right)^{\dagger}\psi^{*}\, dx \tag{322}$$

$$\int \psi\left(\frac{h}{2\pi i}\frac{\partial}{\partial x}\right)^{\dagger}\psi^{*}\, dx = \int \psi\left(\frac{h}{2\pi i}\right)^{\dagger}\left(\frac{\partial}{\partial x}\right)^{\dagger}\psi^{*}\, dx \tag{323}$$

$$\int \psi\left(\frac{h}{2\pi i}\frac{\partial}{\partial x}\right)^{\dagger}\psi^{*}\, dx = \int \psi\left(-\frac{h}{2\pi i}\right)\left(-\frac{\partial}{\partial x}\right)\psi^{*}\, dx \tag{324}$$

$$\int \psi\left(\frac{h}{2\pi i}\frac{\partial}{\partial x}\right)^{\dagger}\psi^{*}\, dx = \int \psi\left(\frac{h}{2\pi i}\frac{\partial}{\partial x}\right)\psi^{*}\, dx \tag{325}$$

Therefore, we can say that the Hermitian conjugate of the linear momentum operator is itself, and hence it is a Hermitian operator. Now from the most primitive definition of Hermitian operators, that all operators which correspond to observable quantities, we can say that the Hermitian conjugates of the following operator are themselves.

| Operator | Hermitian conjugate |
|---|---|
| $\hat{x}$ | $\hat{x}$ |
| $\hat{x}^{2}$ | $\hat{x}^{2}$ |
| $\hat{p}_{x}$ | $\hat{p}_{x}$ |
| $\hat{p}_{x}^{2}$ | $\hat{p}_{x}^{2}$ |
| $\hat{T}_{x}$ | $\hat{T}_{x}$ |
| $\hat{V}(x)$ | $\hat{V}(x)$ |
| $\hat{H}$ | $\hat{H}$ |

*ii) Hermitian conjugates of a constant operator:*

There are some operators which are complex numbers. The Hermitian conjugates of such operators are actually their complex conjugates. Let we have the operator $A$

$$\hat{A} = a + ib \tag{326}$$

and since the definition of Hermitian operator is

$$\langle \varphi | A\psi \rangle = \langle A^\dagger \varphi | \psi \rangle \tag{327}$$

gives the integer as

$$\langle \varphi | (a + ib)\psi \rangle = \langle (a - ib)\varphi | \psi \rangle = (a + ib)\langle \varphi | \psi \rangle \tag{328}$$

Hence, the Hermitian conjugates of constant operators are their complex conjugates. The Hermitian conjugates of some operators are given below.

| Operator | Hermitian conjugate |
|---|---|
| $(a + ib)$ | $(a + ib)^\dagger = (a - ib)$ |
| $(+ib)$ | $(+ib)^\dagger = (-ib)$ |
| $\left(+\dfrac{i}{4}\right)$ | $\left(+\dfrac{i}{4}\right)^\dagger = \left(-\dfrac{i}{4}\right)$ |

*iii) Hermitian conjugates of a mathematical operator:*

The Hermitian conjugates of mathematical operators can be obtained by obtaining their respective integrals as discussed below. Let we have a mathematical operator $A$

$$\hat{A} = \frac{d}{dx} \tag{326}$$

We use the following integral to derive the result

$$\left\langle \varphi \left| \frac{d}{dx} \psi \right\rangle = \int\limits_{-\infty}^{+\infty} \varphi^*(x) \frac{d\psi(x)}{dx} dx \tag{327}$$

Integrating the above equation by part, we get

$$\left\langle \varphi \left| \frac{d}{dx} \psi \right\rangle = [\varphi^*(x)\psi(x)] - \int\limits_{-\infty}^{+\infty} \frac{d\varphi^*(x)}{dx} \psi(x)\, dx \tag{328}$$

$$= 0 - \left\langle \frac{d}{dx} \varphi \middle| \psi \right\rangle \tag{329}$$

$$= - \left\langle \frac{d}{dx} \varphi \middle| \psi \right\rangle \tag{330}$$

Hence, the Hermitian conjugate of $d/dx$ operator is $-d/dx$. Similarly, we can prove that the Hermitian conjugate of $d^2/dx^2$ is $d^2/dx^2$.

**2. Properties of Hermitian conjugates:** From the definition and properties of scalar product, adjoints or Hermitian conjugate show the following properties.

*i*) Let $C$ a constant and $A$ as an operator.

$$(CA)^\dagger = C^*A^\dagger \tag{331}$$

For example

$$\left(\frac{i}{4}\frac{\partial}{\partial x}\right)^\dagger = \left(\frac{i}{4}\right)^\dagger \left(\frac{\partial}{\partial x}\right)^\dagger \tag{332}$$

$$\left(\frac{i}{4}\frac{\partial}{\partial x}\right)^\dagger = \left(-\frac{i}{4}\right)\left(-\frac{\partial}{\partial x}\right) \tag{333}$$

$$\left(\frac{i}{4}\frac{\partial}{\partial x}\right)^\dagger = \frac{i}{4}\frac{\partial}{\partial x} \tag{334}$$

*ii*) Let $A$ and $B$ as two operators.

$$(A + B)^\dagger = A^\dagger + B^\dagger \tag{335}$$

For example

$$\left(\frac{\partial}{\partial x} + \frac{\partial^2}{\partial x^2}\right)^\dagger = \left(\frac{\partial}{\partial x}\right)^\dagger + \left(\frac{\partial^2}{\partial x^2}\right)^\dagger \tag{336}$$

$$\left(\frac{\partial}{\partial x} + \frac{\partial^2}{\partial x^2}\right)^\dagger = \left(-\frac{\partial}{\partial x}\right) + \left(\frac{\partial^2}{\partial x^2}\right) \tag{337}$$

$$\left(\frac{\partial}{\partial x} + \frac{\partial^2}{\partial x^2}\right)^\dagger = \left(-\frac{\partial}{\partial x} + \frac{\partial^2}{\partial x^2}\right) \tag{338}$$

*iii*) Let $A$ and $B$ as two operators, then

$$(AB)^\dagger = A^\dagger B^\dagger \tag{339}$$

For example

$$\left(\frac{\partial}{\partial x}\frac{\partial^2}{\partial x^2}\right)^\dagger = \left(\frac{\partial}{\partial x}\right)^\dagger \left(\frac{\partial^2}{\partial x^2}\right)^\dagger \tag{340}$$

$$\left(\frac{\partial}{\partial x}\frac{\partial^2}{\partial x^2}\right)^\dagger = \left(-\frac{\partial}{\partial x}\right)\left(\frac{\partial^2}{\partial x^2}\right) \tag{341}$$

$$\left(\frac{\partial}{\partial x}\frac{\partial^2}{\partial x^2}\right)^{\dagger} = \left(-\frac{\partial^3}{\partial x^3}\right) \tag{342}$$

*iv*) Let $A$ be the operators, then

$$\left(A^{\dagger}\right)^{\dagger} = A \tag{343}$$

For example

$$\left[\left(\frac{\partial}{\partial x}\right)^{\dagger}\right]^{\dagger} = \left(\frac{\partial}{\partial x}\right) \tag{344}$$

It should also be noted that the multiplication to an anti-hermitian operator by $i$ makes it Hermitian, while the vice-versa is also equally true for adjoints.

*v*) For any operator $A$ and its adjoint, the product $(AA^{\dagger})$ is Hermitian. For instance

$$\left(\frac{\partial}{\partial x}\right)\left(-\frac{\partial}{\partial x}\right) = -\frac{\partial^2}{\partial x^2} \tag{343}$$

*vi*) For any operator $A$ and its adjoint, the sum $(A+A^{\dagger})$ is Hermitian. For instance

$$\left(x + x^{\dagger}\right) = 2x \tag{343}$$

*vii*) For any operator $A$ and its adjoint, then $AA^{\dagger}+A^{\dagger}A$ is Hermitian. For instance

$$(i3)(i3)^{\dagger} + (i3)^{\dagger}(i3) = (i3)(-i3) + (-i3)(i3) = 9 + 9 = 18 \tag{343}$$

**3. Characterization of Hermitian operator:** We know that the average value of any operator (say $\hat{A}$) in quantum mechanics is calculated by the equation given below.

$$\bar{A} = \int \psi^* \hat{A} \psi\, dx \tag{344}$$

Where $\psi$ is the wave function representing any quantum mechanical state and $\psi^*$ is its complexes conjugate. Now because of the fact that the average value of any physical observable must be a real value, we can say that the operator used in equation (344) must follow the following condition.

$$\bar{A} = \bar{A}^* \tag{345}$$

$$\int \psi^* \hat{A} \psi\, dx = \left[\int \psi^* \hat{A} \psi\, dx\right]^* \tag{346}$$

$$\int \psi^* \hat{A} \psi\, dx = \int (\psi^*)^* (\hat{A}\psi)^*\, dx \tag{347}$$

$$\int \psi^* \hat{A}\psi\, dx = \int \psi(\hat{A}\psi)^* dx \tag{348}$$

Every linear operator that satisfies the equation (348) for all quantum-mechanically acceptable wave functions is called the Hermitian operator.

Besides the form given by equation (348), one more popular definition of a Hermitian operator is also given below.

$$\int f^* \hat{A} g\, dx = \int g(\hat{A}f)^* dx \tag{349}$$

From the equation, we can state that a Hermitian operator must fulfill the condition for the well-behaved functions $f$ and $g$. It can be clearly seen that on the left side of the equation (349), $\hat{A}$ is operated over the function $g$; while on the right side, the $\hat{A}$ is operated over the function $f$. However, if we put $f = g$, the equation (349) is also reduced to equation (348); indicating that both definitions are correct.

**4. Properties of Hermitian operators:** The important properties of Hermitian operators are discussed below.

*i) The eigenvalues of Hermitian operators are always real:*

Let $\hat{A}$ be a Hermitian operator with a well-behaved wavefunction $\psi$ representing a quantum mechanical state, then we can say that

$$\hat{A}\psi = a\psi \tag{350}$$

Each side of equation (350) can be expressed as an imaginary and a real part as well; with left-hand real part equal to the right-hand real part, while left side imaginary part equal to right imaginary one. After taking the complex conjugate of equation (350), the imaginary parts would reverse sign but still holding the condition of equivalence.

$$\hat{A}^*\psi^* = a^*\psi^* \tag{351}$$

Multiplying the equation (350) by $\psi^*$ and integrating over the whole configurational space, we get

$$\int \psi^* \hat{A}\psi\, dx = a \int \psi^*\psi\, dx \tag{352}$$

Similarly, multiplying the equation (351) by $\psi$ and integrating over the whole configurational space, we get

$$\int \psi \hat{A}^*\psi^*\, dx = a^* \int \psi\psi^*\, dx \tag{353}$$

Now because left-hand sides of equation (352) and (353) are equal to each other (owing to the Hermitian nature of the operator), the right-hand sides are also equivalent; therefore, we can say that

$$a^* \int \psi\psi^*\, dx = a \int \psi^*\psi\, dx \tag{354}$$

$$0 = (a - a^*) \int \psi^* \psi \, dx \tag{355}$$

Since the wave function is a square-integrable, the integral part of the equation (355) cannot be zero and left us with the only possibility given blow.

$$(a - a^*) = 0 \tag{356}$$

$$a = a^* \tag{357}$$

The physical interpretation of the result given by equation (357) is that $a$ must be real in order to yield zero from equation (356).

*ii) Non-degenerate eigenfunctions of Hermitian operators are always orthogonal to each other:*

Let $\psi_m$ and $\psi_n$ be two square-integrable eigenfunctions of a Hermitian operator $\hat{A}$; therefore, we say

$$\hat{A}\psi_m = a_1 \psi_m \tag{358}$$

also

$$\hat{A}^* \psi_n^* = a_2 \psi_n^* \tag{359}$$

Multiplying the equation (358) by $\psi_n^*$ and integrating over the whole configurational space, we get

$$\int \psi_n^* \hat{A} \psi_m \, dx = a_1 \int \psi_n^* \psi_m \, dx \tag{360}$$

Similarly, multiplying the equation (359) by $\psi_m$ and integrating over the whole configurational space, we get

$$\int \psi_m \hat{A}^* \psi_n^* \, dx = a_2 \int \psi_m \psi_n^* \, dx \tag{361}$$

Now because left-hand sides of equation (360) and (361) are equal to each other (owing to the Hermitian nature of the operator), the right-hand sides are also equivalent; therefore, we can say that

$$a_1 \int \psi_n^* \psi_m \, dx = a_2 \int \psi_m \psi_n^* \, dx \tag{362}$$

$$(a_1 - a_2) \int \psi_m \psi_n^* \, dx = 0 \tag{363}$$

Since the wave functions used are non-degenerate i.e. $a_1 \neq a_2$; the only possibility we are left with for the equation to be true is given below.

$$\int \psi_m \psi_n^* \, dx = 0 \tag{364}$$

Hence, we can say that $\psi_m$ and $\psi_n$ are definitely orthogonal to each other.

*iii) If two Hermitian operators commute, their product is also a Hermitian operator:*

Let $\psi_1$ and $\psi_2$ be two well-behaved functions; while $\hat{A}$ and $\hat{B}$ as two Hermitian operators. Therefore, we can say that

$$\int \psi_1^* \hat{A}\hat{B}\psi_2\, dx \tag{365}$$

Since $\hat{A}$ is Hermitian, we can say that

$$\int \psi_1^* \hat{A}\hat{B}\psi_2\, dx = \int \psi_1^* \hat{A}\left(\hat{B}\psi_2\right) dx \tag{366}$$

$$\int \hat{A}^* \psi_1^* \hat{B}\psi_2\, dx = \int \psi_1^* \hat{A}\left(\hat{B}\psi_2\right) dx \tag{367}$$

Since $\hat{B}$ is also Hermitian, therefore

$$\int \left(\hat{A}^*\psi_1^*\right)\hat{B}\psi_2\, dx = \int \hat{B}^*\hat{A}^*\psi_1^*\psi_2\, dx \tag{368}$$

From equation (366) and (368), we get

$$\int \psi_1^* \hat{A}\hat{B}\psi_2\, dx = \int \hat{B}^*\hat{A}^*\psi_1^*\psi_2\, dx \tag{369}$$

If the operator $\hat{A}$ and $\hat{B}$ commute with each other, we have

$$\hat{A}\hat{B} = \hat{B}\hat{A} \quad or \quad \hat{A}^*\hat{B}^* = \hat{B}^*\hat{A}^* \tag{370}$$

Therefore, equation (369) becomes

$$\int \psi_1^* \hat{A}\hat{B}\psi_2\, dx = \int \hat{A}^*\hat{B}^*\psi_1^*\psi_2\, dx \tag{371}$$

Which is the condition for the product operator to act as Hermitian.

*iv) If two Hermitian operators do not commute, their commutator operator is anti-Hermitian in nature:*

Let $\hat{A}$ and $\hat{B}$ as two Hermitian operators; therefore, we can say that their commutation must follow the following condition.

$$\left[\hat{A},\hat{B}\right]^* = \left(\hat{A}\hat{B}\right)^* - \left(\hat{B}\hat{A}\right)^* \tag{372}$$

or

$$\left[\hat{A},\hat{B}\right]^* = \hat{A}^*\hat{B}^* - \hat{B}^*\hat{A}^* = -\left(\hat{B}^*\hat{A}^* - \hat{A}^*\hat{B}^*\right) \tag{373}$$

or

$$[\hat{A}, \hat{B}]^* = -[\hat{B}, \hat{A}]^* \qquad (374)$$

$$[\hat{A}, \hat{B}]^* = -[\hat{B}, \hat{A}]^* \qquad (375)$$

For instance, consider the commutator of position and momentum operator

$$[\hat{x}, \hat{p}_x] = i\frac{h}{2\pi} \qquad (376)$$

The commutator $i\hbar$ is antihermitian in nature.

> #### The Linear Momentum Operator as Hermitian

In order to prove the linear momentum operator as the Hermitian, we must find its Hermitian conjugate first. The general expression of linear momentum operator is

$$\hat{p}_x = \frac{h}{2\pi i}\frac{\partial}{\partial x} \qquad (377)$$

Let $\hat{p}_x^{\dagger}$ be the Hermitian conjugate which can be calculated as follows:

$$\hat{p}_x^{\dagger} = \left(\frac{h}{2\pi i}\right)^{\dagger}\left(\frac{\partial}{\partial x}\right)^{\dagger} \qquad (378)$$

or

$$\hat{p}_x^{\dagger} = \left(-\frac{h}{2\pi i}\right)\left(-\frac{\partial}{\partial x}\right) \qquad (379)$$

or

$$\hat{p}_x^{\dagger} = \left(\frac{h}{2\pi i}\right)\left(\frac{\partial}{\partial x}\right) \qquad (380)$$

Comparing equation (377) and (380), we can see that the Hermitian conjugate of linear momentum operator is exactly equal to the linear momentum operator i.e. $\hat{p}_x^{\dagger} = \hat{p}_x$; proving that it is defiantly a Hermitian operator.

> #### The Angular Momentum Operator as Hermitian

In order to prove the angular momentum operator as Hermitian, we must find its Hermitian conjugate first. The general expression of the angular momentum operator is

$$\hat{L} = \frac{h}{2\pi i}\left[\left(y\frac{\partial}{\partial z} - z\frac{\partial}{\partial y}\right) + \left(z\frac{\partial}{\partial x} - x\frac{\partial}{\partial z}\right) + \left(x\frac{\partial}{\partial y} - y\frac{\partial}{\partial x}\right)\right] \qquad (381)$$

Let $\hat{L}_x^{\dagger}$ be the Hermitian conjugate which can be calculated as follows:

$$\hat{L}_x^{\dagger} = \left[ \frac{h}{2\pi i} \left[ \left( y\frac{\partial}{\partial z} - z\frac{\partial}{\partial y} \right) + \left( z\frac{\partial}{\partial x} - x\frac{\partial}{\partial z} \right) + \left( x\frac{\partial}{\partial y} - y\frac{\partial}{\partial x} \right) \right] \right]^{\dagger} \tag{382}$$

$$= \left[ \frac{h}{2\pi i} y\frac{\partial}{\partial z} - \frac{h}{2\pi i} z\frac{\partial}{\partial y} + \frac{h}{2\pi i} z\frac{\partial}{\partial x} - \frac{h}{2\pi i} x\frac{\partial}{\partial z} + \frac{h}{2\pi i} x\frac{\partial}{\partial y} - \frac{h}{2\pi i} y\frac{\partial}{\partial x} \right]^{\dagger}$$

or

$$\hat{L}_x^{\dagger} = \left( \frac{h}{2\pi i} \right)^{\dagger} (y)^{\dagger} \left( \frac{\partial}{\partial z} \right)^{\dagger} - \left( \frac{h}{2\pi i} \right)^{\dagger} (z)^{\dagger} \left( \frac{\partial}{\partial y} \right)^{\dagger} + \left( \frac{h}{2\pi i} \right)^{\dagger} (z)^{\dagger} \left( \frac{\partial}{\partial x} \right)^{\dagger} \tag{383}$$

$$- \left( \frac{h}{2\pi i} \right)^{\dagger} (x)^{\dagger} \left( \frac{\partial}{\partial z} \right)^{\dagger} + \left( \frac{h}{2\pi i} \right)^{\dagger} (x)^{\dagger} \left( \frac{\partial}{\partial y} \right)^{\dagger} - \left( \frac{h}{2\pi i} \right)^{\dagger} (y)^{\dagger} \left( \frac{\partial}{\partial x} \right)^{\dagger}$$

or

$$\hat{L}_x^{\dagger} = \left( -\frac{h}{2\pi i} \right) (y) \left( -\frac{\partial}{\partial z} \right) - \left( -\frac{h}{2\pi i} \right) (z) \left( -\frac{\partial}{\partial y} \right) + \left( -\frac{h}{2\pi i} \right) (z) \left( -\frac{\partial}{\partial x} \right) \tag{384}$$

$$- \left( -\frac{h}{2\pi i} \right) (x) \left( -\frac{\partial}{\partial z} \right) + \left( -\frac{h}{2\pi i} \right) (x) \left( -\frac{\partial}{\partial y} \right)$$

$$- \left( -\frac{h}{2\pi i} \right) (y) \left( -\frac{\partial}{\partial x} \right)$$

or

$$\hat{L}^{\dagger} = \frac{h}{2\pi i} y\frac{\partial}{\partial z} - \frac{h}{2\pi i} z\frac{\partial}{\partial y} + \frac{h}{2\pi i} z\frac{\partial}{\partial x} - \frac{h}{2\pi i} x\frac{\partial}{\partial z} + \frac{h}{2\pi i} x\frac{\partial}{\partial y} - \frac{h}{2\pi i} y\frac{\partial}{\partial x} \tag{385}$$

$$\hat{L}^{\dagger} = \frac{h}{2\pi i} \left[ \left( y\frac{\partial}{\partial z} - z\frac{\partial}{\partial y} \right) + \left( z\frac{\partial}{\partial x} - x\frac{\partial}{\partial z} \right) + \left( x\frac{\partial}{\partial y} - y\frac{\partial}{\partial x} \right) \right] \tag{386}$$

Comparing equation (381) and (386), we can see that the Hermitian conjugate of the angular momentum operator is exactly equal to the angular momentum operator i.e. $L^{\dagger} = L$; proving that it is defiantly a Hermitian operator.

> ➤ *The Hamiltonian or Energy Operator as Hermitian*

In order to prove the energy operator as Hermitian, we must find its Hermitian conjugate first. The general expression of the energy operator is

$$\hat{H} = \frac{-h^2}{8\pi^2 m}\frac{\partial^2}{\partial x^2} + V(x) \tag{387}$$

Let $\hat{H}^{\dagger}$ be the Hermitian conjugate which can be calculated as follows:

$$\hat{H}^{\dagger} = \left[\frac{-h^2}{8\pi^2 m}\frac{\partial^2}{\partial x^2} + V(x)\right]^{\dagger} \tag{388}$$

or

$$\hat{H}^{\dagger} = \left[\frac{-h^2}{8\pi^2 m}\frac{\partial}{\partial x}\frac{\partial}{\partial x} + V(x)\right]^{\dagger} \tag{389}$$

$$\hat{H}^{\dagger} = \left(\frac{-h^2}{8\pi^2 m}\right)^{\dagger}\left(\frac{\partial}{\partial x}\right)^{\dagger}\left(\frac{\partial}{\partial x}\right)^{\dagger} + (V(x))^{\dagger} \tag{390}$$

or

$$\hat{H}^{\dagger} = \left(\frac{-h^2}{8\pi^2 m}\right)\left(-\frac{\partial}{\partial x}\right)\left(-\frac{\partial}{\partial x}\right) + (V(x)) \tag{391}$$

$$\hat{H}^{\dagger} = \frac{-h^2}{8\pi^2 m}\frac{\partial^2}{\partial x^2} + V(x) \tag{392}$$

Comparing equation (387) and (392), we can see that the Hermitian conjugate of energy operator is exactly equal to the energy operator i.e. $\hat{H}^{\dagger} = \hat{H}$; proving that it is defiantly a Hermitian operator.

## ❖ The Average Value of the Square of Hermitian Operators

The expectation value of the square of every Hermitian operator is always positive. In other words, we can say that if $A$ is a Hermitian operator, then

$$\langle A^2 \rangle > 0 \tag{393}$$

This can be proved by taking a well-behaved function ψ as discussed below.

$$\langle A^2 \rangle = \frac{\int \psi^* A^2 \,\psi d\tau}{\int \psi^* \psi d\tau} \tag{394}$$

The right-hand side of equation (394) will be positive only if the numerator as well as denominator, both are either positive or negative. Since the wave-function is well-behaved (normalized), the value of denominator is

$$\int \psi^* \,\psi d\tau = 1 \tag{395}$$

Since the denominator is positive, the numerator must also be positive. Now owing to the Hermitian nature of operator $A$, we can evaluate the numerator as given below.

$$\int \psi^* A^2 \, \psi d\tau = \int \psi^* A A^* \psi \, d\tau \tag{396}$$

$$= \int (\psi^* A^*) A \psi \, d\tau \tag{397}$$

or

$$= \int |A\psi|^2 \, d\tau \tag{398}$$

Hence, the value of numerator given by equation (398) is greater than zero i.e. positive, making the average value of the square of the Hermitian operator ($A$) also positive.

## ❖ Commuting Operators and Uncertainty Principle (*x* & *p*; *E* & *t*)

One of the most important properties of operator multiplication is the commutation relation or the commutation rule. Two operators, A and B, are said to be commutating or non-commutating depending upon the magnitude of their commutator.

$$[\hat{A},\hat{B}] = \hat{A}\hat{B} - \hat{B}\hat{A} = 0 \; \rightarrow Commutating \tag{399}$$

and

$$[\hat{A},\hat{B}] = \hat{A}\hat{B} - \hat{B}\hat{A} \neq 0 \; \rightarrow Non\text{-}commutating \tag{400}$$

The physical significance of the commutation relations implies in the fact that when two operators commute, they possess simultaneous set of eigenfunctions; and their respective physical properties can be evaluated simultaneously and accurately. However, if the commutator is non-zero, the respective physical properties cannot be obtained simultaneously and accurately; which is actually the popular uncertainty principal. Two of the most common uncertainty systems; position-momentum and energy-time; can also be proved from commutation relations.

> ### Position-Momentum Uncertainty (*x* & *p*)

The position-momentum uncertainty can be justified only if the commutation of their operators is non-zero. Therefore, we need to find the following.

$$[\hat{x}, \hat{p}_x] \tag{401}$$

Let it be operated over a function $\psi$. We have

$$[\hat{x}, \hat{p}_x]\psi = \hat{x}\,\hat{p}_x\,\psi - \hat{p}_x\,\hat{x}\,\psi \tag{402}$$

or

$$[\hat{x}, \hat{p}_x]\psi = x\frac{h}{2\pi i}\frac{\partial}{\partial x}\psi - \frac{h}{2\pi i}\frac{\partial}{\partial x}x\,\psi \tag{403}$$

$$[\hat{x}, \hat{p}_x]\psi = \frac{h}{2\pi i}x\frac{\partial\psi}{\partial x} - \frac{h}{2\pi i}x\frac{\partial\psi}{\partial x} - \frac{h}{2\pi i}\psi\frac{\partial x}{\partial x} \tag{404}$$

$$[\hat{x}, \hat{p}_x]\psi = -\frac{h}{2\pi i}\psi \tag{405}$$

$$[\hat{x}, \hat{p}_x] = -\frac{h}{2\pi i} = \frac{hi}{2\pi} = i\hbar \tag{406}$$

Equation (406) proves that we cannot determine the position and momentum of a particle along one axis simultaneously and accurately.

> ➤ *Energy-Time Uncertainty (E & t)*

The energy-time uncertainty can be justified only if the commutation of their operators is non-zero. Therefore, we need to find the following.

$$\left[\hat{t}, \hat{E}\right] \tag{407}$$

Let it be operated over a function ψ(*t*). We have

$$\left[\hat{t}, \hat{E}\right]\psi = \hat{t}\,\hat{E}\,\psi - \hat{E}\hat{t}\,\psi \tag{408}$$

or

$$\left[\hat{t}, \hat{E}\right]\psi = t\frac{h}{2\pi i}\frac{\partial}{\partial t}\psi - \frac{h}{2\pi i}\frac{\partial}{\partial t}t\,\psi \tag{409}$$

$$\left[\hat{t}, \hat{E}\right]\psi = \frac{h}{2\pi i}t\frac{\partial\psi}{\partial t} - \frac{h}{2\pi i}t\frac{\partial\psi}{\partial t} - \frac{h}{2\pi i}\psi\frac{\partial t}{\partial t} \tag{410}$$

$$\left[\hat{t}, \hat{E}\right]\psi = -\frac{h}{2\pi i}\psi \tag{411}$$

$$\left[\hat{t}, \hat{E}\right] = -\frac{h}{2\pi i} \tag{412}$$

$$\left[\hat{t}, \hat{E}\right] = \frac{hi}{2\pi} \tag{413}$$

$$\left[\hat{t}, \hat{E}\right] = i\hbar \tag{414}$$

The equation (412) proves that higher the lifetime of the state lower will be energy fluctuation i.e. uncertainty ΔE, and the vice-versa is also true.

## ❖ Schrodinger Wave Equation for a Particle in One Dimensional Box

In the first section of this chapter, we discussed the postulates of quantum mechanics i.e. the step-by-step procedure to solve a quantum mechanical problem. Now it's the time to implement those rules to the simplest quantum mechanical problem i.e. particle in a one-dimensional box. Consider a particle trapped in a one-dimensional box of length "$a$", which means that this particle can travel in only one direction only, say along $x$-axis. The potential inside the box is V, while outside to the box it is infinite.

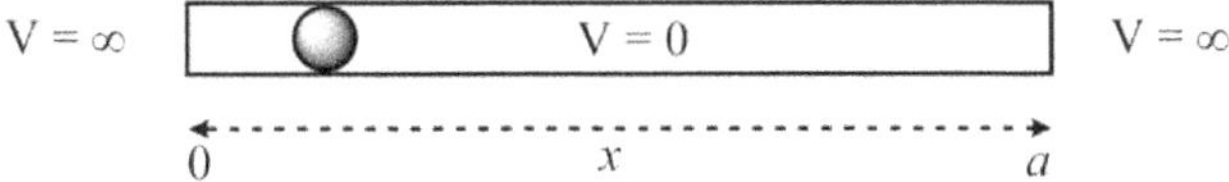

Figure 7. The particle in a one-dimensional box.

One other popular depiction of the particle in a one-dimensional box is also given in which the potential is shown vertically while the displacement is projected along the horizontal line.

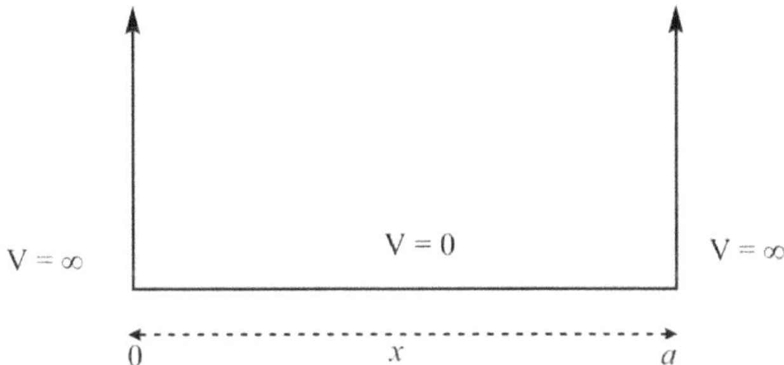

Figure 8. The second representation particle in a one-dimensional box.

So far we have considered a quantum mechanical system of a particle trapped in a one-dimensional box. Now suppose that we need to find various physical properties associated with different states of this system. Had it been a classical system, we would use simple formulas from classical mechanics to determine the value of different physical properties. However, being a quantum mechanical system, we cannot use those expressions because they would give irrational results. Therefore, we need to use the postulates of quantum mechanics to evaluate various physical properties.

Let $\psi$ be the function that describes all the states of the particle in a one-dimensional box. At this point we have no information about the exact mathematical expression of $\psi$; nevertheless, we know that there is one operator that does not need the absolute expression of wave function but uses the symbolic form only, the Hamiltonian operator. The operation of Hamiltonian operator over this symbolic form can be rearranged to give to construct the Schrodinger wave equation; and we all know that the wave function as well the energy, both are the obtained as this second-order differential equation is solved. Mathematically, we can say that



**D** **DALAL**
**INSTITUTE**

$$\hat{H}\psi = E\psi \tag{415}$$

After putting the value of one-dimensional Hamiltonian in equation (415), we get

$$\left[\frac{-h^2}{8\pi^2 m}\frac{\partial^2}{\partial x^2} + V\right]\psi = E\psi \tag{416}$$

or

$$\frac{-h^2}{8\pi^2 m}\frac{\partial^2 \psi}{\partial x^2} + V\psi = E\psi \tag{417}$$

$$\frac{-h^2}{8\pi^2 m}\frac{\partial^2 \psi}{\partial x^2} + V\psi - E\psi = 0 \tag{418}$$

$$\frac{\partial^2 \psi}{\partial x^2} + \frac{8\pi^2 m}{h^2}E\psi - \frac{8\pi^2 m}{h^2}V\psi = 0 \tag{419}$$

or

$$\frac{\partial^2 \psi}{\partial x^2} + \frac{8\pi^2 m}{h^2}(E - V)\psi = 0 \tag{420}$$

The above-mentioned second order differential equation is the Schrodinger wave equation for a particle moving along one dimension only. Since the conditions outside and inside the box are different, the equation (420) must be solved separately for both cases.

**1. The solution of Schrodinger wave equation for outside the box:** After putting the value of potential outside the box in equation (420) i.e. V = ∞, we get

$$\frac{\partial^2 \psi}{\partial x^2} + \frac{8\pi^2 m}{h^2}(E - \infty)\psi = 0 \tag{421}$$

Since $E$ is negligible in comparison to the ∞, the above equation becomes

$$\frac{\partial^2 \psi}{\partial x^2} - \infty\psi = 0 \tag{422}$$

$$\infty\psi = \frac{\partial^2 \psi}{\partial x^2} \tag{423}$$

$$\psi = \frac{1}{\infty}\frac{\partial^2 \psi}{\partial x^2} = 0 \tag{424}$$

The physical significance of the equation (424) is that the particle cannot go outside the box, and is always reflected back when it strikes the boundaries. In other words, as the function describing the existence of particles is zero outside the box, the particle cannot exist outside the box.

**2. Solution of Schrodinger wave equation for inside the box:** After putting the value of potential inside the box in equation (420) i.e. V = 0, we get

$$\frac{\partial^2 \psi}{\partial x^2} + \frac{8\pi^2 m}{h^2}(E - 0)\psi = 0 \tag{425}$$

or

$$\frac{\partial^2 \psi}{\partial x^2} + \frac{8\pi^2 mE}{h^2}\psi = 0 \tag{426}$$

Now consider

$$k^2 = \frac{8\pi^2 mE}{h^2} \tag{427}$$

After using the value from equation (427) in equation (426), we get

$$\frac{\partial^2 \psi}{\partial x^2} + k^2\psi = 0 \tag{428}$$

The general solution of the above equation is

$$\psi = A\,Sin\,kx + B\,Cos\,kx \tag{429}$$

Hence, from just the symbolic form we have obtained some kind of expression for the wave function defining quantum mechanical states. However, the function given by equation (429) cannot be used to find different physical properties or the nature of corresponding quantum mechanical states. The reason is that this expression does have some unknown parameters like $A$, $B$ and $k$. Since the function describing any quantum mechanical state must be single-valued, finite and continuous; the function $\psi$ must also follow these conditions to become a "wave-function". Therefore, these boundary conditions are fulfilled only if the magnitude of $\psi$ is zero at the start and at the end of the box (function outside is zero).

*i) The first boundary condition:* $\psi$ must vanish when $x = 0$ i.e.

$$0 = A\,Sin\,k(0) + B\,Cos\,k(0) \tag{430}$$

$$0 = 0 + B\,Cos\,k(0) \tag{431}$$

$$B = 0 \tag{432}$$

So, the function $\psi$ is acceptable only if the value of the constant $B$ is zero. After putting the value of $B$ in equation (429), we get

$$\psi = A\,Sin\,kx + (0)\,Cos\,kx \tag{433}$$

$$\psi = A\,Sin\,kx \tag{434}$$

*ii) The second boundary condition:* ψ must vanish when $x = a$, i.e.,

$$0 = A\ Sin\ ka \tag{435}$$

$$Sin\ ka = 0 \tag{436}$$

Moreover, as we know that

$$Sin\ 0 = 0 \qquad or \qquad Sin\ 0\pi = 0 \tag{437}$$

$$Sin\ 180 = 0 \qquad or \qquad Sin\ 1\pi = 0 \tag{438}$$

$$Sin\ 360 = 0 \qquad or \qquad Sin\ 2\pi = 0 \tag{439}$$

$$Sin\ 540 = 0 \qquad or \qquad Sin\ 3\pi = 0 \tag{440}$$

or

$$Sin\ n\pi = 0 \tag{441}$$

Where $n = 0, 1, 2, 3, 4, 5 \dots \infty$. Comparing equation (436) and equation (441), we conclude that

$$Sin\ ka = Sin\ n\pi = 0 \tag{442}$$

Which eventually means that

$$ka = n\pi \tag{443}$$

$$k = \frac{n\pi}{a} \tag{444}$$

After putting the value of $k$ in equation (434), we get

$$\psi = A\ Sin\ \frac{n\pi x}{a} \tag{445}$$

The only parameters that is still unknown in equation (445) is $A$, which can also be obtained by the condition of normalization i.e. the function must define the state completely. Therefore, we can say that

$$\int_0^a \psi^2 = A^2 \int_0^a Sin^2\left(\frac{n\pi x}{a}\right) = 1 \tag{446}$$

$$A^2 \cdot \frac{a}{2} = 1 \tag{447}$$

$$A^2 = \frac{2}{a} \quad or \quad A = \sqrt{\frac{2}{a}} \tag{448}$$

After putting the value of $A$ in equation (445), we get

$$\psi = \sqrt{\frac{2}{a}}\ Sin\frac{n\pi x}{a} \tag{449}$$

Since the function ψ also depends upon the discrete variable *n*, it is better to write the above equation given as

$$\psi_n = \sqrt{\frac{2}{a}}\ Sin\frac{n\pi x}{a} \tag{450}$$

The equation (450) represents all the quantum mechanical states of a particle in one-dimensional box. We can obtain functions for individual states just by putting different values of "*n*" allowed by the boundary conditions.

For first quantum mechanical state i.e *n* = 1

$$\psi_1 = \sqrt{\frac{2}{a}}\ Sin\frac{\pi x}{a} \tag{451}$$

For second quantum mechanical state i.e *n* = 2

$$\psi_2 = \sqrt{\frac{2}{a}}\ Sin\frac{2\pi x}{a} \tag{452}$$

For third quantum mechanical state i.e *n* = 3

$$\psi_3 = \sqrt{\frac{2}{a}}\ Sin\frac{3\pi x}{a} \tag{453}$$

Similarly, we can write the expression for $\psi_4$, $\psi_5$, $\psi_6$ and so on. It is also worthy to note that even though the *n* = 0 is permitted by the boundary condition, we still didn't use it in equation (450); which is obviously because it makes the whole function to collapse to zero.

One of the most remarkable results of this procedure that we have not discussed yet is the correlation of equation (427) and equation (444).

$$k^2 = \frac{8\pi^2 mE}{h^2} = \frac{n^2\pi^2}{a^2} \tag{454}$$

$$E_n = \frac{n^2 h^2}{8ma^2} \tag{455}$$

The energy of different quantum mechanical states can be obtained by putting n = 1, 2, 3.... ∞ in equation (455). Hence, we have obtained the wave-function as well as the energy for a particle in one-dimensional box.

## ❖ Evaluation of Average Position, Average Momentum and Determination of Uncertainty in Position and Momentum and Hence Heisenberg's Uncertainty Principle

The third postulate of quantum mechanics states that when the wave-function of a particular quantum mechanical state is multiplied by the operator of an observable quantity, we get a real value multiplied by the wave function itself. However, the value obtained this way can be constant or variable. Mathematically, the constant value of the observable quantity can be reported directly, and the function is called an eigenfunction of the operator under consideration. If the value of the physical property obtained after multiplying the wave function by the corresponding operator is variable i.e. non-eigen, the value can be reported only after averaging it over the whole configurational space.

$$< a > = \frac{\oint \psi^* \hat{O} \psi \, d\tau}{\oint \psi^* \psi \, d\tau} \tag{456}$$

Since the wave function $\psi$ is normalized, the denominator becomes unity; therefore, equation (456) is reduced to the following

$$< a > = \oint \psi^* \hat{O} \psi \, d\tau \tag{457}$$

Since the operation by the Hamiltonian over the symbolic form has already given the absolute expressions for different quantum mechanical states, now we can operate other operators to evaluate their average values. In this section, we will determine the average values of position, position-squared, momentum and momentum-squared; which in turn will be used to prove the Heisenberg's uncertainty finally.

> ### Evaluation of Average Position

The quantum mechanical operator for the position of a particle in one-dimensional is $\hat{x}$; while the general form of wave function is

$$\psi_n = \sqrt{\frac{2}{a}} \, Sin \frac{n\pi x}{a} \tag{458}$$

Using this in equation (457), we get

$$< x > = \oint \psi^* x \, \psi \, d\tau \tag{459}$$

or

$$< x > = \oint x \, \psi^2 \, dx \tag{460}$$

---

$$< x > = \int_0^a x. \frac{2}{a} Sin^2 \left(\frac{n\pi x}{a}\right) dx = \frac{2}{a} \int_0^a x \, Sin^2 \left(\frac{n\pi x}{a}\right) dx \tag{461}$$

$$= \frac{2}{a} \int_0^a x \left[\frac{1 - Cos \left(\frac{2n\pi x}{a}\right)}{2}\right] dx \tag{462}$$

$$= \frac{1}{a} \int_0^a \left(x - x \, Cos \frac{2n\pi x}{a}\right) dx \tag{463}$$

$$= \frac{1}{a} \left[\int_0^a x \, dx - \int_0^a x \, Cos \left(\frac{2n\pi x}{a}\right) dx\right] \tag{464}$$

$$= \frac{1}{a} \left[\frac{a^2}{2} - 0\right] = \frac{a}{2} \tag{465}$$

> ### *Evaluation of Average Position-Squared*

The quantum mechanical operator for the position-squared of a particle in one-dimensional is $\hat{x}^2$; Using this in equation (457), we get

$$< x^2 > = \oint \psi^* \, x^2 \, \psi \, dx \tag{466}$$

$$< x^2 > = \int_0^a x^2. \frac{2}{a} Sin^2 \left(\frac{n\pi x}{a}\right) dx = \frac{2}{a} \int_0^a x^2 \, Sin^2 \left(\frac{n\pi x}{a}\right) dx \tag{467}$$

$$= \frac{2}{a} \int_0^a x^2 \left[\frac{1 - Cos \left(\frac{2n\pi x}{a}\right)}{2}\right] dx \tag{468}$$

$$= \frac{2}{a} \left[\frac{a^3}{6} - \frac{a^3}{4n^2\pi^2}\right] - \frac{1}{a} \left[\frac{a^3}{3} - \frac{a^3}{2n^2\pi^3}\right] \tag{469}$$

$$= \frac{a^2}{3} - \frac{a^2}{2n^2\pi^2} \tag{470}$$

> ### *Evaluation of Average Momentum*

The quantum mechanical operator for the position-squared of a particle in one-dimensional is $\hat{p}_x$; Using this in equation (457), we get

$$< \hat{p}_x > = \oint \psi^* \frac{h}{2\pi i} \frac{\partial}{\partial x} \psi \, dx \tag{471}$$

$$< \hat{p}_x > = \int_0^a \left[ \sqrt{\frac{2}{a}} Sin\left(\frac{n\pi x}{a}\right) \right] \frac{h}{2\pi i} \frac{\partial}{\partial x} \left[ \sqrt{\frac{2}{a}} Sin\left(\frac{n\pi x}{a}\right) \right] dx \tag{472}$$

$$= \frac{h}{2\pi i} \left[\frac{2}{a}\right] \int_0^a Sin\left(\frac{n\pi x}{a}\right) \left(\frac{n\pi}{a}\right) Cos\left(\frac{n\pi x}{a}\right) dx \tag{473}$$

$$= \frac{h}{2\pi i} \left[\frac{2}{a}\right] \left(\frac{n\pi}{a}\right) \int_0^a Sin\left(\frac{n\pi x}{a}\right) Cos\left(\frac{n\pi x}{a}\right) dx \tag{474}$$

$$< \hat{p}_x > = 0 \tag{475}$$

> ### Evaluation of Average Momentum-Squared

The quantum mechanical operator for the position-squared of particle in one-dimensional is $\hat{p}_x^2$; Using this in equation (457), we get

$$< \hat{p}_x^2 > = \oint \psi^* \left( -\frac{h^2}{4\pi^2} \frac{\partial^2}{\partial x^2} \right) \psi \, dx \tag{476}$$

$$< \hat{p}_x^2 > = \int_0^a \left[ \sqrt{\frac{2}{a}} Sin\left(\frac{n\pi x}{a}\right) \right] \left( -\frac{h^2}{4\pi^2} \frac{\partial^2}{\partial x^2} \right) \left[ \sqrt{\frac{2}{a}} Sin\left(\frac{n\pi x}{a}\right) \right] dx \tag{477}$$

$$= -\frac{h^2}{4\pi^2} \left(\frac{2}{a}\right) \int_0^a Sin\left(\frac{n\pi x}{a}\right) \left[ (-)\left(\frac{n\pi}{a}\right)^2 Sin\left(\frac{n\pi x}{a}\right) \right] dx \tag{478}$$

$$= \frac{h^2}{4\pi^2} \left(\frac{2}{a}\right) \left(\frac{n\pi}{a}\right)^2 \int_0^a Sin^2\left(\frac{n\pi x}{a}\right) dx \tag{479}$$

$$= \frac{n^2 h^2}{2a^3} \int_0^a Sin^2\left(\frac{n\pi x}{a}\right) dx \tag{480}$$

$$= \frac{n^2 h^2}{2a^3} \int_0^a \left[ \frac{1 - Cos\left(\frac{2n\pi x}{a}\right)}{2} \right] dx \tag{481}$$

$$= \frac{n^2h^2}{2a^3} \left[ \frac{x - Sin\left(\frac{2n\pi x}{a}\right)}{\frac{2n\pi}{a}} \right]_0^a \tag{482}$$

or

$$= \frac{n^2h^2}{2a^3} \left(\frac{a}{2}\right) \tag{483}$$

or

$$<\hat{p}_x^2> = \frac{n^2h^2}{4a^2} \tag{484}$$

> ### *The Heisenberg's Uncertainty*

In order to prove the Heisenberg's uncertainty principle from for the quantum mechanical system of a particle in one-dimensional box, we first need to find the uncertainties in position and momentum. Once both uncertainties are known, we can simply multiply both to yield final result.

**1. Uncertainty in position:** The uncertainty in position is simply the difference between the square root of the uncertainty in the position-squared. Mathematically, we can say that

$$\Delta x = (< x^2 > -< x >^2)^{1/2} \tag{485}$$

After putting the values of average position and position-squared from equation (465) and (470) in equation (485), we get

$$\Delta x = \left[ \left(\frac{a^2}{3} - \frac{a^2}{2n^2\pi^2}\right) - \left(\frac{a}{2}\right)^2 \right]^{1/2} \tag{486}$$

or

$$\Delta x = \left[ \left(\frac{a^2}{12} - \frac{a^2}{2n^2\pi^2}\right) \right]^{1/2} \tag{487}$$

or

$$\Delta x = a\left(\frac{1}{12} - \frac{1}{2n^2\pi^2}\right)^{1/2} \tag{488}$$

**2. Uncertainty in momentum:** The uncertainty in momentum is simply the square root of the difference between the uncertainty in momentum and uncertainty in the momentum-squared. Mathematically, we can say that

$$\Delta p_x = (< p_x^2 > - < p_x >^2)^{1/2} \tag{489}$$

After putting the values of average position and position-squared from equation (475) and (484) in equation (489), we get

$$\Delta p_x = \left[\left(\frac{n^2 h^2}{4a^2}\right) - (0)^2\right]^{1/2} \tag{490}$$

or

$$\Delta p_x = \frac{nh}{2a} \tag{491}$$

Now multiplying equation (488) and (491), we get

$$\Delta x.\,\Delta p_x = \left[a\left(\frac{1}{12} - \frac{1}{2n^2\pi^2}\right)^{1/2}\right]\left(\frac{nh}{2a}\right) \tag{492}$$

or

$$= \frac{nh}{2}\left(\frac{1}{12} - \frac{1}{2n^2\pi^2}\right)^{1/2} \tag{493}$$

Multiply and divide the above equation by $2n\pi$

$$\Delta x.\,\Delta p_x = \frac{nh}{2}\cdot\frac{2n\pi}{2n\pi}\left(\frac{1}{12} - \frac{1}{2n^2\pi^2}\right)^{1/2} \tag{493}$$

or

$$= \frac{nh}{2}\cdot\frac{1}{2n\pi}\left(\frac{4n^2\pi^2}{12} - \frac{4n^2\pi^2}{2n^2\pi^2}\right)^{1/2} \tag{494}$$

or

$$\Delta x.\,\Delta p_x = \frac{h}{4\pi}\left(\frac{n^2\pi^2}{3} - 2\right)^{1/2} \tag{495}$$

Since $n^2\pi^2/3$ is always greater than 2, we can conclude that

$$\Delta x.\,\Delta p_x > \frac{h}{4\pi} \tag{496}$$

Which is the famous Heisenberg's uncertainty principle.

### ❖ Pictorial Representation of the Wave Equation of a Particle in One Dimensional Box and Its Influence on the Kinetic Energy of the Particle in Each Successive Quantum Level

The solution of the Schrodinger wave equation for a one-dimensional box gives the wave function as well as the energy of the system. The general form of wave-function representing various quantum mechanical states is given below.

$$\psi_n = \sqrt{\frac{2}{a}}\, Sin\,\frac{n\pi x}{a} \tag{497}$$

The energy of the system is given by equation (498) as:

$$E_n = \frac{n^2 h^2}{8ma^2} \tag{498}$$

The general depiction of a particle trapped in a one-dimensional box with zero potential inside, along with the conditions outside, is shown below.

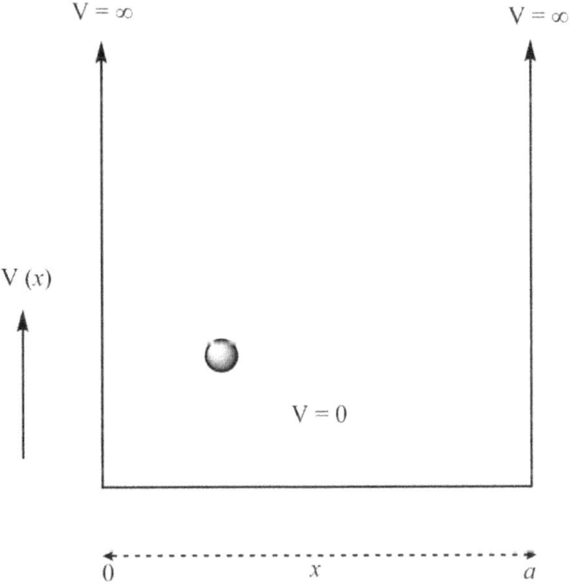


Figure 9. The graphical and pictorial representation of various wave-functions of the particle trapped in a one-dimensional box.

The pictorial representation of the wave-functions in different quantum mechanical states and the corresponding energies are shown below.

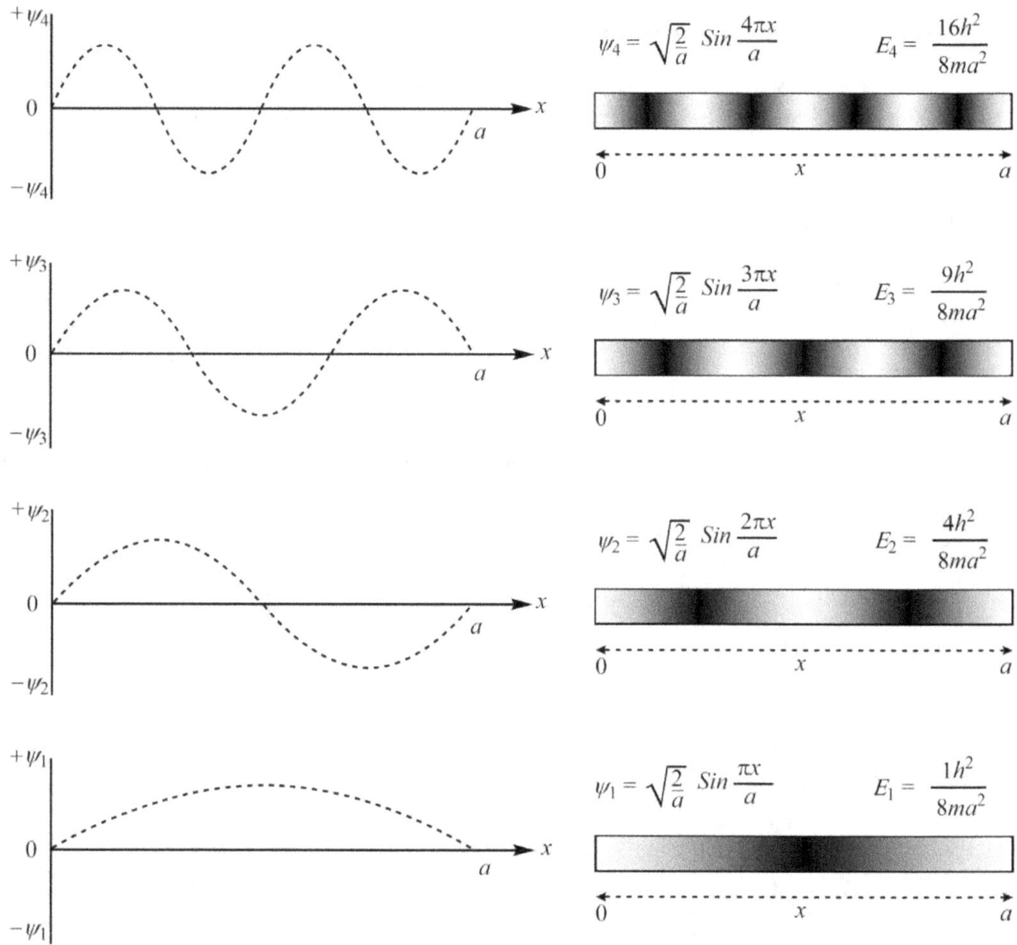

Figure 10. The graphical and pictorial representation of various wave-functions of the particle trapped in a one-dimensional box.

It can be seen clearly from the figure given above that as the number of nodes in wave-function defining a particular quantum mechanical state increases, the energy of the state also increases.

Furthermore, we can also comment on the symmetry of different wave functions w.r.t the center of the box. The symmetry of different states can be classified mainly into two categories as given below.

$$Symmetric \: \rightarrow Even \: function \rightarrow \psi_{odd}$$

and

$$Antisymmetric \: \rightarrow Odd \: function \rightarrow \psi_{even}$$

Hence, function like $\psi_1$, $\psi_3$, $\psi_5$ are symmetric while $\psi_2$, $\psi_4$, $\psi_6$ are antisymmetric. Some of the important results wavefunction and energy analysis for the particle in a one-dimensional box are listed below.

> ### Quantization of Energy

Owing to the discrete domain of $n$ i.e. 1, 2, 3 …. $\infty$; the kinetic energy associated with the particle, that is trapped in a one-dimensional box, can also have discrete or quantized values only. Therefore, the quantized variable is also popularly called as the "quantum number.

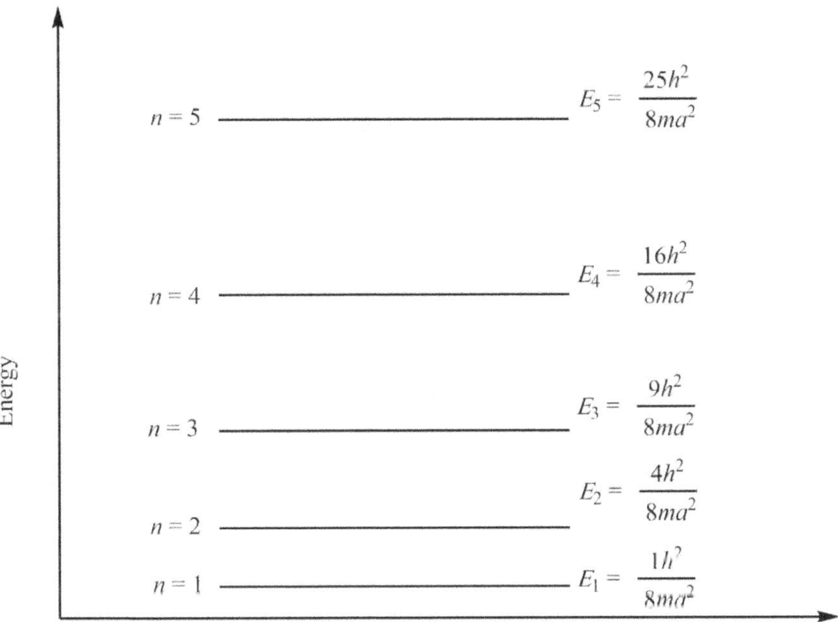


Figure 11. The quantized or discrete energy levels a particle of mass $m$, confined in a one-dimensional box of length $a$.

It is also worthy to note that the energy gap between successive energy levels shows a linear divergence with the increasing value of the quantum number $n$. Moreover, the energy of particle also depends inversely upon the mass and the box length; which eventually means that the energy levels would become continuous if the mass or length of the box becomes very large, proving the Bohr's correspondence principle.

> ### *Non-Quantization of the Energy of the Particle*

If the walls of the box are removed, the boundary conditions will no longer be applicable, and the particle would become free to move. In other words, the constant A, B and $k$ can have any value; and therefore, states of the particle are not quantized anymore. The general expression for the energy of the particle is

$$E_n = \frac{n^2 h^2}{8ma^2} \tag{498}$$

Hence, in such a case, a freely moving particle like an electron has restrictions and gives a continuous energy spectrum.

> ### *Box length and the Wave Function at the Walls*

We have already studied that the magnitude of the wave function at the ends of the box must be equal to zero to maintain its continuity. This is possible only if the length of the box is an integral multiple of half of the wavelength. This can be proved as

$$E_n = \frac{n^2 h^2}{8ma^2} \tag{499}$$

Also

$$E = \frac{1}{2}mv^2 = \frac{m^2 v^2}{2m} = \frac{p^2}{2m} \tag{500}$$

Using the de-Brogli relation ($\lambda = h/p$) in equation (500), we get

$$E = \frac{p^2}{2m} = \frac{(h/\lambda)^2}{2m} = \frac{h^2}{2m\lambda^2} \tag{501}$$

Now from equation (499) and (501), we conclude

$$\frac{n^2 h^2}{8ma^2} = \frac{h^2}{2m\lambda^2} \tag{502}$$

or

$$\frac{n^2}{4a^2} = \frac{1}{\lambda^2} \tag{503}$$

or

$$a = n\left(\frac{\lambda}{2}\right) \tag{504}$$

This result of equation (504) also proves that the number of nodes in $n$th quantum mechanical state are $n-1$.

> ### The Probability Density

The wave density of simply the probability density in the one-dimensional box is not the same at all the points. It is more noticeable when the quantum number defining the state is small. However, it becomes more and more uniform as $n$ increases.

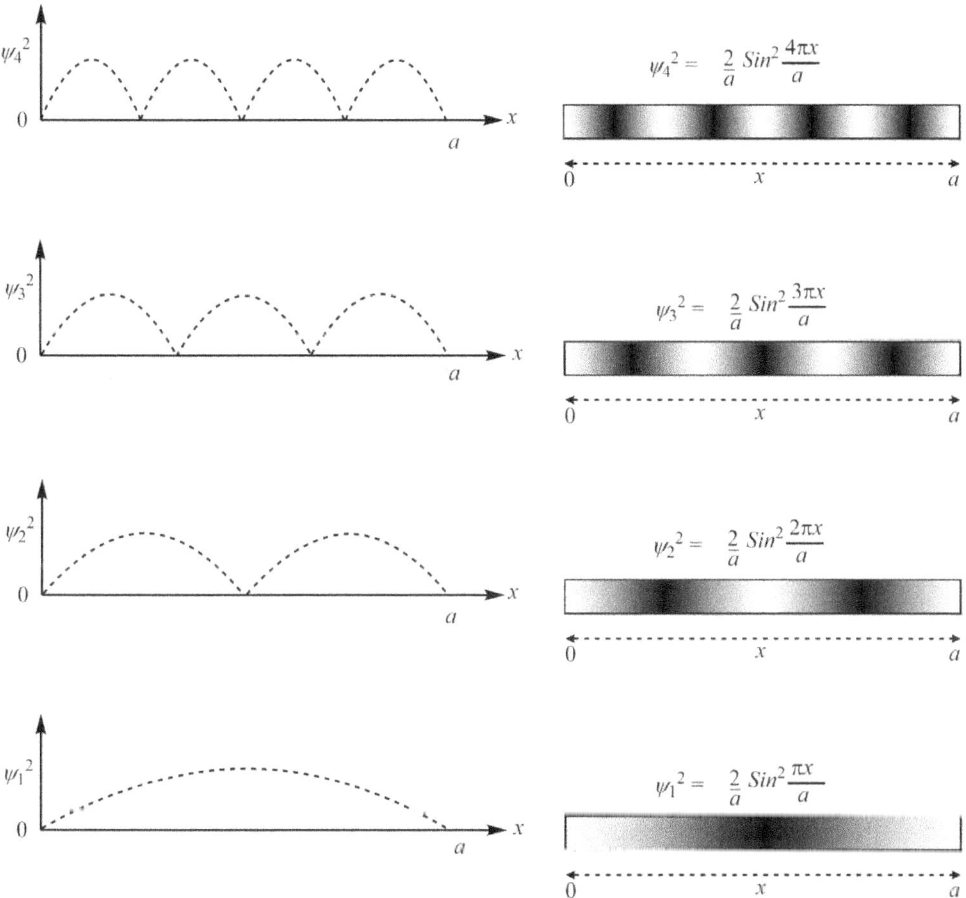


Figure 12. The graphical and pictorial representation of the probability density of a particle with mass $m$ and confined in a one-dimensional box of length $a$.

The increasing uniformity of with increasing value of $n$ is in accordance with the Bohr's correspondence principle which states that the results of quantum mechanics approach classical values at very high quantum numbers.

## ❖ Lowest Energy of the Particle

As we have already discussed that the wave function and energy, both are obtained as the solution of the Schrodinger wave equation for a particle in a one-dimensional box. The general forms of wave-function and energy for various quantum mechanical states are given below.

$$\psi_n = \sqrt{\frac{2}{a}}\ Sin\frac{n\pi x}{a} \qquad and \qquad E_n = \frac{n^2 h^2}{8ma^2} \tag{505}$$

We can write the expressions for $\psi_1$, $\psi_2$, $\psi_3$, $\psi_4$, $\psi_5$, $\psi_6$ and so on; however, it is also worthy to note that even though the $n = 0$ is permitted by the boundary condition, we cannot use it because this would make the whole function to collapse to zero.

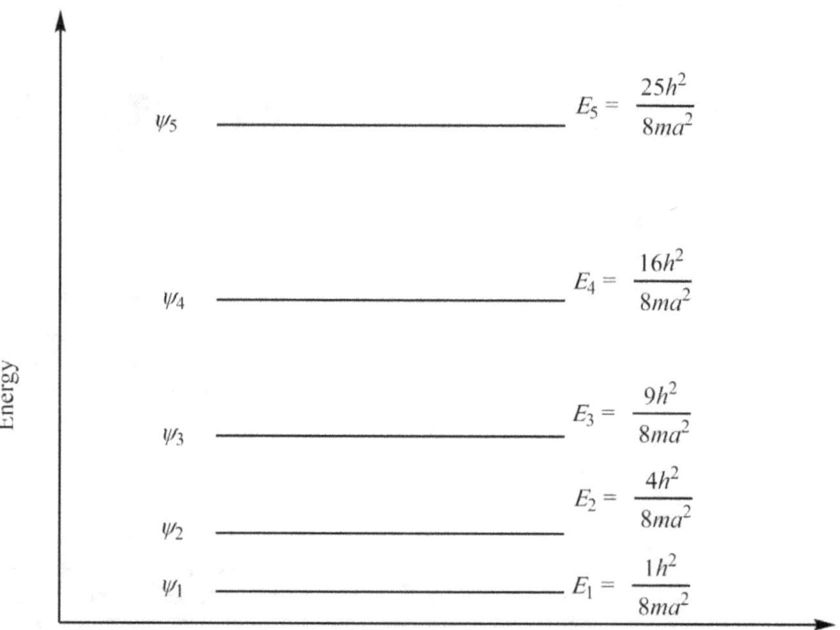


Figure 13. All the energy levels a particle in a one-dimensional box of including the "lowest energy of the particle".

Hence, the minimum acceptable value of the quantum number $n$ is 1 rather than 0; which makes the minimum energy of the particle non-zero.

$$E_1 = \frac{h^2}{8ma^2} \tag{506}$$

This non-zero value is popularly called as the zero-point energy and is a function of the mass of the particle and length of the box.

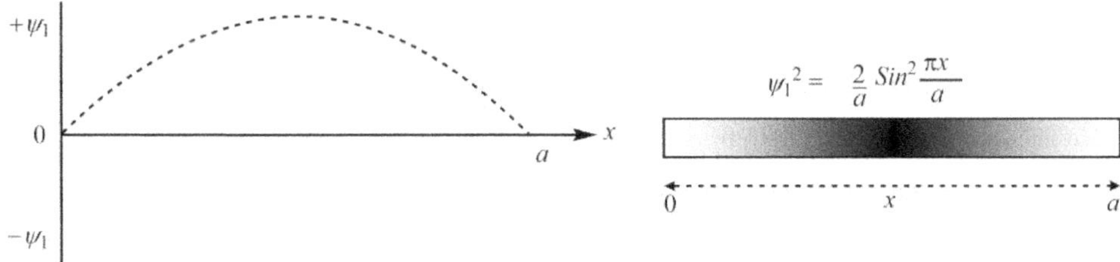



Figure 14. The plot of the wave function (left) and probability for the lowest energy state a particle in one-trapped in dimensional box.

Hence, in order to create the lowest energy, the particle must occupy the whole box without any node, having the highest probability at the center.

### ❖ Problems

Q 1. State and explain the third postulate of quantum mechanics.

Q 2. Why should the function representing a quantum mechanical state be continuous, single-valued and finite?

Q 3. Why don't we report non-eigenvalues directly? What is the need for their expectation values?

Q 4. Derive Schrodinger wave equation from the postulates of quantum mechanics.

Q 5. What is the Max-Born interpretation of "wave function"? Explain in detail by taking the example of one-dimensional systems.

Q 6. What is position-time uncertainty? How would you prove it for the photons passing through a slit of length $d$?

Q 7. What is operator commutation? Evaluate $[\hat{x}^2, \widehat{p_x}]$.

Q 8. Explain the energy-time uncertainty for a particle traveling along $x$-axis. Also, support your argument from the results of operator algebra.

Q 9. What are Hermitian operators? Prove that the operators for linear momentum and angular momentum are Hermitian in nature.

Q 10. Can the average value for the square of the Hermitian operator be negative? If not, explain why?

Q 11. Derive and solve the Schrodinger wave equation for a particle moving in a one-dimensional box.

Q 12. Prove the Heisenberg's uncertainty principle for the particle trapped in a one-dimensional box of length $a$. Also, comment on its validity in other systems.

Q 13. Give the pictorial representation of the first three quantum mechanical states of a particle in a one-dimensional box. Also, formulate the corresponding symmetry and number of nodes.

Q 14. Derive the relation between the box length and the wavelength of the particle in the 1-dimensional box.

Q 15. What is zero-point energy? How is it created by a particle of mass $m$ which is trapped in a one-dimensional box of length $a$.

Q 16. What is the average position? How is it different from the "most probable position"?

Q 17. State and explain the Bohr's correspondence principle.

### ❖ Bibliography

[1] B. R. Puri, L. R. Sharma, M.S. Pathania, *Principles of Physical Chemistry*, Vishal Publications, Jalandhar, India, 2018.

[2] I. N. Levine, *Quantum Chemistry*, Pearson Prentice Hall, New Jersey, USA, 2009.

[3] D. A. McQuarrie, *Quantum Chemistry*, University Science Books, California, USA, 2008.

[4] E. Steiner, *The Chemistry Maths Book*, Oxford University Press, Oxford, UK, 2008.

[5] P. Atkins, J. Paula, *Physical Chemistry*, Oxford University Press, Oxford, UK, 2010.

[6] M. Reed, B. Simon *Functional Analysis*, Elsevier, Amsterdam, Netherlands, 2003.

[7] G. E. Bowman, Essential Quantum Mechanics, Oxford University Press, Oxford, UK, 2008.

[8] W. Heisenberg, *Über den anschaulichen Inhalt der quantentheoretischen Kinematik und Mechanik*, Zeitschrift für Physik, 1927, 172.

[9] E. Schrödinger, *Die gegenwärtige Situation in der Quantenmechanik*, Naturwissenschaften. 1935, 807.

[10] P. Alberto, C. Fiolhaisdag, V. Gil, *Relativistic particle in a box*, European Journal of Physics, 1996, 17.

[11] R. K. Prasad, *Quantum Chemistry*, New Age International Publishers, New Delhi, India, 2010.



**DALAL INSTITUTE**

# CHAPTER 2

# Thermodynamics – I

### ❖ Brief Resume of First and Second Law of Thermodynamics

There are four laws of thermodynamics that define the fundamental physical quantities like temperature, energy, and entropy that characterize thermodynamic systems at thermal equilibrium. These laws describe how these quantities behave under different conditions and rule out the possibility of some phenomena the perpetual motion. The zeroth law of thermodynamics states that If two systems are each in thermal equilibrium with a third system, they are in thermal equilibrium with each other; therefore, this law helps to define the concept of temperature. In this section, we will discuss the elementary ideas and mutual correlation between the first and second laws of thermodynamics.

> #### First Law of Thermodynamics

*The first law of thermodynamics states that the energy can neither be created nor destroyed, but can be converted from one form to another.*

The first law of thermodynamics is obtained on the experimental basis. In other words, we can say that the energy of an isolated system is always constant, which means that whenever some energy disappears from the system, an equal amount of energy in some other form is also produced. In 1847, Helmholtz explained this situation in his famous words, "it is impossible to construct a perpetual machine". The term perpetual machine refers to a device that can work continuously without any energy consumption. Furthermore, we all know that heat is always produced whenever some mechanical work is done. These correlations were studied by Joule (1840 – 1880); and he found that mechanical work is directly proportional to the heat produced. Mathematically, we can say that

$$W \propto Q \ \ or \ W = JQ \tag{1}$$

Where J represents the proportionality constant and is called as "Joule's mechanical equivalent of heat". If $Q = 1$, $W = J$; making $J$ as the amount of mechanical work required to produce one calorie of heat. The experimental value for $J$ was found to be 4.184 joules, which is a very popular relation (1 calorie = 4.184 joule). The first law of thermodynamics can also be deduced from the equivalence of heat and work. Suppose there is now an equivalence between the work and heat; and let $Q$ heat is converted into work. Now when the same amount of work is done to produce the heat $Q'$; considering $Q \neq Q'$, we can say that $Q$ is either greater or less than $Q'$. This would eventually mean that a certain amount of energy has been destroyed or created in this process, which is against the first law of thermodynamics.

The mathematical formulation of the first law thermodynamics can be obtained from the increase in the internal energy of the system. The internal energy of the system can be increased in two ways; one is doing work on the system, and the second one involves the supply of heat. Suppose that the initial internal energy of

---

the system is $E_1$, after supplying heat $q$ and doing work $w$ on the system, the final amount of internal energy can be formulated as:

$$E_2 = E_1 + q + w \tag{2}$$

$$E_2 - E_1 = q + w \tag{3}$$

$$\Delta E = q + w \tag{4}$$

$$q = \Delta E - w \tag{5}$$

If the work is done by the system, putting $w = -P\Delta V$ in equation (5), we get

$$q = \Delta E - (-P\Delta V) \tag{6}$$

$$q = \Delta E + P\Delta V \tag{7}$$

The physical significance of the equation (7) is that heat absorbed by a given system is converted work done by the system and to raise its internal energy.

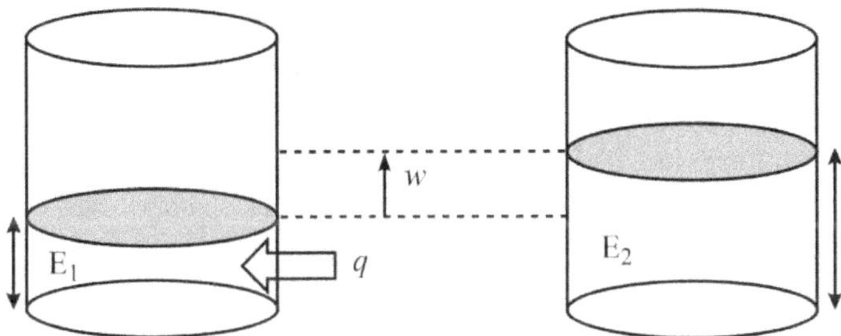


Figure 1. The pictorial representation of the first law of thermodynamics.

It is also worthy to note that the general form of the first law of thermodynamics is applicable only in the case of chemical thermodynamics or physical processes. In 1905, Albert Einstein showed that energy and mass are just the faces of the coin, and can be transformed within each other. In other words, his findings showed that the mass can be converted into energy and the energy can back be converted into mass. Mathematically, the formulation is

$$E - mc^2 \tag{8}$$

Where $m$ is the mass and $c$ is the velocity of light. Since the velocity of light is a very large quantity (so the square), even the small disappearance of mass would generate a huge amount of energy. Such observations are pretty common in case of nuclear reactions and can be neglected here. Therefore, in a broad sense, it is the "law of energy-mass conservation".

> ### *Second Law of Thermodynamics*

*The second law of thermodynamics states that it is impossible to convert the heat completely into work without leaving some effect elsewhere.*

The second law of thermodynamics is actually a rational solution to the limitations of the first law. For instance, the first law talks about the exact equivalence between heat and work, but it is quite far from reality. In 1824, a French scientist Sadi Carnot showed that for every heat engine there is an upper limit to the efficiency of conversion of heat to work. In order to illustrate Carnot's conclusion, consider a locomotive engine that is supplied with a certain amount of heat; however, all of that heat will not be used to move the train but a part of it will always be consumed in some other processes like overcoming the friction. Let $q_2$ be the heat absorbed by the heat engine at temperature $T_2$, and $w$ is the amount of the work done by the system; while $q_1$ is the heat returned to the sink at temperature $T_1$, then the Carnot's formulation can be given as:

$$\eta = \frac{w}{q_2} = \frac{q_2 - q_1}{q_2} = \frac{T_2 - T_1}{T_2} \tag{9}$$

Where $\eta$ is the efficiency of the heat engine and is always less than one. Ideally, $\eta = 1$, which means that such a heat engine would convert 100% of the heat absorbed into work.

One more limitation of the first law is that it does not tell about the feasibility of the process, like whether the heat can flow from cold terminal to the hot one or not. It simply talks if the heat gained or heat lost but not the direction of the process. The second law of thermodynamics states that all the spontaneous processes are thermodynamically irreversible. The word "spontaneous" simply means a process that occurs by itself and external drive is required. In other words, we can also say that heat cannot flow from a cold body to hot, the water cannot uphill without any external drive.

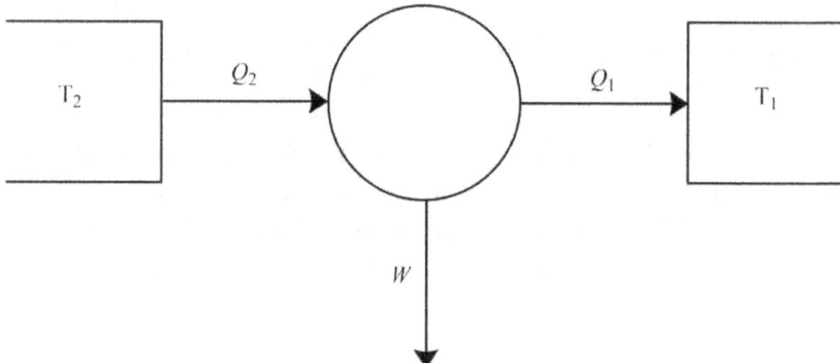


Figure 2. The pictorial representation of the second law of thermodynamics.

The 2nd law of thermodynamics also states that the total entropy of an isolated system can never decline with time; in other words, combined entropy of a system and surroundings remains constant in ideal cases where the system is undergoing a reversible process. In all processes, including spontaneous processes, that occur, the total entropy of the system and surroundings increases and the process is irreversible in the thermodynamic frame. The entropy-increase accounts for the irreversibility of all the natural processes, and the asymmetry between the past and the future. Overall, the 2nd law of thermodynamics can be labeled as an empirical finding that was accepted as a truism of thermodynamic theory. The microscopic origin of the law can be explained by statistical mechanics.

### ❖ Entropy Changes in Reversible and Irreversible Processes

In order to understand the entropy change in reversible and irreversible processes, we need to understand the concept of entropy first. For a Carnot heat engine working at $T_1$ and $T_2$, it has been observed that the heat absorbed ($q_2$) and heat returned ($q_1$) are related as given below.

$$\frac{q_2 - q_1}{q_2} = \frac{T_2 - T_1}{T_2} \tag{10}$$

or

$$1 - \frac{q_1}{q_2} = 1 - \frac{T_1}{T_2} \tag{11}$$

or

$$\frac{q_1}{q_2} = \frac{T_1}{T_2} \tag{12}$$

$$\frac{q_1}{T_1} = \frac{q_2}{T_2} \tag{13}$$

$$\frac{q}{T} = constant \tag{14}$$

Therefore, we can say that for any particular system, the ratio of heat absorbed or lost isothermally and reversibly to the absolute temperature at which this takes place is a constant parameter. If we consider $q_1$ as the heat absorbed at $T_1$, equation (13) becomes

$$\frac{q_2}{T_2} = -\frac{q_1}{T_1} \tag{15}$$

$$\frac{q_2}{T_2} + \frac{q_1}{T_1} = 0 \tag{16}$$

or

$$\sum \frac{q}{T} = 0 \tag{17}$$

Consider a reversible cyclic process, consists of many Carnot cycles. In going from point $A$ to $B$ and then back $A$, all the closed paths cancel each other that results in parent path $ABA$.

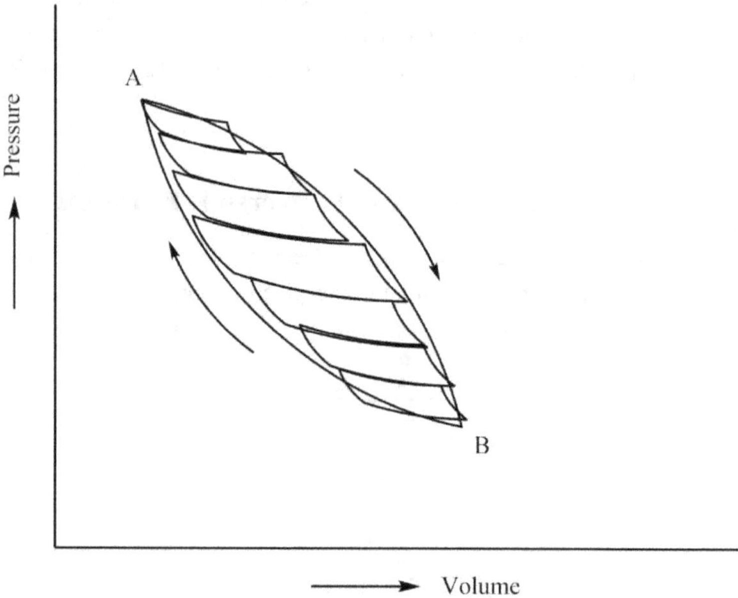

Figure 3. A cyclic process made of many Carnot cycles.

For each cycle, we have equation (17), for an infinite number of cycles

$$\sum \frac{\delta q}{T} = 0 \tag{18}$$

Where $\delta q$ is the extremely small amount of heat absorbed at temperature $T$ during the course of an isothermal and reversible process. Moreover, the total entropy change of the cyclic process $ABA$ can be fragmented into two components as:

$$\sum \frac{\delta q}{T} = \sum_{A \to B} \frac{\delta q}{T} + \sum_{B \to A} \frac{\delta q}{T} = 0 \tag{19}$$

$$\sum_{A \to B} \frac{\delta q}{T} = - \sum_{B \to A} \frac{\delta q}{T} \tag{20}$$

or

$$\left(\frac{q}{T}\right)_{A \to B} = - \left(\frac{q}{T}\right)_{B \to A} \tag{21}$$

The physical significance of the equation (21) is that the value of $q/T$ for the path $A{\rightarrow}B$ is the same as the path $B{\rightarrow}A$; which eventually means that the quantity $q/T$ is actually a state function. This quantity i.e. $q/T$ is called as entropy, and is generally labeled as $S$. If $S_A$ and $S_B$ are the entropies at point A and B, respectively; then we can say that

$$S_A - S_B = \frac{q}{T} \tag{22}$$

$$\Delta S = \frac{q}{T} \tag{23}$$

Hence, the entropy change may be defined as the amount of the heat absorbed isothermally and reversibly divided by the temperature at which the heat is absorbed. Being a state function, the change in entropy always depends upon the initial and final state and not upon the path followed. Moreover, since the heat is absorbed reversibly, it is better to use $q_{rev}$ in equation (23) instead of simply $q$, therefore

$$\Delta S - \frac{q_{rev}}{T} \tag{24}$$

For an extremely minute change, the above equation becomes

$$dS = \frac{q_{rev}}{T} \tag{25}$$

The entropy is an extensive property measured in joule per Kelvin per mole ($JK^{-1}mol^{-1}$). The most important significance of entropy is that it can be used to measure the randomness in the system.

> ***Entropy Changes in Reversible Processes***

Suppose that the heat absorbed by the system and heat lost by the surrounding are under completely reversible conditions. In other words, $q_{rev}$ is the heat absorbed and lost by the surrounding at temperature T, then we can say that the entropy change in the system will be given by the following relation.

$$\Delta S_{system} = \frac{q_{rev}}{T} \tag{26}$$

Similarly, the entropy change in the surrounding will be

$$\Delta S_{sorrounding} = -\frac{q_{rev}}{T} \tag{27}$$

Therefore, the total entropy change will be

$$\Delta S_{system} + \Delta S_{sorrounding} = \frac{q_{rev}}{T} - \frac{q_{rev}}{T} = 0 \tag{28}$$

Hence, we can conclude that the entropy change in an isolated system is always zero i.e. the sum of entropy change in system and entropy change in the surrounding is zero under reversible conditions.

> ### *Entropy Changes in Irreversible Processes*

Every reversible process becomes irreversible even if only one part of it becomes irreversible. To understand this, let us suppose that $q_{irrev}$ heat lost by the surrounding. Although this heat would be absorbed by the system, the entropy of the system depends upon the heat absorbed reversibly. Therefore, entropy change of the system at an absolute temperature T will be

$$\Delta S_{system} = \frac{q_{rev}}{T} \tag{29}$$

Similarly, the entropy change of the surrounding will be

$$\Delta S_{sorrounding} = -\frac{q_{irrev}}{T} \tag{30}$$

The total entropy of the isolated system (system + surrounding) will be

$$\Delta S_{system} + \Delta S_{sorrounding} = \frac{q_{rev}}{T} - \frac{q_{irrev}}{T} \tag{31}$$

Furthermore, as we know that the $w_{rev} > w_{irrev}$, and internal energy is a state function that is independent of whether the process is reversible or irreversible. Mathematically, it is

$$\Delta E = q_{rev} - w_{rev} = q_{irrev} - w_{irrev} \tag{32}$$

we can conclude that,

$$q_{rev} > q_{irrev} \tag{33}$$

$$\frac{q_{rev}}{T} > \frac{q_{irrev}}{T} \tag{34}$$

or

$$\frac{q_{rev}}{T} + \frac{q_{irrev}}{T} > 0 \tag{35}$$

After comparing equation (35) with equation (31), we get

$$\Delta S_{system} + \Delta S_{sorrounding} > 0 \tag{36}$$

The physical interpretation of the above equation lies in the fact that all irreversible processes occur via a net increase in entropy. In other words, the total entropy change of an isolated system in any irreversible process is always greater than zero. The results of equation (28) and equation (36) can be combined to give a more generalized form as:

$$\Delta S_{system} + \Delta S_{sorrounding} \geq 0 \tag{37}$$

Where the sign of '=' stands from reversible and '>' stands for irreversible phenomena, respectively.

The entropy change in the irreversible process like the flow of heat from the hot end ($T_2$) to cold end ($T_1$) can be calculated using the following relation.

$$\Delta S_{total} = -\frac{q_{rev}}{T_2} + \frac{q_{rev}}{T_1} = q_{rev}\left(\frac{1}{T_1} - \frac{1}{T_2}\right) \tag{38}$$

Where $q_{rev}$ is the amount of heat transferred from $T_2$ to $T_1$.

Furthermore, the Clausius inequality can also be proved from equation (36) which governs that all the irreversible processes are accompanied by a net increase in the entropy.

$$\Delta S_{system} > \Delta S_{sorrounding} \tag{39}$$

This can be illustrated with the help of the following examples.

*i) Irreversible isothermal expansion of an ideal gas:* Suppose an ideal gas expanding isothermally and irreversibly against vacuum. The condition isothermal means $\Delta T = 0$, which in turn implies that $\Delta U = 0$. Hence from the first law of thermodynamics, we have $dq = 0$ i.e. no heat is transferred from, or to the surrounding giving $dS_{surrounding} = 0$.

However, since the $dS_{system} = R\ ln\ (V_2/V_1)$. Since $V_2$ is definitely greater than $V_1$, therefore

$$dS_{total} = dS_{system} + dS_{sorrounding} = dS_{system} + 0 \geq 0 \tag{40}$$

*ii) Heat flow from hot to the cold end:* Suppose $dq$ is the amount of heat transferred from temperature $T_2$ to $T_1$. The entropy change at the source will be

$$dS_{source} = -\frac{dq}{T_2} \tag{41}$$

Similarly, the entropy change at the sink will be

$$dS_{sink} = \frac{dq}{T_1} \tag{42}$$

The total entropy change of the system can be calculated as

$$dS_{total} = dS_{source} + dS_{sink} = \frac{dq}{T_1} - \frac{dq}{T_2} \tag{43}$$

or

$$dS_{total} = dq\left(\frac{1}{T_1} - \frac{1}{T_2}\right) \tag{44}$$

Now because the source is always at a higher temperature than sink ($T_2 > T_1$), $dS_{total}$ will be positive for sure, proving that transfer of heat from a hot body to the cold body occurs via a net entropy increment.

## ❖ Variation of Entropy with Temperature, Pressure and Volume

Consider one mole of ideal gas filled in a chamber fitted with a weightless and frictionless piston. If the system absorbs a very small amount of heat $\delta q_{rev}$ isothermally and reversibly at temperature T, the total entropy change in the system can be given by the following relation.

$$dS = \frac{\delta q_{rev}}{T} \tag{45}$$

Also, from the first law of thermodynamics

$$\delta q = dE - \delta w \tag{46}$$

The above equation may be written in the following form if the process is carried out under reversible conditions.

$$\delta q_{rev} = dE - \delta w \tag{47}$$

After putting the value of work of expansion $(-PdV)$, the equation (47) becomes

$$\delta q_{rev} = dE - (-PdV) = dE + PdV \tag{48}$$

Using equation (48) in equation (45), we get

$$dS = \frac{dE + PdV}{T} \tag{49}$$

or

$$TdS = dE + PdV \tag{50}$$

Moreover, for one mole of an ideal gas, the value of heat capacity at constant volume $(C_v)$ is

$$C_v = \frac{dE}{dT} \tag{51}$$

$$dE = C_v dT \tag{52}$$

Also

$$PV = RT \tag{53}$$

$$P = \frac{RT}{V} \tag{54}$$

Where R is the gas constant; while P, T and V are the pressure, temperature and volume, respectively. Now putting the value of equation (52) and (54) in equation (50), we get

$$TdS = C_v dT + \frac{RT}{V} dV \tag{55}$$

$$dS = C_v \frac{dT}{T} + R \frac{dV}{V} \tag{56}$$

Now consider that the initial state ($T_1$, $V_1$) is converted to the final state ($T_2$, $V_2$), the total entropy will be calculated by the following relation

$$\int_{S_1}^{S_2} dS = \int_{T_1}^{T_2} C_v \frac{dT}{T} + \int_{V_1}^{V_2} R \frac{dV}{V} \tag{57}$$

$$\int_{S_1}^{S_2} dS = C_v \int_{T_1}^{T_2} \frac{dT}{T} + R \int_{V_1}^{V_2} \frac{dV}{V} \tag{58}$$

or

$$\Delta S = C_v \ln\frac{T_2}{T_1} + R \ln\frac{V_2}{V_1} \tag{59}$$

Since, from the ideal gas equation, the following also true for the initial and final state of the system.

$$P_1 V_1 = RT_1 \tag{60}$$

$$P_2 V_2 = RT_2 \tag{61}$$

Dividing equation (61) by equation (60), we get

$$\frac{P_2 V_2}{P_1 V_1} = \frac{RT_2}{RT_1} \tag{62}$$

or

$$\frac{V_2}{V_1} - \frac{P_1 T_2}{P_2 T_1} \tag{63}$$

After putting the value of (63) in equation (59), we get

$$\Delta S = C_v \ln\frac{T_2}{T_1} + R \ln\frac{P_1 T_2}{P_2 T_1} \tag{64}$$

As we know that $C_p - C_v = R$; therefore, after putting $C_v = C_p - R$ in equation (64), we get

$$\Delta S = (C_p - R) \ln\frac{T_2}{T_1} + R \ln\frac{P_1 T_2}{P_2 T_1} \tag{65}$$

$$= C_p \ln \frac{T_2}{T_1} - R \ln \frac{T_2}{T_1} + R \ln \frac{P_1}{P_2} + R \ln \frac{T_2}{T_1} \tag{66}$$

or

$$\Delta S = C_p \ln \frac{T_2}{T_1} + R \ln \frac{P_1}{P_2} \tag{67}$$

The equation (59) and equation (67) formulate the variation of entropy with temperature-volume and temperature-pressure, respectively.

The variation of entropy with temperature, pressure or volume in isothermal, isobaric and isochoric processes is discussed below.

**1. Entropy change in isothermal process:** Provided that the temperature of the system is kept constant ($T_2 = T_1 = T$), equation (59) and (67) are reduced to the following.

$$\Delta S = R \ln \frac{V_2}{V_1} = R \ln \frac{P_1}{P_2} \tag{68}$$

**2. Entropy change in isochoric process:** Provided that the volume of the system is kept constant ($V_2 = V_1 = V$), equation (59) is reduced to the following.

$$\Delta S = C_v \ln \frac{T_2}{T_1} \tag{69}$$

**3. Entropy change in isobaric process:** Provided that the pressure of the system is kept constant ($P_2 = P_1 = P$), equation (67) is reduced to the following.

$$\Delta S = C_p \ln \frac{T_2}{T_1} \tag{70}$$

It is also worthy to note that if this reversible expansion takes place adiabatically, then $q_{rev} = 0$, and therefore, the change entropy will also be zero.

## ❖ Entropy Concept as a Measure of Unavailable Energy and Criteria for the Spontaneity of Reaction

The entropy change during the course of a process can be used to rationalize the unavailable energy as well as its spontaneity or feasibility. In this section, we will first discuss the entropy concept as a measure of unavailable energy and then we will study the spontaneity of a process in terms of entropy change.

> ### *Entropy Concept as a Measure of Unavailable Energy*

In order to understand the connection between entropy and unavailable energy, consider a heat source at temperature $T_2$ placed in an atmosphere or surrounding at temperature $T_0$. The surrounding plays the role of heat sink here. Now if $Q$ is the heat transferred from the source to a Carnot heat engine working between $T_2$ and $T_0$, then from Carnot's theorem, we have

$$\frac{W_{max,T_2,T_0}}{Q} = \frac{T_2 - T_0}{T_2} \tag{71}$$

$$W_{max,T_2,T_0} = Q\left(1 - \frac{T_0}{T_2}\right) \tag{72}$$

Where $W_{max,T_2,T_0}$ is the maximum work that can be obtained from a heat engine working between $T_2$ and $T_0$ temperatures. It is worthy to note that only a part of the heat transferred can be turned into work i.e. available to be used as work.

Now suppose that the same amount of heat is first transferred directly from the source at $T_2$ to another source at $T_1 < T_2$. In other words, the heat $Q$ is now first transferred from hot source to a cold source without any Carnot bridging. Now the maximum work that can be obtained from a heat engine working between $T_1$ and $T_0$ temperatures will be

$$W_{max,T_1,T_0} = Q\left(1 - \frac{T_0}{T_1}\right) \tag{73}$$

Where $W_{max,T_1,T_0}$ is the maximum work that can be obtained after the direct heat transfer, since the heat engine is now working between $T_1$ and $T_0$ temperatures. Now before the irreversible heat transfer ($T_2 \rightarrow T_1$), the magnitude of energy that could have been transformed into work is:

$$E_{unavailable} = \left(W_{max,T_2,T_0}\right) - \left(W_{max,T_1,T_0}\right) \tag{74}$$

$$= \left[Q\left(1 - \frac{T_0}{T_2}\right)\right] - \left[Q\left(1 - \frac{T_0}{T_1}\right)\right] \tag{75}$$

$$= Q\left[\left(1 - \frac{T_0}{T_2}\right) - \left(1 - \frac{T_0}{T_1}\right)\right] \tag{76}$$

$$= Q\left(\frac{T_0}{T_1} - \frac{T_0}{T_2}\right) \tag{77}$$

$$= T_0\left(\frac{Q}{T_1} - \frac{Q}{T_2}\right) \tag{78}$$

Since $-Q/T_2$ and $Q/T_1$ are the entropy loss and gain of sources at $T_2$ and $T_1$, respectively, the equation (78) reduces to the following.

$$E_{unavailable} = T_0\left(\Delta S_{source\ at\ T_1} + \Delta S_{source\ at\ T_2}\right) \tag{79}$$

$$= T_0\,\Delta S_{irreversible\ heat\ transfer} \tag{80}$$

Where $E_{unavailable}$ is lost work or the energy which is no longer available to do any useful work.

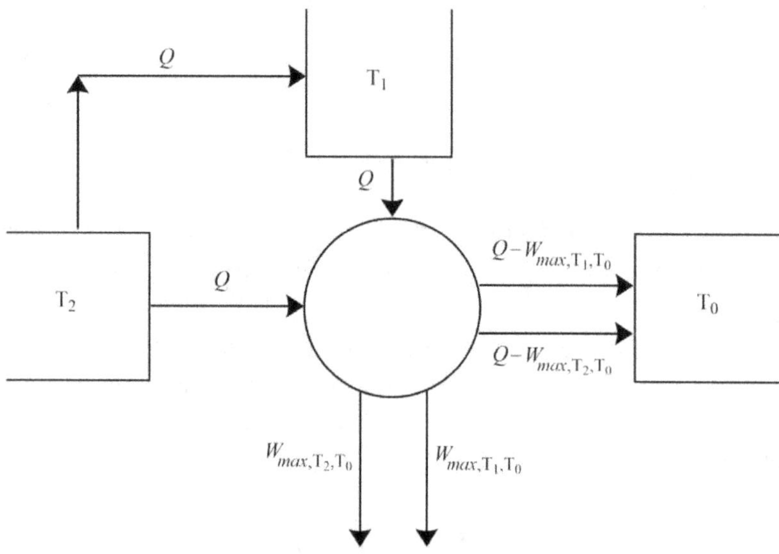

Figure 4. The pictorial representation of the entropy–unavailable-energy.

The physical significance of equation (80) can also be demonstrated using a live example. Let us consider a Carnot engine working between 800K and 200K is supplied with a heat ($Q$) of 5000 joules. The efficiency of such an engine will be

$$\eta = \frac{T_2 - T_0}{T_2} = \frac{800 - 200}{800} = 0.75 \tag{81}$$

The maximum work obtained will be

$$W = \eta Q = 0.75 \times 5000\,J = 3750 \tag{82}$$

Now if this 5000 joules of heat is first transferred to the reservoir at 500K, and then it is fed to a Carnot engine working obviously between "500K" and 200K. The efficiency of such an engine will be

$$\eta = \frac{T_1 - T_0}{T_1} = \frac{500 - 200}{500} = 0.6 \tag{83}$$

The maximum work obtained will be

$$W = \eta Q = 0.6 \times 5000\,J = 3000 \tag{84}$$

Comparing the results of equation (82) and (84), we can say that there is 750 J less work from the same heat transfer in the second process.

> ### *Entropy Concept as the Criteria for the Spontaneity of Reaction*

The second law of thermodynamics states that all the spontaneous processes are thermodynamically irreversible, and accompanied by a net increase in entropy. The entropy change for a reversible process is,

$$\Delta S_{system} + \Delta S_{sorrounding} = 0 \tag{85}$$

And the entropy change for a spontaneous or irreversible process is

$$\Delta S_{system} + \Delta S_{sorrounding} > 0 \tag{86}$$

The results of equation (85) and equation (86) can be combined to give a more generalized form as:

$$\Delta S_{system} + \Delta S_{sorrounding} \geq 0 \tag{87}$$

Where the sign of '=' stands from reversible and '>' stands for irreversible or spontaneous reaction, respectively.

Nevertheless, the criteria from equation (87) is not very much practical as $\Delta S_{system}$ and $\Delta S_{sorrounding}$ are not easy to determine. Therefore, some other criteria must be deduced which requires only the knowledge of the change in some simple thermodynamic parameters. In order to do so, write down the equation (87) for infinitesimally small changes as given below.

$$dS_{system} + dS_{sorrounding} \geq 0 \tag{88}$$

If δq is the heat lost by the surrounding reversibly and isothermally at temperature T, then we can say that

$$dS_{sorrounding} = -\frac{q_{rev}}{T} \tag{89}$$

Also, from the first law of thermodynamics, we have

$$\delta q_{rev} = dU + PdV \tag{90}$$

After putting the value from equation (90) in equation (89), we have

$$dS_{sorrounding} = -\frac{dU + PdV}{T} \tag{91}$$

Now putting the value of $dS_{sorrounding}$ from equation (91) into equation (88), we get

$$dS_{system} - \frac{dU + PdV}{T} \geq 0 \tag{92}$$

$$TdS_{system} - dU - PdV \geq 0 \tag{93}$$

$$TdS \geq dU + PdV \tag{94}$$

Where $dS_{system} = dS$ i.e. change in the entropy of the system. For a spontaneous or irreversible process $TdS$ must be greater the sum of internal energy change and pressure-volume work.

However, the above relation is reduced to the following if the internal energy and volume of the system are kept constant i.e. $dU = 0$ and $dV = 0$.

$$(TdS)_{U,V} \geq 0 \tag{95}$$

or

$$(dS)_{U,V} \geq 0 \tag{96}$$

Here it is worthy to recall that the sign '=' is for reversible reactions whereas '>' condition is applicable to spontaneity or irreversibility of a reaction.

Besides, the spontaneity of equation (94) can also be written in the form of internal energy change,

$$dU \leq TdS - PdV \tag{97}$$

the above relation is reduced to the following if the internal energy and volume of the system are kept constant i.e. $dS = 0$ and $dV = 0$.

$$(dU)_{S,V} \leq 0 \tag{98}$$

Again it should be noted that the sign '=' is for reversible reactions whereas '<' condition is applicable to spontaneity or irreversibility of a reaction.

## ❖ Free Energy, Enthalpy Functions and Their Significance, Criteria for Spontaneity of a Process

The thermodynamic free energy and enthalpies are extremely useful concepts in the field of chemical thermodynamics. The decrease or increase in free energy is the maximum amount of work that a thermodynamic system can perform in a process at a constant temperature. The sign of thermodynamic free energy simply indicates whether a process is thermodynamically forbidden or feasible. The thermodynamic free energy is a state function, like internal energy and enthalpy. In this section, we will discuss the significance of enthalpy and free energy functions, and the corresponding spontaneity of a process.

### ➤ *Enthalpy or Heat Content*

As most of the reactions are carried out in open vessels where the atmospheric pressure remains the same, it must be very interesting to study the heat change that occurs during the course of a chemical reaction. The work-done by gas in the piston-fitted chamber against constant pressure ($P$) with $\Delta V$ volume change is

$$w = -P\Delta V \tag{99}$$

From the first law of thermodynamics, we know that

$$q = \Delta U - w \tag{100}$$

Where $q$ is the heat absorbed and $\Delta U$ is the change in the internal energy of the system respectively. After putting the value of work of expansion from equation (99) in equation (100), we get

$$q_p = \Delta U - (-P\Delta V) \tag{101}$$

$$q_p = \Delta U + P\Delta V \tag{102}$$

The symbol $q_p$ is used instead of $q$ because the whole process is carried out at constant pressure. Now suppose that this heat absorbed increases the internal energy from $U_1$ to $U_2$ and volume from $V_1$ to $V_2$ i.e.

$$\Delta U = U_2 - U_1 \tag{103}$$

$$\Delta V = V_2 - V_1 \tag{104}$$

After putting the values from equation (103) and (104) into equation (102), we get

$$q_p = (U_2 - U_1) + P(V_2 - V_1) \tag{105}$$

or

$$q_p = (U_2 + PV_2) - (U_1 + PV_1) \tag{106}$$

The quantity $U + PV$ is now defined as the enthalpy of the system; and since $U$, $P$ and $V$ all are state functions, the quantity $U + PV$ must also be a state function. Mathematically, the enthalpy can be shown as

$$H = U + PV \tag{107}$$

Therefore, if the enthalpy changes initial state $H_1$ to the final state $H_2$ at constant pressure, we have

$$H_1 = U_1 + PV_1 \tag{108}$$

$$H_2 = U_2 + PV_2 \tag{109}$$

After putting the values from equation (108) and (109) in equation (106), we get

$$q_p = H_2 - H_1 \tag{110}$$

$$q_p = \Delta H \tag{111}$$

Hence, we can conclude that the enthalpy change of a system is simply the amount of heat absorbed at constant pressure. Besides, if we compare equation (102) and equation (111), we have

$$\Delta H = \Delta U + P\Delta V \tag{112}$$

Therefore, the enthalpy change of a reaction may also be defined as the sum of its internal energy change and pressure-volume work done.

**The physical significance of enthalpy:** There is a certain amount of energy stored in every substance or material called internal energy. This energy comprises of many forms like kinetic, rotational, vibrational or inter-particle interactions. Moreover; like pressure, volume and internal energy; the enthalpy function is also an extensive property. Now recalling the expression of enthalpy again

$$H = U + PV \tag{113}$$

We can say that the enthalpy is nothing but the internal energy that is available for the conversion into heat; which is why the enthalpy is also called as "heat content". Furthermore, like internal energy, the heat content or enthalpy is also not obtainable absolutely. However, like some other thermodynamic quantities, the parameter that is needed in the various analysis is enthalpy change which can be derived experimentally.

**Enthalpy as the criteria for the spontaneity of reaction:** The general expression for the enthalpy of a system is given below.

$$H = U + PV \tag{114}$$

After differentiating the equation (114), we get

$$dH = dU + VdP + PdV \tag{115}$$

$$dH - VdP = dU + PdV \tag{116}$$

Also as we know that, for the spontaneity of a process

$$TdS \geq dU + PdV \tag{117}$$

Now after putting the value of $dU + VdP$ from equation (116) in equation (117), we get

$$TdS \geq dH - VdP \tag{118}$$

or

$$dH \leq TdS + VdP \tag{119}$$

For a spontaneous or irreversible process $dH$ must be less than the sum of multiplication of temperature and entropy change, and pressure-volume work.

However, the above relation is reduced to the following if the entropy and pressure of the system are kept constant i.e. $dS = 0$ and $dP = 0$.

$$(dH)_{S,P} \leq 0 \tag{120}$$

Here it is worthy to recall that the sign '=' is for reversible reactions whereas '<' condition is applicable to spontaneity or irreversibility of a reaction.

> ### *Helmholtz Free Energy or Work Function*

The Helmholtz free energy is typically denoted by the symbol '$A$', which is derived from the German word "Arbeit" meaning work. The Helmholtz free energy can be defined mathematically as

$$A = U - TS \tag{121}$$

Where T, S and U are temperature, entropy, and internal energy, respectively. Moreover; like U, T and S; free energy $A$ is also a state function. Since A is independent of its previous state, we can say that

$$A_1 = U_1 - TS_1 \tag{122}$$

$$A_2 = U_2 - TS_2 \tag{123}$$

Where the subscript 1 and 2 represent the initial and final state. The change in Helmholtz free energy in going from initial to final state can obtain by subtracting equation (122) from equation (123) as

$$A_2 - A_1 = (U_2 - TS_2) - (U_1 - TS_1) \tag{124}$$

$$\Delta A = (U_2 - U_1) - T(S_2 - S_1) \tag{125}$$

$$\Delta A = \Delta U - T\Delta S \tag{126}$$

Where $\Delta U$ and $\Delta S$ are the change in internal energy and entropy, respectively. Therefore, the Helmholtz free energy change can be described as the difference of internal energy change and the multiplication of entropy change multiplied with the temperature at which the reaction is actually carried out.

**The physical significance of Helmholtz free energy:** From the definition of entropy change, we know that

$$\Delta S = \frac{q_{rev}}{T} \tag{127}$$

Also, from the first law of thermodynamics, for the work of expansion we have

$$\Delta U = q_{rev} - w_{max} \tag{128}$$

Putting the values of $\Delta S$ and $\Delta U$ from equation (127) and (128) in equation (126), we get

$$\Delta A = (q_{rev} - w_{max}) - T\left(\frac{q_{rev}}{T}\right) \tag{129}$$

$$= q_{rev} - w_{max} - q_{rev}$$

or

$$-\Delta A = w_{max} \tag{130}$$

Hence, the decrease in Helmholtz free energy at constant temperature is equal to the maximum work done by the system; that is why the Helmholtz free energy is also called as work function.

---

**Helmholtz free energy as the criteria for the spontaneity of reaction:** The general expression for the enthalpy of a system is given below.

$$A = U - TS \tag{131}$$

After differentiating the equation (131), we get

$$dA = dU - TdS - SdT \tag{132}$$

$$TdS = dU - SdT - dA \tag{133}$$

Also as we know that, for the spontaneity of a process

$$TdS \geq dU + PdV \tag{134}$$

Now after putting the value of $TdS$ from equation (133) in equation (134), we get

$$dU - SdT - dA \geq dU + PdV \tag{135}$$

or

$$-SdT - dA \geq PdV \tag{136}$$

$$dA \leq -PdV - SdT \tag{137}$$

For a spontaneous or irreversible process, $dA$ must be less than the negative sum of multiplication of pressure and volume change with the multiplication of entropy and temperature change.

However, the above relation is reduced to the following if the volume and temperature of the system are kept constant i.e. $dV = 0$ and $dT = 0$.

$$(dA)_{V,T} \leq 0 \tag{138}$$

Here it is worthy to recall that the sign '=' is for reversible reactions whereas '<' condition is applicable to spontaneity or irreversibility of a reaction.

> ➤ *Gibbs Free Energy or Gibbs Function*

The Gibbs free energy is typically denoted by the symbol '$G$', and can be defined mathematically as

$$G = H - TS \tag{139}$$

Where T, S and H are temperature, entropy, and enthalpy, respectively. Moreover; like H, T and S; the Gibbs free energy $G$ is also a state function. Since G is independent of its previous state, we can say that

$$G_1 = H_1 - TS_1 \tag{140}$$

$$G_2 = H_2 - TS_2 \tag{141}$$

Where the subscript 1 and 2 represent the initial and final state. The change in Gibbs free energy in going from initial to final state can obtain by subtracting equation (141) from equation (140) as

$$G_2 - G_1 = (H_2 - TS_2) - (H_1 - TS_1) \tag{142}$$

$$\Delta G = (H_2 - H_1) - T(S_2 - S_1) \tag{143}$$

$$\Delta G = \Delta H - T\Delta S \tag{144}$$

Where $\Delta H$ and $\Delta S$ are the change in enthalpy and entropy, respectively. Therefore, the Gibbs free energy change can be described as the difference of enthalpy change and the multiplication of entropy change multiplied with the temperature at which the reaction is actually carried out.

**The physical significance of Helmholtz free energy:** From the definition of entropy change, we know that

$$\Delta S = \frac{q_{rev}}{T} \tag{145}$$

Also, at constant pressure, we have

$$\Delta H = \Delta U + P\Delta V \tag{146}$$

Putting the values of $\Delta S$ and $\Delta H$ from equation (145) and (146) in equation (144), we get

$$\Delta G = (\Delta U + P\Delta V) - T\left(\frac{q_{rev}}{T}\right) \tag{147}$$

$$= (\Delta U - q_{rev}) + P\Delta V \tag{148}$$

Also, from the first of thermodynamics, for the work of expansion we have

$$\Delta U - q_{rev} = -w_{max} \tag{149}$$

Now, after putting the value of equation (149) in equation (148), we get

$$\Delta G = -w_{max} + P\Delta V \tag{150}$$

$$-\Delta G = w_{max} - P\Delta V \tag{151}$$

Hence, the decrease in Gibbs free energy for the process occurring at constant pressure and constant temperature is equal to the "maximum net work" that can be obtained from the process. The term "maximum net work" refers to maximum work other than the work of expansion.

**Gibbs free energy as the criteria for the spontaneity of reaction:** The general expression for the enthalpy of a system is given below.

$$G = H - TS \tag{152}$$

Since $H = U + PV$, equation (152) takes the form

$$G = U + PV - TS \tag{153}$$

After differentiating the equation (153), we get

$$dG = dU + PdV + VdP - TdS - SdT \tag{154}$$

$$TdS = dU + PdV + VdP - SdT - dG \tag{155}$$

Also as we know that, for the spontaneity of a process

$$TdS \geq dU + PdV \tag{156}$$

Now after putting the value of $TdS$ from equation (155) in equation (156), we get

$$dU + PdV + VdP - SdT - dG \geq dU + PdV \tag{156}$$

or

$$VdP - SdT - dG \geq 0 \tag{157}$$

$$dG \leq VdP - SdT \tag{158}$$

For a spontaneous or irreversible process, $dG$ must be less than the sum of negative multiplication of volume and pressure change with the multiplication of entropy and temperature change.

However, the above relation is reduced to the following if the pressure and temperature of the system are kept constant i.e. $dP = 0$ and $dT = 0$.

$$(dG)_{P,T} \leq 0 \tag{159}$$

Here it is worthy to recall that the sign '=' is for reversible reactions whereas '<' condition is applicable to spontaneity or irreversibility of a reaction.

## ❖ Partial Molar Quantities (Free Energy, Volume, Heat Concept)

So far we have discussed the variation of various thermodynamic properties with respect to temperature and pressure while the composition of the system was kept constant (closed system). In 1907, G.N. Lewis started the study of open systems i.e. the variation of various thermodynamic properties with respect to the composition of one or more components. In other words, he studied the behavior of a particular thermodynamic property of the system when a component is removed from or added to the system under consideration. Now since a variation like this is observable only for an extensive property, the general definition of partial molar properties can be given as given below.

*A partial molar property may simply be defined as a thermodynamic quantity which indicates how an extensive property of a solution or mixture changes with the variation in the molar composition of the mixture at constant temperature and pressure.*

Basically, it is the partial derivative of the extensive property with respect to the number of moles of the component under consideration. All extensive properties of a mixture have corresponding partial molar properties. In this section, we will discuss some very important partial molar quantities like partial molar free energy ($\overline{G_i}$), partial molar volume ($\overline{V_i}$) and partial molar enthalpy ($\overline{H_i}$).

➢ *Partial Molar Free Energy or Chemical Potential*

In order to derive the expression for partial molar free energy, consider a system that comprises of $n$ types of constituents with $n_1$, $n_2$, $n_3$, $n_4$ … moles. So, being an extensive property, the partial molar free energy depends upon not only the temperature and pressure but also on the number of moles of different components. Mathematically, we can say that

$$G = f(T, P, n_1, n_2, n_3 \dots) \tag{160}$$

Now let us assume a small change in the temperature, pressure and amount of different components, this would impart a variation in partial molar free energy as given below.

$$dG = \left(\frac{\partial G}{\partial T}\right)_{P,n_1,n_2\dots} dT + \left(\frac{\partial G}{\partial P}\right)_{T,n_1,n_2\dots} dP + \left(\frac{\partial G}{\partial n_1}\right)_{T,P,n_2,n_3\dots} dn_1 + \cdots \tag{161}$$

The first term on the right-hand side gives the change in the free energy with temperature at constant pressure and compositions; while the second term gives the change in the free energy with pressure at constant temperature and compositions. The terms afterward represent the variation in free energy with the amount of one component while the temperature, pressure and all other compositions are kept constant.

However, if the temperature and pressure of the system are kept constant i.e. $dT = 0$, $dP = 0$, the equation (161) takes the form

$$(dG)_{T,P} = \left(\frac{\partial G}{\partial n_1}\right)_{T,P,n_2,n_3\dots} dn_1 + \left(\frac{\partial G}{\partial n_2}\right)_{T,P,n_1,n_3\dots} dn_2 + \left(\frac{\partial G}{\partial n_3}\right)_{T,P,n_1,n_2\dots} dn_3 \dots \tag{162}$$

Every term on the right-hand side of the equation (162) is partial molar free energy and is symbolized by a "bar" over it i.e.

$$(dG)_{T,P} - \overline{G_1}\, dn_1 + \overline{G_2}\, dn_2 + \overline{G_3}\, dn_3 \tag{163}$$

For the $i$th component, we can say that

$$\overline{G_i} = \left(\frac{\partial G}{\partial n_i}\right)_{T,P,n_2,n_3\dots} \tag{164}$$

The equation (164) gives the general expression for "partial molar free energy" or the "chemical potential" of the $i$th species.

> ### *Partial Molar Volume*

In order to derive the expression for partial molar volume, consider a system that comprises of $n$ types of constituents with $n_1, n_2, n_3, n_4 \dots$ moles. So, being an extensive property, volume depends upon not only the temperature and pressure but also on the number of moles of different components. Mathematically, we can say that

$$V = f(T, P, n_1, n_2, n_3 \dots) \tag{165}$$

Now let us assume a small change in the temperature, pressure and amount of different components, this would impart a variation in partial molar volume as given below.

$$dV = \left(\frac{\partial V}{\partial T}\right)_{P,n_1,n_2\dots} dT + \left(\frac{\partial V}{\partial P}\right)_{T,n_1,n_2\dots} dP + \left(\frac{\partial V}{\partial n_1}\right)_{T,P,n_2,n_3\dots} dn_1 + \cdots \tag{166}$$

The first term on the right-hand side gives the change in the volume with the temperature at constant pressure and compositions; while the second term gives the change in the volume with pressure at constant temperature and compositions. The terms afterward represent the variation in volume with the amount of one component while the temperature, pressure and all other compositions are kept constant.

However, if the temperature and pressure of the system are kept constant i.e. $dT = 0$, $dP = 0$, the equation (166) takes the form

$$(dV)_{T,P} = \left(\frac{\partial V}{\partial n_1}\right)_{T,P,n_2,n_3\dots} dn_1 + \left(\frac{\partial V}{\partial n_2}\right)_{T,P,n_1,n_3\dots} dn_2 + \left(\frac{\partial V}{\partial n_3}\right)_{T,P,n_1,n_2\dots} dn_3 \dots \tag{167}$$

Every term on the right-hand side of the equation (167) is partial molar volume and is symbolized by a "bar" over it i.e.

$$\overline{V_1} = \left(\frac{\partial V}{\partial n_1}\right)_{T,P,n_2,n_3\dots} \tag{168}$$

$$\overline{V_2} = \left(\frac{\partial V}{\partial n_2}\right)_{T,P,n_1,n_3\dots} \tag{169}$$

After putting the values from equations like (168 – 169) in equation (167), we get

$$(dV)_{T,P} = \overline{V_1}\, dn_1 + \overline{V_2}\, dn_2 + \overline{V_3}\, dn_3 \dots \tag{170}$$

For the $i$th component, we can say that

$$\overline{V_i} = \left(\frac{\partial V}{\partial n_i}\right)_{T,P,n_2,n_3\dots} \tag{171}$$

The equation (171) gives the general expression for the "partial molar volume" of the $i$th species.

> ➤ *Partial Molar Enthalpy or Partial Molar Heat Content*

In order to derive the expression for partial molar enthalpy, consider a system that comprises of $n$ types of constituents with $n_1$, $n_2$, $n_3$, $n_4$ … moles. So, being an extensive property, volume depends upon not only the temperature and pressure but also on the number of moles of different components, i.e.,

$$H = f(T, P, n_1, n_2, n_3 \ldots) \tag{172}$$

Now let us assume a small change in the temperature, pressure and amount of different components, this would impart a variation in molar enthalpy as given below.

$$dH = \left(\frac{\partial H}{\partial T}\right)_{P,n_1,n_2\ldots} dT + \left(\frac{\partial H}{\partial P}\right)_{T,n_1,n_2\ldots} dP + \left(\frac{\partial H}{\partial n_1}\right)_{T,P,n_2,n_3\ldots} dn_1 + \cdots \tag{173}$$

The first term on the right-hand side gives the change in the enthalpy with the temperature at constant pressure and compositions; while the second term gives the change in the enthalpy with pressure at constant temperature and compositions. The terms afterward represent the variation in enthalpy with the amount of one component while the temperature, pressure and all other compositions are kept constant.

However, if the temperature and pressure of the system are kept constant i.e. $dT = 0$, $dP = 0$, the equation (173) takes the form

$$(dH)_{T,P} = \left(\frac{\partial H}{\partial n_1}\right)_{T,P,n_2,n_3\ldots} dn_1 + \left(\frac{\partial H}{\partial n_2}\right)_{T,P,n_1,n_3\ldots} dn_2 + \left(\frac{\partial H}{\partial n_3}\right)_{T,P,n_1,n_2\ldots} dn_3 \ldots \tag{174}$$

Every term on the right-hand side of the equation (174) is partial molar enthalpy and is symbolized by a "bar" over it i.e.

$$\overline{H_1} = \left(\frac{\partial H}{\partial n_1}\right)_{T,P,n_2,n_3\ldots} \tag{175}$$

$$\overline{H_2} = \left(\frac{\partial H}{\partial n_2}\right)_{T,P,n_1,n_3\ldots} \tag{176}$$

After putting the values from equations like (175 – 176) in equation (174), we get

$$(dH)_{T,P} = \overline{H_1}\, dn_1 + \overline{H_2}\, dn_2 + \overline{H_3}\, dn_3 \ldots \tag{177}$$

For the $i$th component, we can say that

$$\overline{H_i} = \left(\frac{\partial H}{\partial n_i}\right)_{T,P,n_2,n_3\ldots} \tag{178}$$

The equation (178) gives the general expression for the "partial molar enthalpy" of the $i$th species.

## ❖ Gibb's-Duhem Equation

In order to derive the expression for Gibbs's-Duhem equation, consider a system that comprises of $n$ types of constituents with $n_1, n_2, n_3, n_4 \dots$ moles. So, being an extensive property, the partial molar free energy depends upon not only the temperature and pressure but also on the number of moles of different components. Mathematically, we can say that

$$G = f(T, P, n_1, n_2, n_3 \dots) \tag{179}$$

Now let us assume a small change in the temperature, pressure and amount of different components, this would impart a variation in partial molar free energy as given below.

$$dG = \left(\frac{\partial G}{\partial T}\right)_{P,n_1,n_2\dots} dT + \left(\frac{\partial G}{\partial P}\right)_{T,n_1,n_2\dots} dP + \left(\frac{\partial G}{\partial n_1}\right)_{T,P,n_2,n_3\dots} dn_1 + \cdots \tag{180}$$

The first term on the right-hand side gives the change in the free energy with the temperature at constant pressure and compositions; while the second term gives the change in the free energy with pressure at constant temperature and compositions. The terms afterward represent the variation in free energy with the amount of one component while the temperature, pressure and all other compositions are kept constant.

However, if the temperature and pressure of the system are kept constant, i.e., $dT = 0$, $dP = 0$, the equation (180) takes the form

$$(dG)_{T,P} = \left(\frac{\partial G}{\partial n_1}\right)_{T,P,n_2,n_3\dots} dn_1 + \left(\frac{\partial G}{\partial n_2}\right)_{T,P,n_1,n_3\dots} dn_2 + \left(\frac{\partial G}{\partial n_3}\right)_{T,P,n_1,n_2\dots} dn_3 \dots \tag{181}$$

Every term on the right-hand side of the equation (181) is partial molar free energy or simply the "chemical potential" i.e.

$$\mu_1 = \left(\frac{\partial G}{\partial n_1}\right)_{T,P,n_2,n_3\dots} \tag{182}$$

$$\mu_2 = \left(\frac{\partial G}{\partial n_2}\right)_{T,P,n_1,n_3\dots} \tag{183}$$

$$\mu_3 = \left(\frac{\partial G}{\partial n_3}\right)_{T,P,n_1,n_2\dots} \tag{184}$$

$$\mu_4 = \left(\frac{\partial G}{\partial n_4}\right)_{T,P,n_1,n_2\dots} \tag{185}$$

$$\mu_5 = \left(\frac{\partial G}{\partial n_5}\right)_{T,P,n_1,n_2\dots} \tag{186}$$

After putting the values from equations like (182 – 186) in equation (181), we get

$$(dG)_{T,P} = \mu_1\, dn_1 + \mu_2\, dn_2 + \mu_3\, dn_3 + \mu_4\, dn_4 + \mu_5\, dn_5 \ldots \tag{187}$$

Now, if the system composition is definite, the integration of the above equation gives

$$G_{T,P,N} = \mu_1 n_1 + \mu_2 n_2 + \mu_3 n_3 + \mu_4\, dn_4 + \mu_5\, dn_5 \ldots \tag{188}$$

The differentiation of equation (188) at constant temperature and constant pressure but changing composition gives the following relation

$$(dG)_{T,P} = (n_1 d\mu_1 + \mu_1 dn_1) + (n_2 d\mu_2 + \mu_2 dn_2) + (n_3 d\mu_3 + \mu_3 dn_3) \tag{189}$$
$$+ (n_4 d\mu_4 + \mu_4 dn_4) + (n_5 d\mu_5 + \mu_5 dn_5) \ldots$$

or

$$(dG)_{T,P} = (\mu_1 dn_1 + \mu_2 dn_2 + \mu_3 dn_3 + \mu_4 dn_4 + \mu_5 dn_5 \ldots) \tag{190}$$
$$+ (n_1 d\mu_1 + n_2 d\mu_2 + n_3 d\mu_3 + n_4 d\mu_4 + n_5 d\mu_5 \ldots)$$

After comparing the equation (187) and (190), we can conclude that the content included in the second bracket must be equal to zero. Mathematically, we can say that

$$n_1 d\mu_1 + n_2 d\mu_2 + n_3 d\mu_3 \ldots = 0 \tag{191}$$

or

$$\sum n_i d\mu_i = 0 \tag{192}$$

Which is the popular Gibbs-Duhem equation, and is applicable to the systems under constant temperature-pressure conditions.

The physical significance of the Gibbs-Duhem equation can be understood by taking the example of binary solutions i.e. a system of two components only. The Gibbs-Duhem equation for such systems is

$$n_1 d\mu_1 + n_2 d\mu_2 = 0 \tag{193}$$

or

$$n_1 d\mu_1 = -n_2 d\mu_2 \tag{194}$$

or

$$d\mu_1 = -\frac{n_2}{n_1}\, d\mu_2 \tag{195}$$

Hence, the chemical potential of one constituent is not independent of another component in binary solutions. In other words, the chemical potentials or partial molar free energies of two components of the binary system are mutually dependent; and if the one increases the other one decreases.

It is also worthy to note that if the number of moles of different constituents remains constant (closed system), i.e., $dn_i = 0$; equation (180) reduces to the following.

$$dG = \left(\frac{\partial G}{\partial T}\right)_{P,N} dT + \left(\frac{\partial G}{\partial P}\right)_{T,N} dP \tag{196}$$

Also, for a closed system, we know that

$$dG = VdP - SdT \tag{197}$$

Which means that

$$\left(\frac{\partial G}{\partial T}\right)_{P,N} = -S \quad and \quad \left(\frac{\partial G}{\partial P}\right)_{T,N} = V \tag{198}$$

Which is the variation of free energy with temperature and pressure in closed systems. These two relations can further be used to deduce the variations of chemical potentials with temperature and volume.

## ❖ Problems

Q 1. Define the first law of thermodynamics with pictorial description.

Q 2. Discuss the limitations of the first law of thermodynamics and the need for the second law.

Q 3. What is entropy? How does it behave in reversible and irreversible reactions?

Q 4. Comment on the statement "all spontaneous processes are thermodynamically irreversible".

Q 5. Derive the relation showing the variation of entropy with temperature and pressure.

Q 6. What do you mean by "unavailable energy" or the "lost work"?

Q 7. Discuss the spontaneity of a process in terms of entropy change.

Q 8. What is Gibbs free energy? What are its significances?

Q 9. Define the work function and discuss the related spontaneity science.

Q 10. What are partial molar quantities? Discuss with the special reference of chemical potential.

Q 11. Derive Gibbs-Duhem equation.

### ❖ Bibliography

[1] G. Raj, *Thermodynamics*, Krishna Prakashan, Meerut, India, 2004.

[2] B. R. Puri, L. R. Sharma, M. S. Pathania, *Principles of Physical Chemistry*, Vishal Publications, Jalandhar, India, 2008.

[3] P. Atkins, J. Paula, *Physical Chemistry*, Oxford University Press, Oxford, UK, 2010.

[4] E. Steiner, *The Chemistry Maths Book*, Oxford University Press, Oxford, UK, 2008.

[5] S. Carnot, R. Fox, *Reflections on the Motive Power of Fire: a Critical Edition with the Surviving Scientific Manuscripts*, Manchester University Press, New York, USA, 1986.

[6] P. A. Rock, *Chemical Thermodynamics*, University Science Books, California, USA, 1983.

[7] E. B. Smith, *Basic Chemical Thermodynamics*, Imperial College Press, London, UK, 2014.

# CHAPTER 3

# Chemical Dynamics – I

### ❖ Effect of Temperature on Reaction Rates

The temperature of the system shows a very marked effect on the overall rate of the reaction. In fact, it has been observed that the rate of a chemical reaction typically gets doubled with every 10°C rise in the temperature. However, this ratio may differ considerably and may reach up to 3 for different reactions. Besides, this ratio also varies as the temperature of the reaction increases gradually. The ratio of rate constant at two different temperatures is called as "temperature coefficient" of the reaction. Although we can determine the temperature coefficient between any two temperatures for any chemical reaction, generally it is calculated for 10°C difference.

$$Temperature\ Coefficient = \frac{k_{T+10}}{k_T} = 2 - 3 \tag{1}$$

Where $k_T$ and $k_{T+10}$ are rate constants at temperature $T$ and $T+10$, respectively. Now, if once the temperature coefficient is known, you can determine the relative increase or decrease in the overall reaction-rate by using the following relation.

$$\frac{R_2}{R_1} = \frac{k_2}{k_1} = (Temperature\ Coefficient)^{\frac{T_2-T_1}{\Delta T_{tc}}} \tag{2}$$

Where $R_2$ and $R_1$ are the reaction-rates at temperatures $T_2$ and $T_1$, respectively. The $\Delta T_{tc}$ is the temperature range for the temperature coefficient.

In order to illustrate the dominance of the effect of temperature change on the reaction rate, consider a reaction in which the temperature of the system is raised from 310°C to 400°C. Now, if the temperature coefficient for 10°C temperature-rise is 2, the relative increase in the rate constant or rate will be

$$\frac{R_2}{R_1} = \frac{k_2}{k_1} = (2)^{\frac{400-310}{10}} \tag{3}$$

$$\frac{R_2}{R_1} = \frac{k_2}{k_1} = (2)^9 = 512 \tag{4}$$

$$\frac{R_2}{R_1} = 512 \tag{5}$$

Hence, a 90°C rise in temperature increases the rate of reaction 512 times, which is definitely huge. Now the question arises, why is it so? What did the temperature do that made this happen? In this section, we will answer these questions.

## ➤ *Fundamentals of Temperature-Rate Correlation*

Before we discuss the effect of temperature on the reaction rate, we must understand the cause of a reaction itself first. The primary requirement for a reaction to occur is the collision between the reacting molecules. In other words, the reactant molecules must collide with each other to form the product. Therefore, if we assume that every collision results in the formation of the product, the rate of reaction should simply be equal to the collision frequency of the reacting system.

For a reaction between $A$ and $B$, the collision frequency $(Z)$ is the number of collisions between $A$ and $B$ occurring in the container per unit volume per unit time.

$$Z = n_A\, n_B\, \Theta_{AB} \sqrt{\frac{8k_B T}{\pi \mu_{AB}}} \qquad (6)$$

Where $n_A$ and $n_A$ are the number densities (in the units of $m^{-3}$) of particles $A$ and $B$, respectively. The term $\Theta_{AB}$ is the reaction cross-section (in $m^2$) when particle A with radius $r_A$ and B with radius $r_B$ collide with each other i.e. $\Theta_{AB} = \pi\,(\sigma_{AB})^2 = \pi(r_A + r_B)^2 = \pi(\sigma_A/2 + \sigma_B/2)^2$. $k_B$ is the Boltzmann's constant ($m^2\ kg\ s^{-2}\ K^{-1}$). $T$ represents the temperature of the system. The term $\mu_{AB}$ represents the reduced mass of the reactants A and B i.e. $\mu_{AB} = m_A m_B / m_A + m_B$. From equation (6), it follows that when we heat the substance, the particles collide more frequently and hence increase the collision frequency. Now one may think that this collision frequency would result in a larger rate of reaction, and therefore, the mystery is solved. However, this isn't sufficient to rationalize the experimental observations. For instance, if we increase the temperature from 300 K to 310 K, the relative increase in the collision frequency ($n_Z$), and hence in reaction rate, from equation (6) can be determined as given below.

$$n_Z = \sqrt{\frac{310}{300}} = 1.0165 \qquad (7)$$

This is only 1.65% increase for a 10° rise in temperature. This is pretty far from the reality i.e. reaction-rate almost gets double (100% increase). So, the actual mechanism is still behind the scene and must be understood.

At this point we must introduce the concept of activation energy, otherwise, the concept cannot be discussed further. The collision of reacting molecules would result in the chemical reaction only if they possess a certain amount of minimum energy i.e. threshold energy. Since every molecule does have some energy, the energy it needs to reach the threshold is less than the actual threshold energy. The energy required by reactant molecules to cross the barrier is called the activation energy or the enthalpy of activation for the reaction. A simple equation can be used to deduce their relationships as given below.

Activation energy = Threshold energy – Energy actually possessed by the molecules

The rate of a chemical reaction is inversely proportional to the magnitude of the activation energy i.e. larger the activation energy, slower will be the reaction and vice-versa.

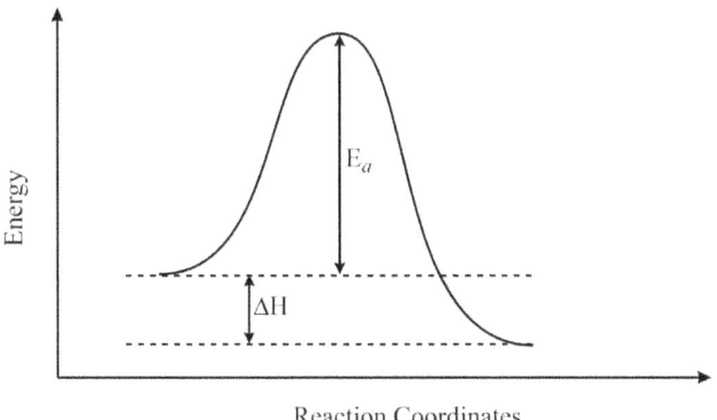



Figure 1. The reaction coordinate diagram for a typical chemical reaction.

Hence, we can say that only effective collisions would result in the chemical reactions, but how can we find the number of molecules having energy high enough to react with each other. For this, we need to go into the basics of energy distribution among a large number of particles i.e. Maxwell's distribution of energies.

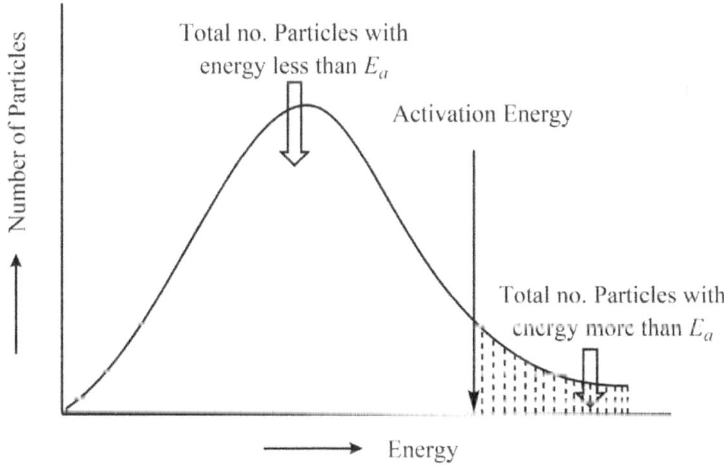



Figure 2. The Maxwell-Boltzmann distribution of energies at temperature T.

After marking the activation energy on the Maxwell-Boltzmann distribution curve, the particle with sufficient energy to react can easily be found from the area under the corresponding curve i.e. dashed area. The undashed area at a particular temperature is quite large, and therefore, represents the number particle whose collision would not result in any reaction chemical change.  It can be clearly seen that most of the particles don't have enough energy, and hence, are unable to yield the product. The reaction-rate will be very small If there are very few particles with enough energy at any time.

However, if the temperature is raised, the maxima of the Maxwell-Boltzmann distribution curve shifts towards higher energy. This makes the number of "efficient particles" to increase and thereby increases the number of effective collisions too. Consider the Maxwell-Boltzmann energy distributions at temperature T and T+10.

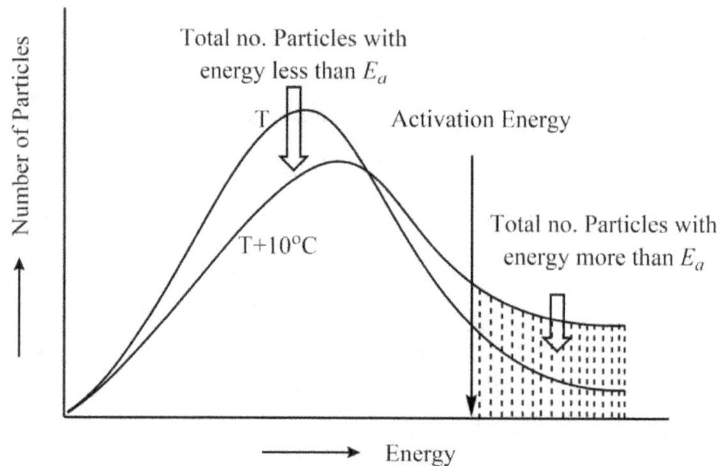


Figure 3. The Maxwell-Boltzmann distribution of energies at temperature T and T+10°C.

Now although the area under the whole curve remains the same, the dashed area is doubled. Thus, the primary reason for almost 100% rise in the overall rate of reaction for every 10°C is the 100% increase in the number of effective collisions.

> ### The Arrhenius Equation

In 1884, the famous Dutch chemist Jacobus Henricus Van't Hoff realized that his equation (Van't Hoff equation) could also be used to suggests a formula for the rates of both forward and backward reactions. In 1889, Svante Arrhenius immediately noticed the importance of this invention and proposed an empirical equation based on Van't Hoff's work. This equation is extremely useful in the modeling of the temperature variation of many chemical reactions. The equation proposed by Arrhenius is

$$k = A\, e^{-E_a/RT} \tag{8}$$

Where the symbol $k$, $R$ and T represent rate constant, gas constant and temperature, respectively. $A$ is popularly known as the pre-exponential factor or Arrhenius constant with the units identical to those of the rate constant used, and therefore, will vary depending on the order of the reaction. The term $E_a$ represents the activation energy measured in joule mole$^{-1}$.

Another popular form of the Arrhenius equation is

$$k = A\,e^{-E_a/k_B T} \tag{9}$$

The only difference in the equation (8) and equation (9) is the energy units of $E_a$; the former one uses energy per mole, which is more common in chemistry, while the latter form uses energy per molecule directly, which is common in physics. The different units are accounted for in using either the gas constant, $R$, or the Boltzmann constant, $k_B$, as the multiplier of temperature T. If the reaction is first order, $A$ will have the units of $s^{-1}$ and can be called as collision frequency or frequency factor.

The physical significance of $k$ is that it represents the number of collisions that result in a reaction per second; A is the number of collisions (leading to a reaction or not) per second occurring with the proper orientation to react. The exponential factor is the probability that any given collision will result in a reaction. It can also be seen that either increasing the temperature or decreasing the activation energy (for example through the use of catalysts) will result in an increase in the rate of reaction. Taking the natural logarithm of both side of equation (8), we get

$$\ln k = \ln A + \ln e^{-E_a/RT} \tag{10}$$

$$\ln k = \ln A - \frac{E_a}{RT} \tag{11}$$

Rearrange the above equation, we get

$$\ln k = -\frac{E_a}{RT} + \ln A \tag{12}$$

The equation (12) has the same form as the equation of straight line i.e. $y = mx + c$; which means that if we plot "$\ln k$" vs 1/T, the slope and intercept will yield "$-E_a/R$" and "$\ln A$", respectively.

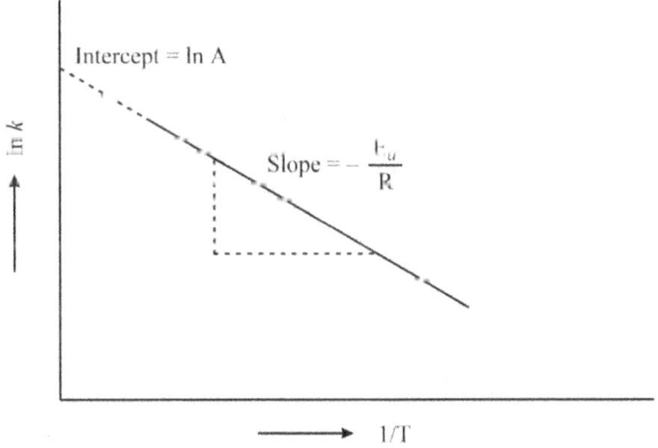


Figure 4. The Arrhenius plot of ln $k$ vs 1/T.

In addition to the equation (12), one of the more popular forms of the Arrhenius equation can be derived by converting it to the common logarithm as given below.

$$2.303 \log k = -\frac{E_a}{RT} + 2.303 \log A \tag{13}$$

or

$$\log k = -\frac{E_a}{2.303\, RT} + \log A \tag{14}$$

The equation (14) also has the same form as the equation of straight line i.e. $y = mx + c$; which means that if we plot "log $k$" vs 1/T, the slope and intercept will yield "$-E_a/2.303R$" and "log A", respectively.

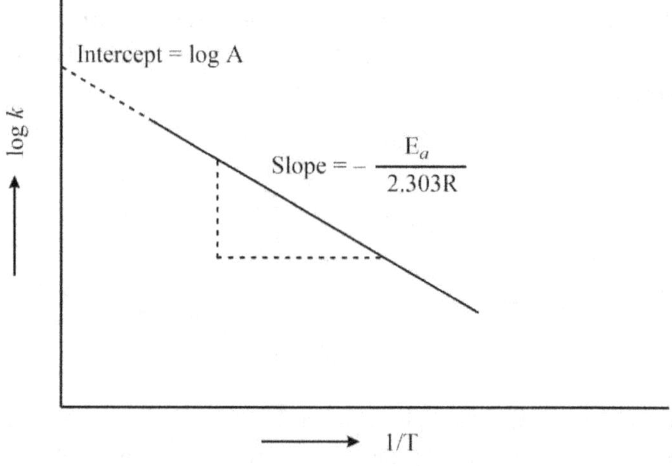


Figure 5. The Arrhenius plot of log $k$ vs 1/T.

To obtain the integrated form of the Arrhenius equation, differentiate the equation (12) as given below.

$$\frac{d \ln k}{dT} = \frac{E_a}{RT^2} \tag{15}$$

Now integrating the above equation between temperature $T_1$ and $T_2$, we get

$$\ln \frac{k_2}{k_1} = \frac{E_a}{R}\left[\frac{1}{T_1} - \frac{1}{T_2}\right] \quad or \quad \log \frac{k_2}{k_1} = \frac{E_a}{2.303R}\left[\frac{1}{T_1} - \frac{1}{T_2}\right] \tag{16}$$

The above equation can be used to find the activation energy if rate constants are known at two different temperatures.

## ❖ Rate Law for Opposing Reactions of Ist Order and IInd Order

*A reaction will be called as the opposing or reversible reaction if the reactants react together to form a product and the products also react to yield the reactants simultaneously under the same conditions.*

In a simple context, we can say these reactions proceed not only in the forward direction but also in the backward direction. These reactions can be classified into the following categories based upon the kinetic order of the reactions involved.

➢ *First Order Opposed by First Order*

In order to understand the kinetic profile of first-order reactions opposed by the first order, consider a general reaction in which the reactant A forms product B i.e.

$$A \underset{k_b}{\overset{k_f}{\rightleftharpoons}} B \tag{17}$$

Now, if $k_f \ggg k_b$, $k_b$ can be neglected. However, $k_f$ and $k_b$ have comparable values, a rate law depending upon both the constants can be written. To do so, suppose that $a$ is the initial concentration of the reactant $A$ and $x$ is the decrease in the concentration of $A$ after '$t$' time. The concentration of the product after the same time would also be equal to $x$. Hence, the rates of forward reaction ($R_f$) and backward reaction ($R_b$) can be given as:

$$R_f = k_f[A] = k_f(a - x) \tag{17}$$

$$R_b = k_b[B] = k_b x \tag{18}$$

The net reaction rate i.e. rate of formation of the product can be given as

$$\frac{dx}{dt} = k_f(a - x) - k_b x \tag{19}$$

However, when the equilibrium is attained, the rate of forward reaction will be equal to the rate of backward reaction i.e. $R_f = R_b$. Therefore, the will take the form

$$k_f(a - x_{eq}) = k_b x_{eq} \tag{20}$$

Where $x_{eq}$ is the concentration of product $B$ or the decrease in the concentration of reactant $A$ at equilibrium. Now putting the value of $k_b$ from equation (20) into equation (19), we get

$$\frac{dx}{dt} = k_f(a - x) - \frac{k_f(a - x_{eq})}{x_{eq}} x \tag{21}$$

$$\frac{dx}{dt} = k_f\left[(a - x) - \frac{(a - x_{eq})}{x_{eq}} x\right] \tag{22}$$

or

$$\frac{dx}{dt} = k_f \left[ \frac{(a - x)x_{eq} - (a - x_{eq})x}{x_{eq}} \right] \tag{23}$$

$$\frac{dx}{dt} = k_f \frac{ax_{eq} - ax}{x_{eq}} = \frac{k_f a(x_{eq} - x)}{x_{eq}} \tag{24}$$

or

$$\frac{x_{eq}}{(x_{eq} - x)} dx = k_f a\, dt \tag{25}$$

Integrating equation (25), we get

$$-x_{eq}\ln(x_{eq} - x) = k_f at + C \tag{26}$$

Where $C$ is the constant of integration. When $t = 0$, $x = 0$; putting these values in equation (26), we get

$$-x_{eq}\ln x_{eq} = C \tag{27}$$

Using the value of C form equation (27) in (26)

$$-x_{eq}\ln(x_{eq} - x) = k_f at - x_{eq}\ln x_{eq} \tag{28}$$

or

$$x_{eq}\ln x_{eq} - x_{eq}\ln(x_{eq} - x) = k_f at \tag{29}$$

$$k_f = \frac{x_{eq}}{at}\ln \frac{x_{eq}}{x_{eq} - x} \tag{30}$$

Using equation (30), the rate constant for the forward reaction can easily be determined by measuring simple quantities like $a$, $t$, $x_{eq}$ and $x$. Now rearranging equation (20) for $k_b$

$$k_b = \frac{k_f(a - x_{eq})}{x_{eq}} \tag{31}$$

$$k_b = \frac{k_f(a - x_{eq})}{x_{eq}} = k_f\left(\frac{a}{x_{eq}} - \frac{x_{eq}}{x_{eq}}\right) = k_f\left(\frac{a}{x_{eq}} - 1\right) \tag{32}$$

Now putting the value $k_f$ from equation (30) in equation (32), we get

$$k_b = \left(\frac{1}{t}\ln \frac{x_{eq}}{x_{eq} - x}\right) - \frac{x_{eq}}{at}\ln \frac{x_{eq}}{x_{eq} - x} \tag{33}$$

Hence, the value of the rate constant for backward reaction can also be obtained just by measuring $t$, $x_{eq}$ and $x$; or in other words, the $k_f$ eventually yields the $k_b$ also.

Alternatively, the values of $k_f$ and $k_b$ can also be obtained a slightly different route. Rearranging equation (30), we get

$$\frac{a\,k_f}{x_{eq}} = \frac{1}{t}\ln\frac{x_{eq}}{x_{eq}-x} \tag{34}$$

Also rearranging equation (20), we have

$$\frac{k_f a}{x_{eq}} = k_b + k_f \tag{35}$$

Equating the right-hand sides of equation (34) and (35), we get

$$\frac{1}{t}\ln\frac{x_{eq}}{x_{eq}-x} = k_b + k_f \tag{36}$$

$$\ln\frac{x_{eq}}{x_{eq}-x} = \left(k_b + k_f\right)t \tag{37}$$

or

$$\log\frac{x_{eq}}{x_{eq}-x} = \frac{\left(k_b + k_f\right)}{2.303}t \tag{38}$$

Equation (37) and (38) are the equations of straight line ($y = mx + c$) with zero intercept.

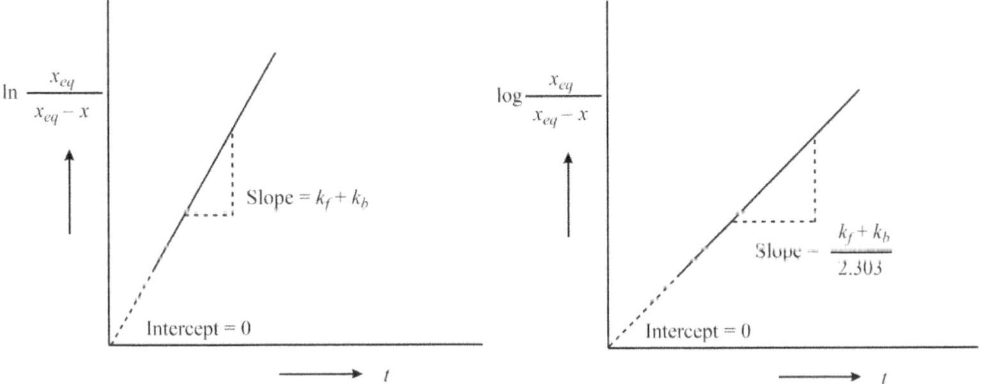


Figure 6. The plot of $\ln\frac{x_{eq}}{x_{eq}-x}$ and $\log\frac{x_{eq}}{x_{eq}-x}$ vs time for first order opposed by the first order.

Now, finding $k_f + k_b$ from slope and $k_f/k_b$ from equilibrium constant, $k_f$ and $k_b$ can easily be obtained from the elimination method. It should also be noted that if plot $\ln\frac{x_{eq}-x}{x_{eq}}$ and $\log\frac{x_{eq}-x}{x_{eq}}$, the slopes will become negative.

➤ *First Order Opposed by Second Order*

In order to understand the kinetic profile of first-order reactions opposed by second-order, consider a general reaction in which the reactant $A$ forms product $B$ and $C$ i.e.

$$A \underset{k_b}{\overset{k_f}{\rightleftharpoons}} B + C \tag{39}$$

Now, if $k_f \ggg k_b$, $k_b$ can be neglected. However, $k_f$ and $k_b$ have comparable values, a rate law depending upon both the constants can be written. To do so, suppose that $a$ is the initial concentration of the reactant $A$ and $x$ is the decrease in the concentration of $A$ after '$t$' time. The concentration of both the products after same time would also be equal to $x$. Hence, the rates of forward reaction ($R_f$) and backward reaction ($R_b$) can be given as:

$$R_f = k_f[A] = k_f(a - x) \tag{40}$$

$$R_b = k_b[B][C] = k_b x^2 \tag{41}$$

The net reaction rate i.e. rate of formation of the product can be given as

$$\frac{dx}{dt} = k_f(a - x) - k_b x^2 \tag{42}$$

However, when the equilibrium is attained, the rate of forward reaction will be equal to the rate of backward reaction i.e. $R_f = R_b$. Therefore, the equation (42) will take the form

$$k_f\left(a - x_{eq}\right) = k_b x_{eq}^2 \tag{43}$$

Where $x_{eq}$ is the concentration of product $B$ and $C$ or the decrease in the concentration of reactant $A$ at equilibrium. Now putting the value of $k_b$ from equation (43) into equation (42), we get

$$\frac{dx}{dt} = k_f(a - x) - \frac{k_f\left(a - x_{eq}\right)}{x_{eq}^2} x^2 \tag{44}$$

$$\frac{dx}{dt} = k_f\left[(a - x) - \frac{\left(a - x_{eq}\right)}{x_{eq}^2} x^2\right] \tag{45}$$

or

$$\frac{dx}{dt} = k_f\left[\frac{(a - x)x_{eq}^2 - \left(a - x_{eq}\right)x^2}{x_{eq}^2}\right] \tag{46}$$

$$\frac{dx}{dt} = k_f\left[\frac{ax_{eq}^2 - xx_{eq}^2 - ax^2 + x_{eq}x^2}{x_{eq}^2}\right] \tag{47}$$

or

$$\frac{x_{eq}^2}{\left(ax_{eq}^2 - xx_{eq}^2 - ax^2 + x_{eq}x^2\right)}\, dx = k_f dt \tag{48}$$

Integrating equation (48), and then rearranging

$$k_f = \frac{x_{eq}}{t(2a - x_{eq})}\ln\frac{ax_{eq} + x_{eq}(a - x_{eq})}{a(x_{eq} - x)} \tag{49}$$

Using equation (49), the rate constant for the forward reaction can easily be determined by measuring simple quantities like $a$, $t$, $x_{eq}$ and $x$. Now, we know that the equilibrium constant for the first order opposed by second-order will be

$$K = \frac{[B][C]}{[A]} \tag{50}$$

also

$$K = \frac{k_f}{k_b} \tag{51}$$

Now putting the value of $k_f$ from equation (49) in equation (51) and the rearranging for $k_b$, we get

$$k_b = \frac{x_{eq}}{t(2a - x_{eq})K}\ln\frac{ax_{eq} + x_{eq}(a - x_{eq})}{a(x_{eq} - x)} \tag{52}$$

Hence, the value of the rate constant for backward reaction can also be obtained just by measuring $t$, $x_{eq}$ and $x$ and the equilibrium constant from equation (50).

> ➢ *Second Order Opposed by First Order*

In order to understand the kinetic profile of second-order reactions opposed by first order, consider a general reaction in which two reactants $A$ and $B$ form product $C$ i.e.

$$A + D \underset{k_b}{\overset{k_f}{\rightleftharpoons}} C \tag{53}$$

Now, if $k_f \ggg k_b$, $k_b$ can be neglected. However, $k_f$ and $k_b$ have comparable values, a rate law depending upon both the constants can be written. To do so, suppose that $a$ is the initial concentration of both the reactant $A$ and $B$; while $x$ is the decrease in the concentrations of both reactants after '$t$' time. The concentration of the product after the same time would also be equal to $x$. Hence, the rates of forward reaction ($R_f$) and backward reaction ($R_b$) can be given as:

$$R_f = k_f[A][B] = k_f(a - x)^2 \tag{54}$$

$$R_b = k_b[C] = k_b x \tag{55}$$

The net reaction rate i.e. rate of formation of the product can be given as

$$\frac{dx}{dt} = k_f(a - x)^2 - k_b x \tag{56}$$

However, when the equilibrium is attained, the rate of forward reaction will be equal to the rate of backward reaction i.e. $R_f = R_b$. Therefore, the equation (56) will take the form

$$k_f\left(a - x_{eq}\right)^2 = k_b x_{eq} \tag{57}$$

Where $x_{eq}$ is the concentration of the product $C$ or the decrease in the concentration of reactant $A$ or $B$ at equilibrium. Now putting the value of $k_b$ from equation (57) into equation (56), we get

$$\frac{dx}{dt} = k_f(a - x)^2 - \frac{k_f\left(a - x_{eq}\right)^2}{x_{eq}} x \tag{58}$$

$$\frac{dx}{dt} = k_f\left[(a - x)^2 - \frac{\left(a - x_{eq}\right)^2}{x_{eq}} x\right] \tag{59}$$

or

$$\frac{dx}{dt} = k_f\left[\frac{(a - x)^2 x_{eq} - \left(a - x_{eq}\right)^2 x}{x_{eq}}\right] \tag{60}$$

$$\frac{dx}{dt} = k_f\left[\frac{(a^2 + x^2 - 2ax)x_{eq} - \left(a^2 + x_{eq}^2 - 2ax_{eq}\right)x}{x_{eq}}\right] \tag{61}$$

$$\frac{dx}{dt} = k_f\left[\frac{a^2 x_{eq} + x^2 x_{eq} - 2axx_{eq} - a^2 x - x_{eq}^2 x + 2ax_{eq}x}{x_{eq}}\right] \tag{62}$$

$$\frac{dx}{dt} = k_f\left[\frac{a^2 x_{eq} + x^2 x_{eq} - a^2 x - x_{eq}^2 x}{x_{eq}}\right] \tag{63}$$

or

$$\frac{x_{eq}}{\left(a^2 x_{eq} + x^2 x_{eq} - a^2 x - x_{eq}^2 x\right)} dx = k_f dt \tag{64}$$

Integrating equation (64), and then rearranging

$$k_f = \frac{x_{eq}}{t(a^2 - x_{eq}^2)} \ln \frac{x_{eq}(a^2 - x_{eq}x)}{a^2(x_{eq} - x)} \tag{65}$$

Using equation (66), the rate constant for the forward reaction can easily be determined by measuring simple quantities like $a$, $t$, $x_{eq}$ and $x$. Now, we know that the equilibrium constant for a second-order reaction opposed by first order will be

$$K = \frac{[C]}{[A][B]} \tag{66}$$

$$K = \frac{k_f}{k_b} \tag{67}$$

Now putting the value of $k_f$ from equation (65) in equation (67) and the rearranging for $k_b$, we get

$$k_b = \frac{x_{eq}}{t(a^2 - x_{eq}^2)K} \ln \frac{x_{eq}(a^2 - x_{eq}x)}{a^2(x_{eq} - x)} \tag{68}$$

Hence, the value of the rate constant for backward reaction can also be obtained just by measuring $t$, $x_{eq}$ and $x$ and the equilibrium constant from equation (66).

> ➤ ***Second Order Opposed by Second Order***

In order to understand the kinetic profile of second-order reactions opposed by second-order, consider a general reaction in which two reactants $A$ and $B$ form product $C$ and $D$ i.e.

$$A + B \;\underset{k_b}{\overset{k_f}{\rightleftharpoons}}\; C + D \tag{69}$$

Now, if $k_f \ggg k_b$, $k_b$ can be neglected. However, $k_f$ and $k_b$ have comparable values, a rate law depending upon both the constants can be written. To do so, suppose that $a$ is the initial concentration of both the reactant $A$ and $B$; while $x$ is the decrease in the concentrations of both reactants after '$t$' time. The concentration of the products after the same time would also be equal to $x$. Hence, the rates of forward reaction ($R_f$) and backward reaction ($R_b$) can be given as:

$$R_f = k_f[A][B] = k_f(a - x)^2 \tag{70}$$

$$R_b = k_b[C][D] = k_b x^2 \tag{71}$$

The net reaction rate i.e. rate of formation of the product can be given as

$$\frac{dx}{dt} = k_f(a - x)^2 - k_b x^2 \tag{72}$$

However, when the equilibrium is attained, the rate of forward reaction will be equal to the rate of backward reaction i.e. $R_f = R_b$. Therefore, the equation (72) will take the form

$$k_f(a - x_{eq})^2 = k_b x_{eq}^2 \tag{73}$$

Where $x_{eq}$ is the concentration of the product $C$ and $D$ or the decrease in the concentration of reacant $A$ or $B$ at equilibrium. Now putting the value of $k_b$ from equation (73) into equation (72), we get

$$\frac{dx}{dt} = k_f(a-x)^2 - \frac{k_f(a-x_{eq})^2}{x_{eq}^2}x^2 \tag{74}$$

$$\frac{dx}{dt} = k_f\left[(a-x)^2 - \frac{(a-x_{eq})^2}{x_{eq}^2}x^2\right] \tag{75}$$

or

$$\frac{dx}{dt} = k_f\left[\frac{(a-x)^2x_{eq}^2 - (a-x_{eq})^2x^2}{x_{eq}^2}\right] \tag{76}$$

$$\frac{dx}{dt} = k_f\left[\frac{a^2x_{eq}^2 + x^2x_{eq}^2 - 2axx_{eq}^2 - a^2x^2 - x_{eq}^2x^2 + 2ax_{eq}x^2}{x_{eq}^2}\right] \tag{78}$$

or

$$\frac{x_{eq}^2}{\left(a^2x_{eq}^2 - 2axx_{eq}^2 - a^2x^2 + 2ax_{eq}x^2\right)}dx = k_f dt \tag{80}$$

Integrating equation (80), and then rearranging

$$k_f = \frac{x_{eq}}{2at(a-x_{eq})}\ln\frac{x(a-2x_{eq}) + ax_{eq}}{a(x_{eq}-x)} \tag{81}$$

Using equation (81), the rate constant for the forward reaction can easily be determined by measuring simple quantities like $a$, $t$, $x_{eq}$ and $x$. Now, we know that the equilibrium constant for a second-order reaction opposed by second-order will be

$$K = \frac{[C][D]}{[A][B]} \tag{82}$$

$$K = \frac{k_f}{k_b} \tag{83}$$

Now putting the value of $k_f$ from equation (81) in equation (83) and the rearranging for $k_b$, we get

$$k_b = \frac{x_{eq}}{2at(a-x_{eq})K}\ln\frac{x(a-2x_{eq}) + ax_{eq}}{a(x_{eq}-x)} \tag{84}$$

Hence, the value of the rate constant for backward reaction can also be obtained just by measuring $t$, $x_{eq}$ and $x$ and the equilibrium constant from equation (84).

## ❖ Rate Law for Consecutive & Parallel Reactions of Ist Order Reactions

In addition to the opposing or reversible reactions, two other types of simultaneous reactions are consecutive and parallel reactions. In this section, we will discuss the kinetic profiles of these two types of reactions up to the first order only.

### ➤ *Consecutive Reactions*

In many complex reactions, the order of the reaction has not been found equal to the molecularity noted from the stoichiometry. So, these reactions must take place in multiple steps rather than a single step. These multiple steps are individually labeled as consecutive reactions.

*The consecutive reactions may be defined as the single-step reactions which can be written to represent an overall reaction.*

In order to understand the kinetic profile of consecutive reactions, consider two first-order reactions in which reactant $A$ converts to $B$ which in turn converts to product $C$.

$$A \xrightarrow{\ k_1\ } B \xrightarrow{\ k_2\ } C \tag{85}$$

Where $k_1$ and $k_2$ are the rate constants for the first and second steps, respectively. In other words, $A$ is the reactant, $B$ is simply the intermediate and $C$ is the final product.

However, $k_f$ and $k_b$ have comparable values, a rate law depending upon both the constants can be written. Now suppose that the initial concentrations of reactant $A$ is $C_0$; while the concentrations of $A$, $B$ and $C$ after time $t$ are $C_A$, $C_B$ and $C_C$, respectively. So, we can say that

$$C_0 = C_A + C_B + C_C \tag{86}$$

Now, the rate can be deduced in terms of $C_A$, $C_B$ and $C_C$ as given below.

**1. Rate law in terms of $C_A$:** The rate of disappearance reactant of $A$ in the given reaction can be given by the following relation.

$$-\frac{d[C_A]}{dt} = k_1 C_A \tag{87}$$

or

$$-\frac{d[C_A]}{C_A} = k_1 dt \tag{88}$$

Integrating both sides, we get

$$-\ln C_A = k_1 t + I \tag{89}$$

Where $I$ is the constant of integration. However, when $t = 0$, $C_A = C_0$, the equation (89) takes the form

$$-\ln C_0 = I \tag{90}$$

Using the value of integration constant from equation (90) in equation (89), we get

$$-\ln C_A = k_1 t - \ln C_0 \tag{91}$$

or

$$\ln C_0 - \ln C_A = k_1 t \tag{92}$$

or

$$-\ln \frac{C_A}{C_0} = k_1 t \tag{93}$$

or

$$\ln \frac{C_A}{C_0} = -k_1 t \tag{94}$$

or

$$\frac{C_A}{C_0} = e^{-k_1 t} \tag{95}$$

or

$$C_A = C_0 e^{-k_1 t} \tag{96}$$

**2. Rate law in terms of $C_B$:** The rate of formation of intermediate $B$ can be given by the following relation.

$$\frac{d[C_B]}{dt} = -k_2 C_B + k_1 C_A \tag{97}$$

or

$$\frac{d[C_B]}{dt} = k_1 C_A - k_2 C_B \tag{98}$$

After putting the value of $C_A$ from equation (96) in equation (98), we get a linear differential equation of first order i.e.

$$\frac{d[C_B]}{dt} = k_1 C_0 e^{-k_1 t} - k_2 C_B \tag{99}$$

Integrating and then rearranging equation (99), both side, we get

$$[C_B] = C_0 \left( \frac{k_1}{k_2 - k_1} \right) \left( e^{-k_1 t} - e^{-k_2 t} \right) \tag{100}$$

**3. Rate law in terms of $C_C$:** The overall rate of formation of the product $C$ in the given reaction can be given by the following relation.

$$\frac{d[C_C]}{dt} = k_2 C_B \tag{101}$$

After putting the value of $C_A$ and $C_B$ from equation (96) and equation (100) into equation (86), we get the following result.

$$C_C = C_0 - C_0 e^{-k_1 t} - C_0 \left(\frac{k_1}{k_2 - k_1}\right)\left(e^{-k_1 t} - e^{-k_2 t}\right) \tag{102}$$

or

$$= C_0\left[1 - e^{-k_1 t} - \left(\frac{k_1}{k_2 - k_1}\right)\left(e^{-k_1 t} - e^{-k_2 t}\right)\right] \tag{103}$$

or

$$= C_0\left[1 - e^{-k_1 t} - \frac{k_1 e^{-k_1 t}}{k_2 - k_1} + \frac{k_1 e^{-k_2 t}}{k_2 - k_1}\right] \tag{104}$$

or

$$= C_0\left[1 - \left(e^{-k_1 t} + \frac{k_1 e^{-k_1 t}}{k_2 - k_1} - \frac{k_1 e^{-k_2 t}}{k_2 - k_1}\right)\right] \tag{105}$$

or

$$= C_0\left[1 - \left(\frac{(k_2 - k_1)e^{-k_1 t} + k_1 e^{-k_1 t} - k_1 e^{-k_2 t}}{k_2 - k_1}\right)\right] \tag{106}$$

$$= C_0\left[1 - \left(\frac{k_2 e^{-k_1 t} - k_1 e^{-k_1 t} + k_1 e^{-k_1 t} - k_1 e^{-k_2 t}}{k_2 - k_1}\right)\right] \tag{107}$$

$$- C_0\left[1 - \left(\frac{k_2 e^{-k_1 t} - k_1 e^{-k_2 t}}{k_2 - k_1}\right)\right] \tag{108}$$

or

$$C_C = C_0\left[1 - \frac{1}{(k_2 - k_1)}\left(k_2 e^{-k_1 t} - k_1 e^{-k_2 t}\right)\right] \tag{109}$$

The equation (96), (100) and (109) can be used to plot the time-dependent variation of $C_A$, $C_B$ and $C_C$, respectively.

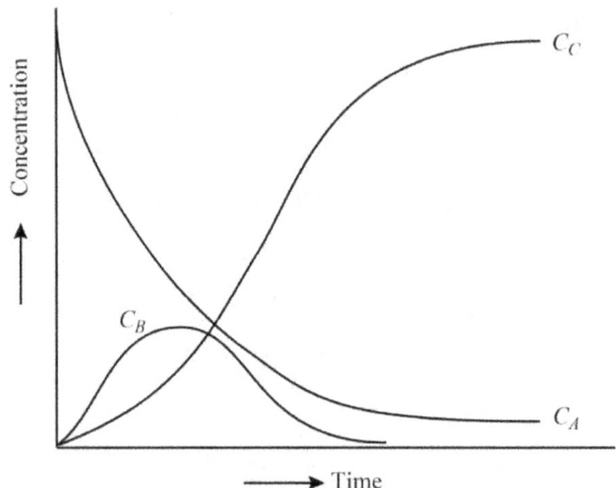


Figure 7. The plot $C_A$, $C_B$ and $C_C$ as a function of time in a typical consecutive reaction.

It can be clearly seen that the concentration of $A$ decreases exponentially, while the concentration of $B$ increases first and then declines. The concentration of $C$ increases continuously and finally becomes equal to the concentration of $A$.

**Maxima in the concentration of $B$:** In addition to the time-dependent concentration variation of different species, one more important parameter to measure is the maximum $B$ concentration. Since, for this the value of $[dC_B]/dt = 0$, the differentiation of equation (100) and then putting equal to zero gives

$$\frac{d[C_B]}{dt} = C_0 \left(\frac{k_1}{k_2 - k_1}\right)\left(-k_1 e^{-k_1 t} + k_2 e^{-k_2 t}\right) = 0 \tag{110}$$

$$-k_1 e^{-k_1 t} + k_2 e^{-k_2 t} = 0 \tag{111}$$

or

$$k_1 e^{-k_1 t} = k_2 e^{-k_2 t} \tag{112}$$

or

$$\frac{k_1}{k_2} = \frac{e^{-k_2 t}}{e^{-k_1 t}} \tag{113}$$

$$\frac{k_1}{k_2} = e^{(k_1 - k_2)t} \tag{114}$$

Taking logarithm both side, we get

$$\ln\frac{k_1}{k_2} = \ln e^{(k_1-k_2)t} \tag{115}$$

$$\ln\frac{k_1}{k_2} = (k_1 - k_2)t \tag{116}$$

$$t_{max} = \frac{1}{(k_1 - k_2)}\ln\frac{k_1}{k_2} \tag{117}$$

Now putting the value of $t$ from equation (117) in equation (100), we get

$$[C_B] = C_0\left(\frac{k_1}{k_2 - k_1}\right)\left[\exp\left\{\frac{-k_1\ln(k_1/k_2)}{k_1 - k_2}\right\} - \exp\left\{\frac{-k_2\ln(k_1/k_2)}{k_1 - k_2}\right\}\right] \tag{118}$$

Simplifying and then rearranging the above equation, we get

$$[C_B]_{max} = C_0\left(\frac{k_2}{k_1}\right)e^{k_2/(k_1-k_2)} \tag{119}$$

**Rate law in special cases:** In addition to the typical consecutive reaction i.e. $k_1 = k_2$, two special cases also arise from the nature of the step reactions discussed below.

*i) When $k_2 \ggg k_1$:* In these types of reactions, the value of $k_1$ can be neglected. Therefore, the equation (109) takes the form

$$C_C = C_0\left(1 - e^{-k_1t}\right) \tag{120}$$

Graphically,

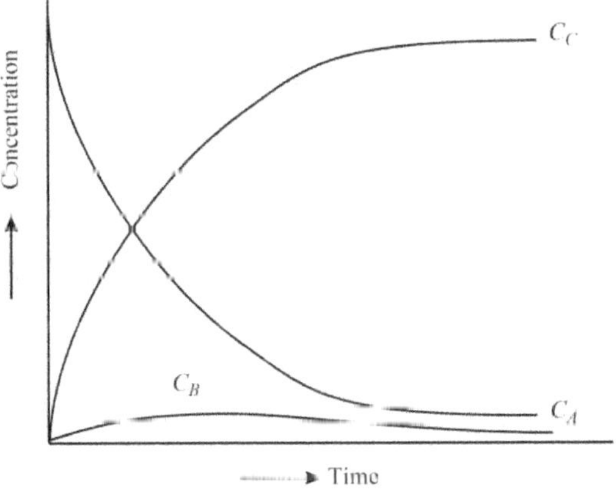


Figure 8. The plot $C_A$, $C_B$ and $C_C$ vs time in a typical consecutive reaction when $k_2 \ggg k_1$.

It can be clearly seen that the concentration of the intermediate practically remains constant, and therefore, the steady-state approximation can be applied in this case.

*ii) When $k_1 \ggg k_2$:* In these types of reactions, the value of $k_2$ can be neglected. Therefore, the equation (109) takes the form

$$C_C = C_0\left(1 - e^{-k_2 t}\right) \tag{121}$$

Graphically,

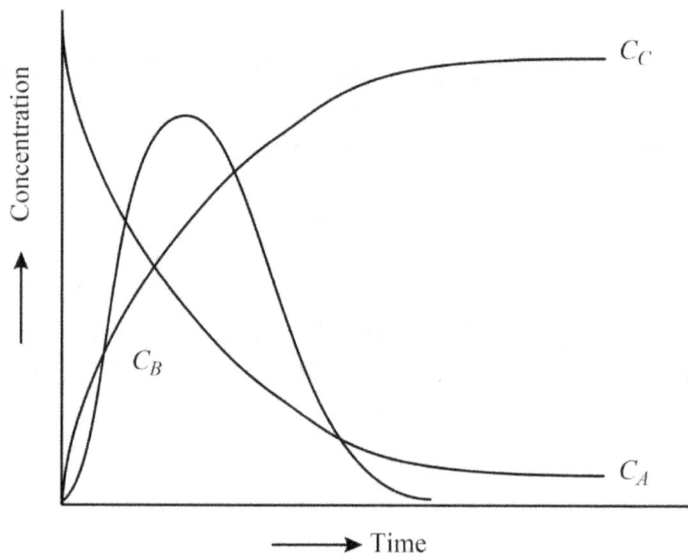


Figure 9. The plot $C_A$, $C_B$ and $C_C$ vs time in a typical consecutive reaction when $k_1 \ggg k_2$.

> ➢ *Parallel Reactions*

In many reactions, the reactant reacts to form more than one product simultaneously. If the amount of one the reaction product is very large in comparison to the others, then we can simply neglect these other reactions. However, if the amount of the product formed by other reactions are significant, we must refine the overall rate equation to represent this.

*The parallel or side reactions may simply be defined as the reactions in which initial species react to give multiple products simultaneously.*

In order to understand the kinetics of parallel reactions of the first order, suppose that a reactant $A$ reacts to form product $B$ and $C$ simultaneously. A typical depiction of the parallel or side reaction with two pathways is given below.

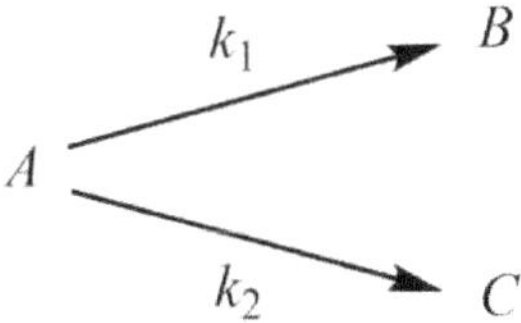


Where $k_1$ and $k_2$ are the rate contents. Now suppose that $a$ is the initial concentration of the reactant $A$, while $x$ is the decrease in the concentrations of the reactant after '$t$' time. Hence, the rates of first ($R_1$) and second reaction ($R_2$) can be given as:

$$R_1 = \frac{d[B]}{dt} = k_1[A] = k_1(a - x) \tag{122}$$

or

$$R_2 = \frac{d[C]}{dt} = k_2[A] = k_2(a - x) \tag{123}$$

The overall reaction rate can be obtained by adding equation (122) and equation (123) as

$$\frac{dx}{dt} = \frac{d[B]}{dt} + \frac{d[C]}{dt} = k_1(a - x) + k_2(a - x) \tag{124}$$

or

$$\frac{dx}{dt} = (k_1 + k_2)(a - x) \tag{125}$$

or

$$\frac{dx}{(a - x)} = (k_1 + k_2)dt \tag{126}$$

Integrating the equation (126) and then rearranging, we get

$$k_1 + k_2 - \frac{1}{t} \ln \frac{a}{(a - x)} \tag{127}$$

Also, dividing equation (122) by (123), we get

$$\frac{R_1}{R_2} = \frac{k_1(a - x)}{k_2(a - x)} \tag{128}$$

Which implies that

$$\frac{R_1}{R_2} = \frac{k_1}{k_2} \tag{129}$$

Hence, the value of rate constants involved, i.e., $k_1$ and $k_2$ can easily be obtained from the use of equation (127) and equation (129).

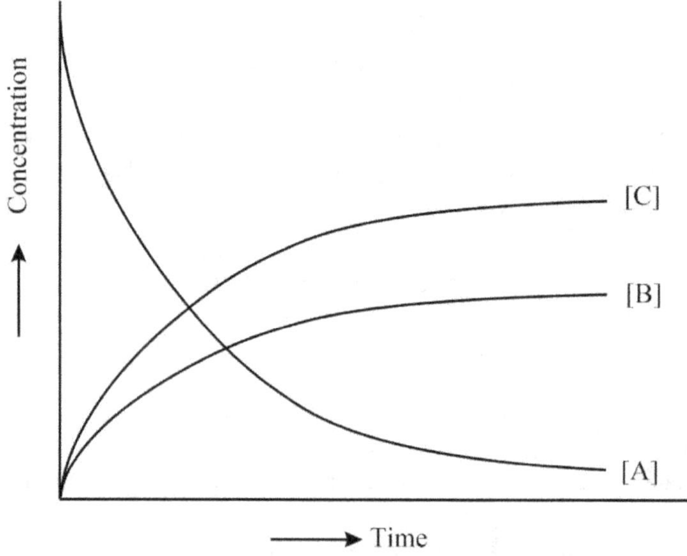

Figure 10. The variation of the concentrations of reactants and products as a function of time in a typical parallel reaction.

It should also be noted from the equation (129) that the ratio of the concentration of products remains the same with time. Furthermore, the percentage of both products can also be obtained from the knowledge of rate constants using the relations given below.

$$Fractional\ Quantum\ Yield\ of\ A = \frac{k_1}{k_1 + k_2} \qquad (130)$$

also

$$Fractional\ Quantum\ Yield\ of\ B = \frac{k_2}{k_1 + k_2} \qquad (131)$$

The percentage is obtained by multiplying corresponding fractional quantum yields by 100. Similarly, parts per thousand (ppt) and parts per million (ppm) can be obtained by multiplying equations (130) and (131) by $10^3$ and $10^6$, respectively.

## ❖ Collision Theory of Reaction Rates and Its Limitations

In 1916, a German chemist Max Trautz proposed a theory based on the collisions of reacting molecules to explain reaction kinetics. Two years later, a British chemist William Lewis published similar results, however, he was completely unaware of Trautz's work. The remarkable work of these two gentlemen was extremely beneficial in explaining the rate of many chemical reactions.

*The collision theory states that when the right reactant particles strike each other, only a definite fraction of the collisions induce any significant or noticeable chemical change; these successful changes are called successful collisions and are possible only if reacting molecules have sufficient energy at the moment of impact to break the pre-existing bonds and form all new bonds.*

The minimum energy required to make a collision successful is called as the activation energy, and these types of collisions result in the products of the reaction. The rise in reactant concentration or increasing the temperature, both result in more collisions and hence more successful collisions, and therefore, increase the reaction rate. Sometimes, a catalyst is involved in the collision between the reactant molecules that decreases the energy required for the chemical change to take place, and so more collisions would have sufficient energy for the reaction to happen. In this section, we will discuss the collision theory of bimolecular and unimolecular reactions in the gaseous phase.

### ➤ *Collision Theory for Bimolecular Reactions*

In order to understand the collision theory for bimolecular reactions, we must understand the cause of a reaction itself first. The primary requirement for a reaction to occur is the collision between the reacting molecules. Therefore, if we assume that every collision results in the formation of the product, the rate of reaction should simply be equal to collision frequency ($Z$) of the reacting system i.e. the number of collisions occurring in the container per unit volume per unit time. Mathematically, we can say that

$$Rate = Z \qquad\qquad (132)$$

However, the actual rate would be much less than what is predicted by the equation (132); which is obviously due to the fact that all the collisions are not effective. Therefore, equation (132) must be modified to represent this factor. If $f$ is the fraction of the molecules which are activated, the rate expression can be written as given below.

$$Rate = Z \times f \qquad\qquad (133)$$

Now, according to the Maxwell-Boltzmann distribution of energies, the fraction of the molecules having energy greater than a particular energy $E$ is

$$f = \frac{\Delta N}{N} = e^{-E/RT} \qquad\qquad (134)$$

Where $N$ is the total number of molecules while $\Delta N$ represents the number of molecules having energy greater than $E$. However, if $E = E_a$, the fraction of activated molecules can be written as

Copyright © *Mandeep Dalal*

$$f = \frac{\Delta N}{N} = e^{-E_a/RT} \tag{135}$$

Where $R$ is the gas constant and T is the reaction temperature. After putting the value of $f$ from equation (135) into equation (133), we get

$$Rate = Ze^{-E_a/RT} \tag{136}$$

At this point, two possibilities arise; one, when the colliding molecules are similar and other, is when the colliding molecules are dissimilar. We will discuss these cases one by one.

**1. Rate of reaction when the colliding molecules are dissimilar:** Consider a bimolecular reaction between different molecules $A$ and $B$ yielding product $P$ as

$$A + B \rightarrow P \tag{137}$$

The number of collisions between $A$ and $B$ occurring in the container per unit volume per unit time can be given by the following relation.

$$Z = n_A \, n_B \, \sigma_{AB}^2 \sqrt{\frac{8\pi k_B T}{\mu_{AB}}} \tag{138}$$

Where $n_A$ and $n_A$ are the number densities (in the units of $m^{-3}$) of particles $A$ and $B$, respectively. The term $\sigma_{AB}$ is simply the average collision diameter i.e. $\sigma_{AB} = (\sigma_A + \sigma_B)/2$. $k_B$ is the Boltzmann's constant ($m^2 \, kg \, s^{-2} \, K^{-1}$). $T$ represents the temperature of the system. The term $\mu_{AB}$ represents the reduced mass of the reactants $A$ and $B$ i.e. $\mu_{AB} = m_A m_B / m_A + m_B$.

The equation (138) can also be expressed in terms of molar masses by putting $m_A = M_A/N$, $m_B = M_B/N$ and $k = R/N$; where $M_A$ and $M_B$ are molar masses of the reactants, $R$ is the gas constant and N represents to Avogadro number. Therefore, equation (138) takes the form

$$Z = \sigma_{AB}^2 \sqrt{\frac{8\pi \left(\frac{R}{N}\right) T \left(\frac{M_A}{N} + \frac{M_B}{N}\right)}{\frac{M_A}{N} \times \frac{M_B}{N}}} \; n_A \, n_B \tag{139}$$

or

$$Z = \sigma_{AB}^2 \sqrt{\frac{8\pi RT (M_A + M_B)}{M_A M_B}} \; n_A \, n_B \tag{140}$$

Also, as we know that the reaction rate can be written in terms of molecules of reactants reacting per cm$^3$ per second as

$$Rate = -\frac{dn_A}{dt} = -\frac{dn_B}{dt} = Ze^{-E_a/RT} \tag{141}$$

or

$$Rate = -\frac{dn_A}{dt} = -\frac{dn_B}{dt} = \sigma_{AB}^2 \sqrt{\frac{8\pi RT(M_A + M_B)}{M_A M_B}} \times n_A\, n_B \times e^{-E_a/RT} \tag{142}$$

Now, in order to express the rate in terms of molar concentrations, we need to recall some typical relations like

$$n_A = \frac{N[A]}{10^3} \quad \text{and} \quad n_B = \frac{N[B]}{10^3} \tag{143}$$

also

$$dn_A = \frac{N}{10^3} d[A] \quad \text{and} \quad dn_B = \frac{N}{10^3} d[B] \tag{144}$$

Using the results of equation (143) and (144) in equation (142), we get

$$Rate = -\frac{N}{10^3}\frac{d[A]}{dt} = -\frac{N}{10^3}\frac{d[B]}{dt} = \sigma_{AB}^2 \sqrt{\frac{8\pi RT(M_A + M_B)}{M_A M_B}} \frac{N[A]}{10^3}\frac{N[B]}{10^3} e^{-E_a/RT} \tag{145}$$

or

$$Rate = -\frac{d[A]}{dt} = -\frac{d[B]}{dt} = \frac{N}{10^3} \sigma_{AB}^2 \sqrt{\frac{8\pi RT(M_A + M_B)}{M_A M_B}} [A][B]\, e^{-E_a/RT} \tag{146}$$

Comparing equation (146) with general rate law expressed in molar concentrations i.e. Rate = $k$[A][B], we get

$$k = \frac{N}{10^3} \sigma_{AB}^2 \sqrt{\frac{8\pi RT(M_A + M_B)}{M_A M_B}} e^{-E_a/RT} \tag{147}$$

Comparing equation (147) with Arrhenius rate constant i.e. $k = Ae^{-E_a/RT}$, we get

$$A = \frac{N}{10^3} \sigma_{AB}^2 \sqrt{\frac{8\pi RT(M_A + M_B)}{M_A M_B}} \tag{148}$$

**2. Rate of reaction when the colliding molecules are similar:** Consider a bimolecular reaction between similar molecules $A$ and $A$ yielding product $P$ as

$$A + A \rightarrow P \tag{149}$$

The number of collisions between $A$ and $A$ occurring in the container per unit volume per unit time can be given by the following relation.

$$Z = n_A^2\, \sigma^2 \sqrt{\frac{4\pi k_B T}{m_A}} \tag{150}$$

Where $n_A$ is the number density (in the units of $m^{-3}$) of particle $A$. The term $\sigma$ is simply the average collision diameter. $k_B$ is the Boltzmann's constant ($m^2\ kg\ s^{-2}\ K^{-1}$). $T$ represents the temperature of the system. The term $m_A$ represents the mass of the reactants $A$.

The equation (150) can also be expressed in terms of molar masses by putting $m_A = M_A/N$ and $k = R/N$; where $M_A$ is the molar mass of the reactant, $R$ is the gas constant and N represents to Avogadro number. Therefore, equation (150) takes the form

$$Z = \sigma^2 \sqrt{\frac{4\pi \left(\frac{R}{N}\right) T}{\frac{M_A}{N}}}\, n_A^2 \tag{151}$$

or

$$Z = \sigma^2 \sqrt{\frac{4\pi RT}{M_A}}\, n_A^2 \tag{152}$$

Also, as we know that the reaction rate can be written in terms of molecules of reactants reacting per $cm^3$ per second as

$$-\frac{1}{2}\frac{dn_A}{dt} = Z e^{-E_a/RT} \tag{153}$$

or

$$Rate = -\frac{dn_A}{dt} = 2\left( \sigma^2 \sqrt{\frac{4\pi RT}{M_A}} \times n_A^2 \times e^{-E_a/RT} \right) \tag{154}$$

Now, in order to express the rate in terms of molar concentrations, we need to recall some typical relations like

$$n_A = \frac{N[A]}{10^3} \tag{155}$$

also

$$dn_A = \frac{N}{10^3} d[A] \tag{156}$$

Using the results of equation (155) and (156) in equation (154), we get

$$Rate = -\frac{N}{10^3}\frac{d[A]}{dt} = 2\left(\sigma^2\sqrt{\frac{4\pi RT}{M_A}} \times \frac{N^2[A]^2}{10^6} \times e^{-E_a/RT}\right) \tag{157}$$

or

$$Rate = -\frac{d[A]}{dt} = \frac{4N}{10^3}\sigma^2\sqrt{\frac{\pi RT}{M_A}} \times [A]^2 \times e^{-E_a/RT} \tag{158}$$

Comparing equation (158) with general rate law expressed in molar concentrations i.e. Rate = $k[A]^2$, we get

$$k = \frac{4N}{10^3}\sigma^2\sqrt{\frac{\pi RT}{M_A}}\, e^{-E_a/RT} \tag{159}$$

Comparing equation (159) with Arrhenius rate constant i.e. $k = Ae^{-E_a/RT}$, we get

$$A = \frac{4N}{10^3}\sigma^2\sqrt{\frac{\pi RT}{M_A}} \tag{160}$$

> ➢ *Collision Theory for Unimolecular Reactions*

In order to understand the collision theory for unimolecular reactions, we must understand the root cause of these reactions. In a typical unimolecular reaction, a single molecule converts into the product by simply rearranging itself. However, the question that arises here is how these molecules get activated. The mystery was solved by a British physicist, Frederick Alexander Lindemann, who proposed a time-leg between activation and actual reaction. In other words, when ordinary molecules collide with each other, some of them get activated, and the rate depends only upon these molecules but not the ordinary ones i.e.

$$A + A \underset{k_2}{\overset{k_1}{\rightleftharpoons}} A^* + A \xrightarrow{k_3} P$$

Where $k_1$, $k_2$ and $k_3$ are the rate constants for the different processes; while $A$ and $A^*$ are the ordinary and activated molecule. The overall rate of formation of the product can be given as

$$\frac{d[P]}{dt} = k_3[A^*] \tag{161}$$

Now, since the molar concentration of $[A^*]$ is unknown we must apply the steady-state approximation on $[A^*]$.

At steady state

$$\text{Rate of formation of } [A^*] = \text{Rate of disappearance of } [A^*]$$

$$k_1[A]^2 = k_2[A^*][A] + k_3[A^*] \tag{162}$$

or

$$k_1[A]^2 = (k_2[A] + k_3)[A^*] \tag{163}$$

$$[A^*] = \frac{k_1[A]^2}{k_2[A] + k_3} \tag{164}$$

Using the value of $[A^*]$ from equation (164) in equation (161), we get

$$Rate = \frac{d[P]}{dt} = \frac{k_3 k_1[A]^2}{k_2[A] + k_3} \tag{165}$$

Equation (165) gives rise to two possibilities discussed below.

**1. If the concentration of reactant A is very high:** In this situation, $k_2[A] \ggg k_3$, and $k_3$ can be neglected, therefore, the equation (165) takes the form

$$Rate = \frac{d[P]}{dt} = \frac{k_3 k_1[A]^2}{k_2[A]} \tag{166}$$

or

$$\frac{d[P]}{dt} = \frac{k_3 k_1}{k_2}[A] \tag{167}$$

$$\frac{d[P]}{dt} = k_o[A] \tag{168}$$

Where $k_o$ is the overall rate constant. It is clear from the above result that the unimolecular reactions follow first-order kinetics in such cases.

**2. If the concentration of reactant A is very low:** In this situation, $k_3 \ggg k_2[A]$ and $k_2[A]$ can be neglected, therefore, the equation (165) takes the form

$$Rate = \frac{d[P]}{dt} = \frac{k_3 k_1[A]^2}{k_3} \tag{169}$$

$$\frac{d[P]}{dt} = k_1[A]^2 \tag{170}$$

Where $k_1$ is the overall rate constant. It is clear from the above result that the unimolecular reactions follow second-order kinetics in such cases.

> *Limitations of Collision Theory*

The collision theory of reaction rate is extremely successful in rationalizing the kinetics of many reactions, however, it does suffer from some serious limitations discussed below.

1. This theory finds application only to reactions occurring in the gas phase and solution having simple reactant molecules.

2. The rate constants obtained by employing collision theory are found to be comparable to what has been obtained from the Arrhenius equation only for the simple reactions but not for complex reactions.

3. This theory tells nothing about the exact mechanism behind the chemical reaction i.e. making and breaking of chemical bonds.

4. The collision theory considers only the kinetic energy of reacting molecules and just ignored rotational and vibrational energy which also plays an important role in reaction rate.

5. This theory did not consider the steric factor at all i.e. the proper orientation of the colliding molecules needed to result in the chemical change.

## ❖ Steric Factor

One of the most glaring limitations of the collision is that the predicted values of rate constants for many reactions were found to be considerably different from the values obtained experimentally. Moreover, it was also noticed that more the complexity, the higher was the deviation. This happened because the collision theory supposed that the particles participating in the chemical reaction are completely spherical, and thus, are able to react in every direction. However, this is far from the truth since the orientation of the collisions is not always appropriate to result in the chemical change. For instance, in the hydrogenation of ethylene, the dihydrogen molecule must approach the bonding zone between the atoms, and not all the possible collisions would be able to satisfy this requirement. For more clear view, consider the formation of $CO_2$ as shown below.

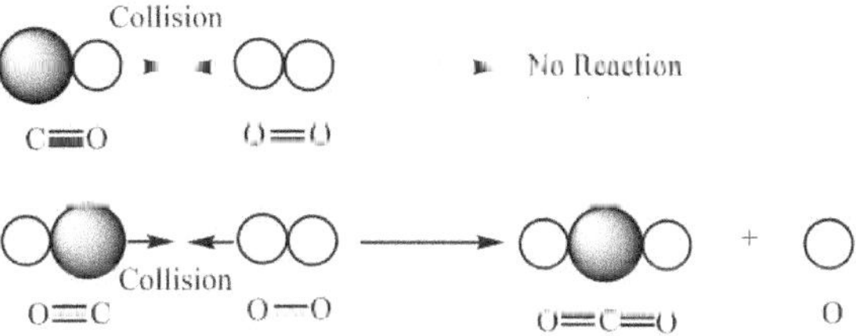


To solve this problem, the concept of steric factor ($\rho$) was introduced, which is simply the ratio of experimental value to the predicted value of the rate constant. In other words, the steric factor may be defined as the ratio between the frequency factor and the collision frequency i.e.

$$\rho = \frac{A_{observed}}{Z_{calculated}} \tag{171}$$

It is worthy to note that the value of the steric factor most of the cases is less than unity. Typically, it has been seen that more the complex the reactant molecules are, the lower is the steric factor. However, some reactions do have steric factors higher than unity; for instance, the harpoon reactions in which atoms involved exchange electrons generating ions. The deviation from unity may arise due to different reasons such as non-spherical shape of reacting molecules, or the partial delivery of kinetic energy, the presence of a solvent when applied to solutions.

In order to derive the expression for modified collision theory that does consider the reactants steric, recall the rate constant calculated from simple collision theory first i.e.

$$Rate = Ze^{-E_a/RT} \tag{172}$$

For experimental rate, multiply the equation (172) by the probability or steric factor i.e.

$$Rate = \rho Ze^{-E_a/RT} \tag{173}$$

Now considering both the possibilities i.e. whether the reacting species are the same or different, we can simplify the above equation in two ways:

*i) For dissimilar molecules*:

If the colliding molecules are not the same, the exponential part in equation (173) takes the form

$$A = \rho \frac{N}{10^3} \, \sigma_{AB}^2 \sqrt{\frac{8\pi RT(M_A + M_B)}{M_A M_B}} \tag{174}$$

Substituting the values of different constants, we get

$$A = 2.753 \times 10^{29} \times \rho \times \sigma_{AB}^2 \sqrt{\frac{T(M_A + M_B)}{M_A M_B}} \tag{175}$$

*ii) For similar molecules:*

If the colliding molecules are the same, the exponential part in equation (173) takes the form

$$A = \rho \frac{N}{10^3} \, 4\sigma^2 \sqrt{\frac{\pi RT}{M_A}} \tag{176}$$

$$A = 3.893 \times 10^{29} \times \rho \times \sigma^2 \sqrt{\frac{T}{M_A}} \tag{177}$$

This modified collision theory can account for probability factors up to $10^{-4}$ but not less than that. This limitation can be overcome by "transition state theory" discussed in the next section.

## ❖ Activated Complex Theory

In 1935, an American chemist Henry Eyring; alongside two British chemists, Meredith Gwynne Evans and Michael Polanyi; proposed a new theory to rationalize the rate of different chemical reactions which was based upon the formation of an activated intermediate complex. This theory is also known as the "transition state theory", "theory of absolute reaction rates", and "absolute-rate theory".

*The activated complex theory states that the rates of various elementary chemical reactions can be explained by assuming a special type of chemical equilibria (quasi-equilibrium) between reactants and activated complexes.*

Before the development of activated complex theory, the Arrhenius rate law was popularly used to determine energies for the potential barrier. However, the Arrhenius equation was based on empirical observations rather than mechanistic investigations as if one or more intermediates are involved in the conversion or not. For that reason, more development was essential to know the two factors present in the Arrhenius equation, the activation energy ($E_a$) and the pre-exponential factor ($A$). The Eyring equation from transition state theory successfully addresses these two issues and therefore contributed significantly to the conceptual understanding of reaction kinetics.

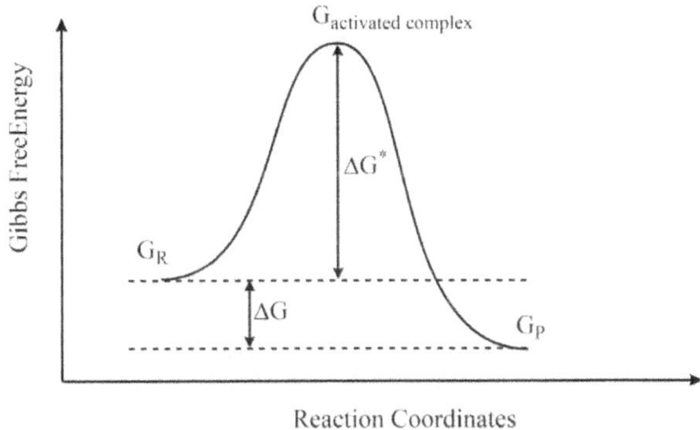


Figure 11. The variation of free energy as the reaction proceeds.

To explore the concept mathematically, consider a reaction between reactant $A$ and $B$ forming a product $P$ via the activated complex $X^*$ as:

$$A + B \;\overset{K^*}{\rightleftharpoons}\; X^* \;\xrightarrow{k}\; P \tag{178}$$

The equilibrium constant for reactants to activated complex conversion is

$$K^* = \frac{[X^*]}{[A][B]} \tag{179}$$

Copyright © *Mandeep Dalal*                                    **DALAL INSTITUTE**

Henry Eyring showed that the rate constant for a chemical reaction with any order or molecularity can be given by the following relation.

$$k = \frac{RT}{Nh} K^* \tag{180}$$

Where $K^*$ is the equilibrium constant for reactants to activated complex conversion at temperature T. Whereas, $R$, $N$ and $h$ represent the gas constant, Avogadro number and Planck's constant, respectively. Now, as we know from thermodynamics

$$\Delta G^* = -RT \ln K^* \tag{181}$$

$$-\frac{\Delta G^*}{RT} = \ln K^* \tag{182}$$

$$K^* = e^{-\frac{\Delta G^*}{RT}} \tag{183}$$

Where $\Delta G^*$ is the free energy of activation for reactants to activated complex conversion step. Using the value of $K^*$ from equation (183) into equation (180), we get

$$k = \frac{RT}{Nh} e^{-\frac{\Delta G^*}{RT}} \tag{184}$$

If we put the thermodynamic value of free energy i.e. $\Delta G^* = \Delta H^* - T\Delta S^*$ in equation (184), we get

$$k = \frac{RT}{Nh} e^{-\frac{(\Delta H^* - T\Delta S^*)}{RT}} \tag{185}$$

or

$$k = \frac{RT}{Nh} \times e^{\frac{\Delta S^*}{R}} \times e^{-\frac{\Delta H^*}{RT}} \tag{186}$$

Where $\Delta H^*$ and $\Delta S^*$ are enthalpy change and the entropy change of the activation step. Equation (186) is popularly known as the Eyring equation. Now, since the equation (186) contains very fundamental factors of the reacting species, that is why this theory got its name of "theory of absolute reaction rates".

> ➤ *Significance of Entropy of Activation and Enthalpy of Activation*

As far as the equation (184) is concerned, it can easily be seen that as the free energy change of the activation step increases, the rate constant would decrease. However, if we look at the simplified form i.e. equation (186), we find three factors; one is $RT/Nh$ which is constant if the temperature is kept constant. The second factor involves $\Delta S^*$, and therefore, we can conclude that the reaction rate would show exponential increase if the entropy of activation increases. The third factor includes $\Delta H^*$, and therefore, we can conclude that the reaction rate would show exponential decrease if the enthalpy of activation increases. It is also worthy to note that the first two terms collectively make the frequency factor.

> ### *Comparison with Arrhenius Rate Constant*

Like the collision theory, the validity of the "activated complex theory" must also be checked against the results of the Arrhenius rate equation. In order to do so, recall the Arrhenius equation i.e.

$$k = A\, e^{-\frac{E_a}{RT}} \tag{187}$$

By looking at the analogy with equation (187), we get

$$A = \frac{RT}{Nh} \times e^{\frac{\Delta S^*}{R}} \tag{188}$$

and

$$e^{-\frac{E_a}{RT}} = e^{-\frac{\Delta H^*}{RT}} \tag{189}$$

Now we can use equation (188) and (189) to determine the value of entropy of activation and enthalpy of activation.

**1. Calculation of entropy of activation:** In order to determine the entropy change of the activation step, take the natural logarithm of the equation (188) i.e.

$$\ln A = \ln \frac{RT}{Nh} + \ln e^{\frac{\Delta S^*}{R}} \tag{190}$$

$$\ln A = \ln \frac{RT}{Nh} + \frac{\Delta S^*}{R} \tag{191}$$

or

$$\frac{\Delta S^*}{R} = \ln A - \ln \frac{RT}{Nh} \tag{192}$$

$$\Delta S^* = R \ln A - R \ln \frac{RT}{Nh} \tag{193}$$

Thus, higher is the value of the frequency factor, larger will be the entropy of activation.

**2. Calculation of enthalpy of activation:** In order to determine the entropy change of the activation step, we must look at the equation (189) which suggests that the enthalpy change of the activation step as exactly equal to the activation energy of the reaction dictating the rate considerably i.e. $\Delta H^* = E_a$. However, it has been observed that the exact value of activation energy is slightly different than the enthalpy of activation. In order to prove the aforementioned statement, rewrite the equation (186) as:

$$k = \frac{R}{Nh} e^{\frac{\Delta S^*}{R}} \times T \times e^{-\frac{\Delta H^*}{RT}} \tag{194}$$

or

$$k = C \times T \times e^{-\frac{\Delta H^*}{RT}} \tag{195}$$

Where $C$ is another constant. Now taking natural logarithm both side, equation (195) takes the form

$$\ln k = \ln C + \ln T + \ln e^{-\frac{\Delta H^*}{RT}} \tag{196}$$

or

$$\ln k = \ln C + \ln T - \frac{\Delta H^*}{RT} \tag{197}$$

Now, differentiating both side w.r.t temperature, we get

$$\frac{d \ln k}{dt} = \frac{1}{T} + \frac{\Delta H^*}{RT^2} \tag{198}$$

We also know from the Arrhenius equation that

$$k = A\, e^{-\frac{E_a}{RT}} \tag{199}$$

Now taking natural logarithm both side, equation (199) takes the form

$$\ln k = \ln A + \ln e^{-\frac{E_a}{RT}} \tag{200}$$

or

$$\ln k = \ln A - \frac{E_a}{RT} \tag{201}$$

Now, differentiating both side w.r.t temperature, we get

$$\frac{d \ln k}{dt} = \frac{E_a}{RT^2} \tag{202}$$

Comparing (198) and (202), we get

$$\frac{1}{T} + \frac{\Delta H^*}{RT^2} = \frac{E_a}{RT^2} \tag{203}$$

$$\frac{RT + \Delta H^*}{RT^2} = \frac{E_a}{RT^2} \tag{204}$$

$$\Delta H^* = E_a - RT \tag{205}$$

If $n$ is the change in the number of moles of gas in going from reactant to activated complex, the result of equation (205) takes the form $\Delta H^* = E_a - \Delta n_g RT$.

## ❖ Ionic Reactions: Single and Double Sphere Models

It is a quite well-known fact that the rate of ionic reactions is generally small, which is obviously due to the larger magnitude of activation energies arising from the very strong nature of electrostatic interactions. The magnitude of the frequency factor in ionic reactions is a function of ionic charges. The frequency factors have larger values if the charges on the participating ions are opposite, while smaller values are obtained in the case of like-charged ions. This behavior can be explained in terms of the kinetic theory of gases; which suggests that oppositely charged ions are more prone to collision due to attraction than the ions colliding with same charges (repulsive forces). Besides the collision theory, the activated complex theory also provides an alternate explanation for the ionic reactions. In this section, we will discuss the rationalization of ionic reactions on the basis of the single-sphere model and the double-sphere model in detail.

### ➤ *Double Sphere Model*

Before we discuss the double sphere model of the ionic reactions, a simplified surrounding must be assumed. Although it would be an oversimplification of the actual situation, it is highly beneficial as far as conceptual and quantitative understanding is concerned. To do so, the solvent is considered as continuous surrounding with a $\varepsilon$ as the dielectric constant.

According to this model, two ions, which can same or opposite charges, combine together to form an activated complex. In the initial state, the ions are considered as discrete; while in the final state, they assumed to form a dumbbell like coordination with '$r$' as the distance of separation between their centers.

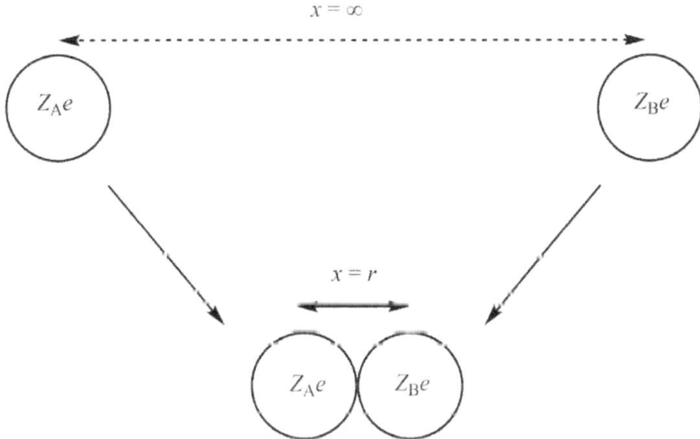


Figure 12. The pictorial depiction of the double-sphere model of ionic reactions.

Now, if $Z_A$ and $Z_B$ are the charge numbers of the participating ions and $x$ as the distance of separation, the force of electrostatic interaction ($F_{AB}$) between them can be given from the Coulomb's law as:

 **DALAL INSTITUTE**

$$F_{AB} = \frac{Z_A Z_B e^2}{4\pi\varepsilon_0\varepsilon x^2} \tag{206}$$

Where $\varepsilon_0$ and $\varepsilon$ are permittivities of the vacuum ($8.854 \times 10^{-12}\ C^2\ N^{-1}\ m^{-2}$) and the dielectric constant of the solvent used, respectively. The symbol $e$ represents the elementary charge and has a value equal to $1.6 \times 10^{-19}$ C. The value of parameter varies from $\infty$ to $r$ with the mutual approach of two ions. The amount of work done in moving the two ions closer by an extant $dx$ will be

$$work = force \times displacement \tag{207}$$

$$dw = F_{AB} \times dx \tag{208}$$

$$dw = -\frac{Z_A Z_B e^2}{4\pi\varepsilon_0\varepsilon x^2}\, dx \tag{209}$$

The negative sign is an indicator of decreasing separation i.e. distance is reduced by $dx$. The total amount of work done in moving the two ions from $x = \infty$ to $x = r$ will be

$$w = -\int_{\infty}^{r} \frac{Z_A Z_B e^2}{4\pi\varepsilon_0\varepsilon x^2}\, dx \tag{210}$$

$$w = \frac{Z_A Z_B e^2}{4\pi\varepsilon_0\varepsilon\, r} \tag{211}$$

The work given the above equation is actually the potential energy of the system which would have a negative sign for oppositely charged ions and positive sign if the ions have same charges. Furthermore, we can also say that this work is the free energy change due to electrostatic interactions, therefore, multiplying it by Avogadro number ($N$) would give the value of the corresponding molar free energy change ($\Delta G_{EI}^*$) i.e.

$$\Delta G_{EI}^* = \frac{N Z_A Z_B e^2}{4\pi\varepsilon_0\varepsilon\, r} \tag{212}$$

Correcting the above equation for non-electrostatic contribution $\Delta G_{NEI}^*$, the total molar free energy change for the whole process can be given by the following relation.

$$\Delta G^* = \Delta G_{NEI}^* + \Delta G_{EI}^* = \Delta G_{NEI}^* + \frac{N Z_A Z_B e^2}{4\pi\varepsilon_0\varepsilon\, r} \tag{213}$$

Also, from the activated complex theory, we know that

$$k = \frac{RT}{Nh}e^{-\frac{\Delta G^*}{RT}} \tag{214}$$

After putting the value of $\Delta G^*$ from equation (213) into equation (214), we get

---

$$k = \frac{RT}{Nh} e^{-\left(\frac{\Delta G^*_{NEI}}{RT} + \frac{NZ_AZ_Be^2}{RT4\pi\varepsilon_0\varepsilon\,r}\right)}$$

(215)

$$k = \frac{RT}{Nh} e^{-\frac{\Delta G^*_{NEI}}{RT}} \cdot e^{-\frac{NZ_AZ_Be^2}{RT4\pi\varepsilon_0\varepsilon\,r}}$$

(216)

Taking natural logarithm both side of equation (216), we get

$$\ln k = \ln \frac{RT}{Nh} + \ln e^{-\frac{\Delta G^*_{NEI}}{RT}} + \ln e^{-\frac{NZ_AZ_Be^2}{RT4\pi\varepsilon_0\varepsilon\,r}}$$

(217)

or

$$\ln k = \ln \frac{RT}{Nh} - \frac{\Delta G^*_{NEI}}{RT} - \frac{NZ_AZ_Be^2}{RT4\pi\varepsilon_0\varepsilon\,r}$$

(218)

Which can also be expressed as

$$\ln k = \ln k_0 - \frac{NZ_AZ_Be^2}{RT4\pi\varepsilon_0\varepsilon\,r}$$

(219)

Where $k_0$ represents the magnitude of the rate constant for the ionic reaction carried out in a solvent of infinite dielectric constant so that the electrostatic interactions become zero.

> ### ➢ *Single Sphere Model*

Besides the double-sphere model, another theoretical model that is quite rationalizing is a single-sphere model. Just like the double-sphere model, the solvent is also considered as a continuum with a $\varepsilon$ as the dielectric constant. However, the primary differentiating aspect of this model is that it considers the two ions, which can same or opposite charges, to form a single-sphere activated complex.

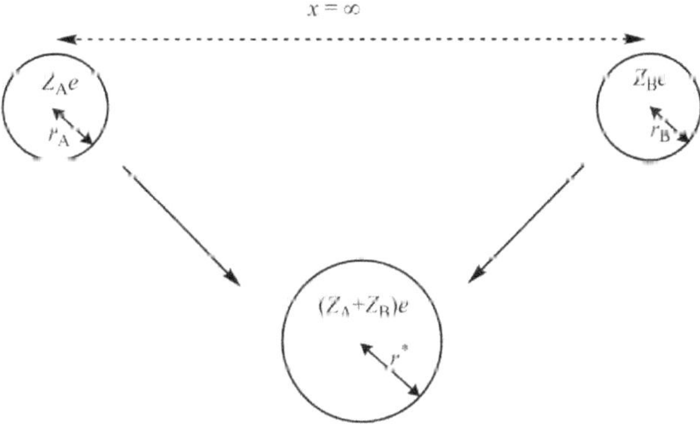


Figure 12. The pictorial depiction of the single-sphere model of ionic reactions.

In the initial state, the ions are considered as discrete; while in the final state, they assumed to form a single-sphere activated complex with '$r^*$' as the overall radius. The rate law for this case was derived by Born by considering the energy required to charge an ion in solution. Now suppose that we need to charge a conducting sphere of radius $r$ from an initial value of zero to the final value $Ze$. This can be visualized as a process in which a very small charge is $e.d\lambda$ ($\lambda = 0 - Z$) is carried from infinite to this sphere.

Now, if $Z_A$ and $Z_B$ are the charge numbers of the participating ions and $x$ as the distance of separation between the sphere and the "increment" at any time, the force of electrostatic interaction ($dF$) between them can be given from the Coulomb's law as:

$$dF = \frac{\lambda e^2 d\lambda}{4\pi\varepsilon_0\varepsilon x^2} \tag{220}$$

Where $\varepsilon_0$ and $\varepsilon$ are permittivities of the vacuum ($8.854 \times 10^{-12}$ $C^2$ $N^{-1}$ $m^{-2}$) and the dielectric constant of the solvent used, respectively. The symbol $e$ represents the elementary charge and has a value equal to $1.6 \times 10^{-19}$ $C$. The amount of work done in moving the "increment" closer by an extant $dx$ will be

$$dw = dF \times dx \tag{221}$$

$$dw = \frac{\lambda e^2 d\lambda}{4\pi\varepsilon_0\varepsilon x^2} dx \tag{222}$$

The total amount of work done can be obtained by carrying out the double integration with respect to $x = \infty - r$ and $\lambda = 0 - Z$ i.e.

$$w = \frac{e^2}{4\pi\varepsilon_0\varepsilon} \int_0^Z \int_\infty^r \frac{\lambda}{x^2} d\lambda\, dx \tag{223}$$

$$w = \frac{Z^2 e^2}{8\pi\varepsilon_0\varepsilon\, r} \tag{224}$$

The work given the above equation is actually the contribution of the electrostatic interactions to the Gibbs energy of the ion i.e.

$$G_{EI} = \frac{Z^2 e^2}{8\pi\varepsilon_0\varepsilon\, r} \tag{225}$$

In the light of the above correlation, the electrostatic contribution to the Gibbs free energy of discrete ions and activated complex can be written as

$$G_{EI}(A) = \frac{Z_A^2 e^2}{8\pi\varepsilon_0\varepsilon\, r_A} \tag{225}$$

$$G_{EI}(B) = \frac{Z_B^2 e^2}{8\pi\varepsilon_0\varepsilon\, r_B} \tag{226}$$

$$G_{EI}^* = \frac{(Z_B + Z_B)^2 e^2}{8\pi\varepsilon_0\varepsilon\, r^*} \tag{227}$$

Hence, the change in electrostatic contribution can be obtained simply by subtracting the sum of individual contributions from the overall contribution i.e.

$$\Delta G_{EI}^* = \frac{(Z_B + Z_B)^2 e^2}{8\pi\varepsilon_0\varepsilon\, r^*} - \frac{Z_A^2 e^2}{8\pi\varepsilon_0\varepsilon\, r_A} - \frac{Z_B^2 e^2}{8\pi\varepsilon_0\varepsilon\, r_B} \tag{228}$$

or

$$\Delta G_{EI}^* = \frac{e^2}{8\pi\varepsilon_0\varepsilon}\left[\frac{(Z_B + Z_B)^2}{r^*} - \frac{Z_A^2}{r_A} - \frac{Z_B^2}{r_B}\right] \tag{229}$$

Correcting the above equation for non-electrostatic contribution $\Delta G_{NEI}^*$, the total molar free energy change for the whole process can be given by the following relation.

$$\Delta G^* = \Delta G_{NEI}^* + \Delta G_{EI}^* = \Delta G_{NEI}^* + \frac{Ne^2}{8\pi\varepsilon_0\varepsilon}\left[\frac{(Z_B + Z_B)^2}{r^*} - \frac{Z_A^2}{r_A} - \frac{Z_B^2}{r_B}\right] \tag{230}$$

Also, from the activated complex theory, we know that

$$k = \frac{RT}{Nh}e^{-\frac{\Delta G^*}{RT}} \tag{231}$$

After putting the value of $\Delta G^*$ from equation (230) into equation (231), we get

$$k = \frac{RT}{Nh}e^{-\left(\frac{\Delta G_{NEI}^*}{RT} + \frac{Ne^2}{RT8\pi\varepsilon_0\varepsilon}\left[\frac{(Z_B+Z_B)^2}{r^*} - \frac{Z_A^2}{r_A} - \frac{Z_B^2}{r_B}\right]\right)} \tag{232}$$

Taking natural logarithm both side of equation (232) and rearranging, we get

$$\ln k = \ln\frac{RT}{Nh} - \frac{\Delta G_{NEI}^*}{RT} - \frac{Ne^2}{RT8\pi\varepsilon_0\varepsilon}\left[\frac{(Z_B + Z_B)^2}{r^*} - \frac{Z_A^2}{r_A} - \frac{Z_B^2}{r_B}\right] \tag{233}$$

Which can also be expressed as

$$\ln k = \ln k_0 - \frac{Ne^2}{RT8\pi\varepsilon_0\varepsilon}\left[\frac{(Z_B + Z_B)^2}{r^*} - \frac{Z_A^2}{r_A} - \frac{Z_B^2}{r_B}\right] \tag{234}$$

Where $k_0$ represents the magnitude of the rate constant for the ionic reaction carried out in a solvent of infinite dielectric constant so that the electrostatic interactions become zero.

## ❖ Influence of Solvent and Ionic Strength

The rate of reaction in the case of ionic reactions is strongly dependent upon the nature of the solvent used and the ionic strength. The single and double-sphere treatment of these reactions enables us to study their effect in detail. In this section, we discuss the application and validity of solvent influence and ionic strength on the reaction rate.

### ➢ *Influence of the Solvent*

In order to study the influence of solvent on the rate of ionic reactions, recall the rate equation derived using the double sphere model i.e.

$$\ln k = \ln k_0 \; - \frac{NZ_AZ_Be^2}{RT4\pi\varepsilon_0\varepsilon\, r} \tag{235}$$

Where $\varepsilon_0$ and $\varepsilon$ are permittivities of the vacuum ($8.854 \times 10^{-12}$ $C^2$ $N^{-1}$ $m^{-2}$) and the dielectric constant of the solvent used, respectively. The symbol $e$ represents the elementary charge and has a value equal to $1.6 \times 10^{-19}$ C. $k_0$ represents the magnitude of the rate constant for the ionic reaction carried out in a solvent of infinite dielectric constant so that the electrostatic interactions become zero. $Z_A$ and $Z_B$ are the charge numbers of the participating ions. The symbol $N$ and R represents the Avogadro number and gas constant, respectively. Rearranging equation (235), we get

$$\ln k = -\frac{NZ_AZ_Be^2}{RT4\pi\varepsilon_0\, r}\frac{1}{\varepsilon} + \ln k_0 \tag{236}$$

Which is clearly the equation of the straight line ($y = mx + c$) with a negative slope and positive intercept. Therefore, it is obvious that the logarithm of the rate constant shows a linear variation with the reciprocal of dielectric constant.

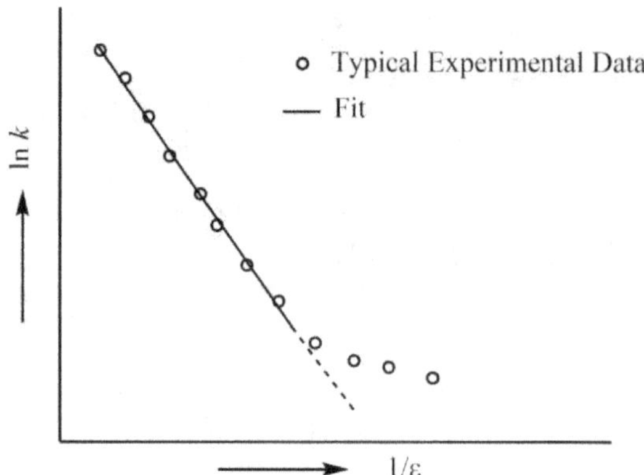


Figure 13. The plot of ln $k$ vs 1/ε for a typical ionic reaction.

It is quite obvious from the plot that equation (236) holds very good over a wide range of dielectric contents; however, as the large deviations are observed at lower values of $\varepsilon$. Moreover, if '$m$' is the experimental slope then from equation (236), we have

$$m = -\frac{NZ_AZ_Be^2}{RT4\pi\varepsilon_0\, r} \tag{237}$$

Every term in the above equation is known apart from $r$, suggesting its straight forward determination from the slope of $\ln k$ vs $1/\varepsilon$. The values of $r$ obtained from equation (237) are found to be quite comparable to other methods, which in turn suggests its practical application.

Besides the calculation of $r$, the influence of dielectric constant of solvent can also be used to explain the entropy of activation. In order to do so, recall from the principles of thermodynamics

$$\left(\frac{\partial G}{\partial T}\right)_P = -S \tag{238}$$

Also, the electrostatic contribution to the Gibbs free energy using the double sphere model is

$$\Delta G^*_{EI} = \frac{NZ_AZ_Be^2}{4\pi\varepsilon_0\varepsilon\, r} \tag{239}$$

However, the only quantity which is temperature-dependent in the above equation is $\varepsilon$. Therefore, differentiating equation (239) with respect to temperature at constant pressure gives

$$\Delta S^*_{EI} = -\left(\frac{\partial(\Delta G^*_{EI})}{\partial T}\right)_P = -\frac{NZ_AZ_Be^2}{4\pi\varepsilon_0\, r}\left(\frac{\partial(1/\varepsilon)}{\partial T}\right)_P \tag{240}$$

or

$$\Delta S^*_{EI} = \frac{NZ_AZ_Be^2}{4\pi\varepsilon_0\varepsilon^2\, r}\left(\frac{\partial\varepsilon}{\partial T}\right)_P \tag{241}$$

or

$$\Delta S^*_{EI} = \frac{NZ_AZ_Be^2}{4\pi\varepsilon_0\varepsilon\, r}\left(\frac{\partial\ln\varepsilon)}{\partial T}\right)_P \tag{242}$$

Therefore, knowing the dielectric constant of the solvent and $r$, the entropy of activation can be obtained. Moreover, it is also worthy to note that the entropy of activation is negative and decreases with an increase in $Z_AZ_B$.

One more factor that affects the entropy of activation is the phenomena of "electrostriction" or the solvent binding. This can be explained by considering the combination of two ions as of same and opposite charges. If the ions forming activated complex are having one-unit positive charge each, the double-sphere will have a total of two-unit positive charge. This would result in a very strong interaction between the activated complex and the surrounding solvent molecules. This would eventually result in a restriction of free movement and hence decreased entropy. On the other hand, If the ions forming activated complex possess opposite charges, the double-sphere would have less charge resulting in decreased electrostriction, and therefore, increased entropy.

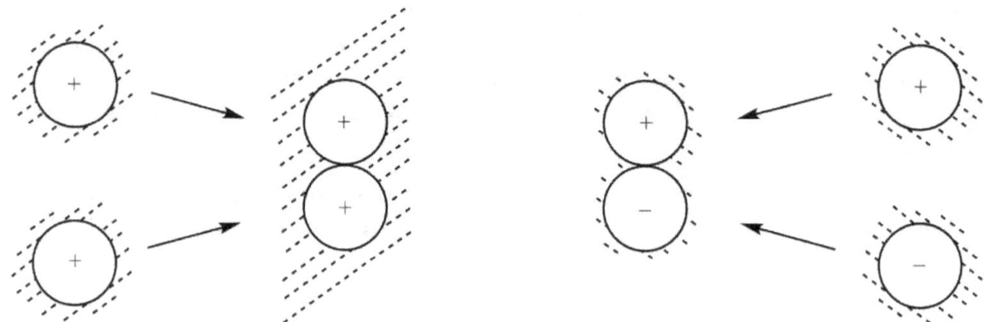

Figure 14. The dependence of entropy of activation on the solvent electrostriction in case of (left) same charges and (right) opposite charges.

> ### Influence of Ionic Strength

In order to study the influence of ionic strength ($I$) on the rate of ionic reactions, we need to recall the quantity itself first i.e.

$$I = \frac{1}{2}\sum_{i=1}^{i=n} m_i z_i^2 \tag{243}$$

Where $m_i$ and $z_i$ are the molarity and charge number of $i$th species, respectively. For instance, the value of $z$ for $Ca^{2+}$ and $Cl^-$ in $CaCl_2$ are $+2$ and $-1$, respectively. It has been found that an increase the ionic strength increases the rate of reaction if charges on the reacting species are of the same sign. On the other hand, the reaction rate has been found to follow a declining trend with increasing ionic strength if reaction ions are of opposite sign. The mathematical treatment of the abovementioned statement is discussed below.

To rationalize the effect of ionic strength of the solution on the rate of reaction in case of ionic reactions, consider a typical case i.e.

$$A^{Z_A} + B^{Z_B} \rightarrow X^{Z_A + Z_B} \rightarrow P \tag{244}$$

A Danish physical chemist, J. N. Brønsted, proposed the rate equation relating reaction-rate ($R$) and activity coefficient as

$$R = k_0 [A][B]\frac{y_A y_B}{y_X} \tag{245}$$

Where $y_A$, $y_B$ and $y_X$ are the activity coefficients for the reactant $A$, $B$ and the activated complex $X$, respectively. Brønsted collectively labeled the term $y_A y_B / y_X$ as the "kinetic activity factor", and it was found be quite accurate with experimental data. Now, rearranging equation (245), we get

$$\frac{R}{[A][B]} = k_0 \frac{y_A y_B}{y_X} \tag{246}$$

Since the left-hand side simply equals to a second-order rate constant, the above equation takes the form

$$k = k_0 \frac{y_A y_B}{y_X} \tag{247}$$

Taking logarithm both side, we get

$$\log k = \log k_0 + \log \frac{y_A y_B}{y_X} \tag{248}$$

or

$$\log k = \log k_0 + \log y_A + \log y_B - \log y_X \tag{249}$$

Now the correlation of mean ionic activity coefficient with the ionic strength is given by famous Debye-Huckel theory i.e.

$$\log y_i = -B z_i^2 \sqrt{I} \tag{250}$$

Where B is the Debye-Huckel constant. Using the concept of equation (250) in equation (249), we get

$$\log k = \log k_0 - B z_A^2 \sqrt{I} - B z_B^2 \sqrt{I} + B z_X^2 \sqrt{I} \tag{251}$$

or

$$\log k = \log k_0 - B z_A^2 \sqrt{I} - B z_B^2 \sqrt{I} + B(z_A + z_B)^2 \sqrt{I} \tag{252}$$

$$\log k = \log k_0 + B[(z_A + z_B)^2 - z_A^2 - z_B^2] \sqrt{I} \tag{253}$$

$$\log k = \log k_0 + B[z_A^2 + z_B^2 + 2 z_A z_B - z_A^2 - z_B^2] \sqrt{I} \tag{254}$$

or simply

$$\log k = \log k_0 + 2 B z_A z_B \sqrt{I} \tag{255}$$

Rearranging the above equation, we get

$$\log k = 2 B z_A z_B \sqrt{I} + \log k_0 \tag{256}$$

Which is clearly the equation of straight line ($y = mx + c$) with a positive slope ($2 B z_A z_B$) and positive intercept ($\log k_0$). For aqueous solutions at 25°C, $2B = 1.02$, equation (256) takes the form

$$\log k - \log k_0 = 1.02\, z_A z_B \sqrt{I} \tag{257}$$

or

$$\log \frac{k}{k_0} = 1.02\, z_A z_B \sqrt{I} \tag{258}$$

Where $k_0$ is the rate constant at zero ionic strength and can be obtained by the extrapolation of log $k_0$ vs square root of the ionic strength.

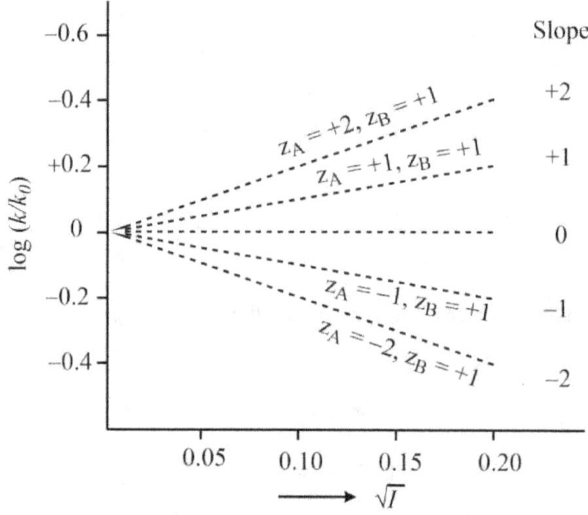


Figure 15. The plot of log ($k/k_0$) vs $(I)^{1/2}$ for different ionic reactions in aqueous solution at 25°C.

The above equation can also be extended to explain the dependence of reaction-rate on ionic strength for third-order reactions. To do so, consider

$$A^{Z_A} + B^{Z_B} + C^{Z_C} \rightarrow X^{Z_A + Z_B + Z_C} \rightarrow P \tag{259}$$

Following the same route as in second-order reactions, we will get

$$\log k = \log k_0 + B[(z_A + z_B + z_C)^2 - z_A^2 - z_B^2 - z_C^2]\sqrt{I} \tag{260}$$

$$\log k = \log k_0 + 2B(z_A z_B + z_B z_C + z_C z_A)\sqrt{I} \tag{261}$$

Therefore, a negative slope will be observed if $(z_A z_B + z_B z_C + z_C z_A)$ is negative while a positive slope is expected for a positive value of $(z_A z_B + z_B z_C + z_C z_A)$.

## ❖ The Comparison of Collision and Activated Complex Theory

After studying the collision as well as the transition state theories in detail, it is time to highlight the key points of similarities and differences between the two. A comparative analysis of both theories is quite beneficial as far as the practicality is concerned.

Table 1. The side-by-side comparison between the collision theory and transition state theory.

| Collision Theory | Activated Complex Theory |
|---|---|
| 1. According to the collision theory, the chemical reactions occur when the reactant molecules collide with a sufficient amount of kinetic energy. | 1. According to the transition state theory, the primary cause of the reaction is actually the formation of an activated complex or the transition state, which in turn, converts to the final product. |
| 2. It is based upon the kinetic theory of gases. | 2. It is derived from the fundamentals of thermodynamics. |
| 3. This theory considers the activation energy as the minimum energy required to make the collision effective. | 3. This theory assumes the activation energy as the difference between the energy of the reacting molecules and the energy of the activated complex. |
| 4. This theory tells nothing about the entropy of activation. | 4. The transition state theory enables us to measure the entropy of activation. |
| 5. Collision theory is applicable to simple chemical reactions and large deviations with experimental results are observed as the complexity increases. | 5. This theory provided reasonable predictions even for the complex reactions. |
| 6. The incorporation of the correction factor in modified collision theory was arbitrary. | 6. The incorporation of correction factor was justified in terms of entropy of activation i.e. $\Delta S^*$. |
| 7. This theory tells nothing about the mechanism involved. | 7. The formation of the activated complex is very much correlated with the actual mechanism going on. |

## ❖ Problems

Q 1. Discuss the fundamental concept of the effect of temperature on the reaction-rate with special reference to Maxwell-Boltzmann distribution of energies.

Q 2. Derive Arrhenius equation for the rate of reactions. How it can be used to determine the activation energy?

Q 3. Deduce the rate expression for the first order opposed by first-order reactions. How it can be used to yield the value of backward reaction too?

Q 4. Derive and discuss the rate law for second-order opposed by second order.

Q 5. What are the consecutive reactions? Discuss the condition required to apply the steady-state approximation to a typical consecutive reaction.

Q 6. Define concurrent or parallel reactions. Derive the rate law for the first-order case.

Q 7. Discuss the collision of the theory bimolecular reaction when the colliding particles are dissimilar.

Q 8. What are unimolecular reactions? How they are treated in the collision framework?

Q 9. Discuss the limitations of collision theory and the incorporation of the steric factor?

Q 10. Derive the rate law in the activated complex framework. How it can be used to determine the entropy of activation?

Q 11. What is the single-sphere model of ionic reactions? How is it different from the double sphere model?

Q 12. Discuss the influence of ionic strength on the rate of ionic reactions of the third order.

Q 13. Give five points of difference between the collision and activated complex theory.

## ❖ Bibliography

[1] B. R. Puri, L. R. Sharma, M. S. Pathania, *Principles of Physical Chemistry*, Vishal Publications, Jalandhar, India, 2008.

[2] P. Atkins, J. Paula, *Physical Chemistry*, Oxford University Press, Oxford, UK, 2010.

[3] E. Steiner, *The Chemistry Maths Book*, Oxford University Press, Oxford, UK, 2008.

[4] S. A. Arrhenius, *Über die Dissociationswärme und den Einfluß der Temperatur auf den Dissociationsgrad der Elektrolyte.* Z. Phys. Chem. 4 (1889) 96–116.

[5] S. A. Arrhenius, *Über die Reaktionsgeschwindigkeit bei der Inversion von Rohrzucker durch Säuren* Z. Phys. Chem. 4 (1889) 226–248.

[6] K. J. Laidler, *Chemical Kinetics*, Harper & Row, New York, USA, 1987.

[7] K. J. Laidler, *The World of Physical Chemistry*, Oxford University Press, Oxford, UK, 1993.

[8] H. Eyring, *The Activated Complex in Chemical Reactions*, J. Chem. Phys. 3 (1935) 107-115.



**D DALAL INSTITUTE**

# CHAPTER 4

# Electrochemistry – I: Ion-Ion Interactions

## ❖ The Debye-Huckel Theory of Ion-Ion Interactions

When an ion goes into the solution, it interacts not only with the solvent molecules but also with other ions that are produced as a result of electrolytic dissociation. The interaction of this ion with its solvent-surrounding is quite important as it helps the ion to get stabilized in that medium. Nevertheless, the other types of interactions, i.e., the ion-ion interactions, are also of very much significant value to the overall understanding of the system because these interactions affect a number of properties of these solutions such as partial molar free energy change or the electrolytic conductance. Now, since the phenomenon of ion-ion interactions is a function of interionic separations, which in turn depends upon the population density of these ions; the first step to know this phenomenon must be the rationalization of ionic population density.

In the case of weak electrolytes, the degree of dissociation is very small at high concentrations yielding a very low population density of charge carriers. This would result in almost zero ion-ion interactions at high concentrations. Now, although the degree of dissociation increases with dilution, which in turn also increases the total number of charge carriers, the population density remains almost unchanged since the extra water has also been added for these extra ions. Thus, we can conclude that there are no ion-ion interactions in weak electrolytes, neither at higher nor at the lower concentrations. On the other hand, in the case of strong electrolytes, the degree of dissociation is a hundred percent even at high concentrations yielding a very high population density of charge carriers. This would result in very strong ion-ion interactions at high concentrations. Now, when more and more solvent is added, the total number of charge carriers remains the same but the population density decreases continuously, creating large interionic separations. This would result in a decrease in the magnitude of ion-ion interaction with increasing dilution.

In 1913, a British physicist Samuel Milner proposed the first quantitative explanation of ion-ion interactions in a statistical framework. After that, in 1918, an Indian Chemist, Sir Jnan Chandra Ghosh proposed a model to compute the energy of Coulombic interaction of the ions with an assumption that the ions in the solution are fixed just like in the crystals. The Milner's model was quite complicated mathematically, while Ghosh's model did not consider the distorting effect of thermal motion on the ionic distribution in the solution phase of electrolyte under consideration.

In 1923, Peter Debye and Erich Huckel proposed an extremely important idea to quantify the ion-ion interactions in the solutions of strong electrolytes. In this model, the solute, i.e., the strong electrolyte, is assumed to dissociate completely. They considered the ions as perfect spheres that cannot be polarized by the surrounding electric field while the solvation of ions was ignored. The solvent plays no role other than providing a medium of the constant relative permittivity (dielectric constant). A reference ion was thought to be suspended in solvent-continuum of dielectric constant $\varepsilon$ with zero electrostriction, and this ion is surrounded by oppositely charged ionic cloud.

---

## ➢ *Fundamental Concept*

It is quite interesting to recall the fact that like many other coupling effects, the phenomena of ion-ion interaction can also be quantified in terms of free energy change. In this case, it can be achieved by assuming an initial state with no ion-ion interaction and a final state in which the ion-ion interactions do exist; and then calculating the free energy change in going from initial to the final state.

The initial state, i.e., the state with no ion-ion interaction, can be created hypothetically by taking ions in the vacuum. However, since the final state possesses ions in solution, the initial state should also have ions in solution and not in the vacuum, otherwise, two states would differ not only by ion-ion interaction but also in respect of ion-solvent interaction. Conversely, if we imagine some discharged ions in the solution (discharged ions mean no ion-ion interaction) and then charge them up fully, free energy change of ion-ion interaction would be the amount of work done in doing so.

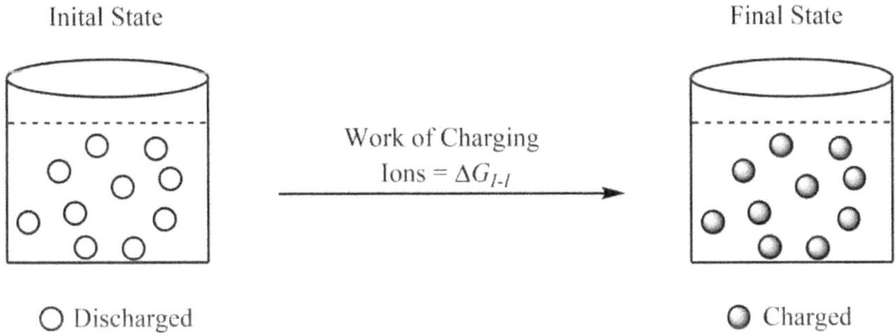


Figure 1. The pictorial representation of the fundamental idea for calculating the free energy of ion-ion in the Debye-Huckel model of ionic interactions.

It should also be noted that the procedure depicted in 'Figure 1' gives the work of charging (i.e., the free energy of ion-ion interaction) for all the ions present in the solution whether they are positive or negative. However, the scientific analysis of electrolytic solutions typically needs the free energy of ionic interactions for individual species. This partial free energy change is called as chemical potential change and is typically denoted as $\Delta\mu_{i\text{-}I}$. The value of $\Delta\mu_{i\text{-}I}$ for $i$th species of radius $r_i$ can be obtained by computing the work of charging this sphere from a neutral state to its full charge of $Z_ie_0$. The work of charging ($w$) a spherical conductor can be given by the following relation (from electrostatics):

$$w = \frac{1}{2}\,[Charge\ on\ the\ conductor \times Electrostatic\ Potential\ of\ conductor] \tag{1}$$

Since the Debye-Huckel model also considers the ions as spherical conductors, the work of charging an ion of radius $r$ will be

$$w = \frac{1}{2}[Z_i e_0 \times \psi] \tag{2}$$

Where $\psi$ represents the electrostatic potential of the reference ion due to the influence on it by the Coulombic forces of the neighboring ions. Multiplying equation (2) by Avogadro number, we can obtain the amount of work done required to charge one mole of such ions i.e. the partial molar free energy change.

$$\Delta\mu_{i-I} = N_A w = \frac{N_A}{2}[Z_i e_0 \times \psi] \tag{3}$$

The only thing that is unknown in the above expression is $\psi$. Therefore, the partial molar free energy change due to the interaction of $i$th species with the rest of the solution requires the determination of electrostatic potential exerted at the reference ion by its ionic-surrounding. Now, from the law of superposition of potentials, we know that the potential at any point due to an assembly of charges is just the sum of the individual potentials exerted at that point by the individual charges. Since the electrostatic potentials depend upon the distances between the charges under consideration, the time-averaged distribution of all ions around must be known (distances from the reference ion are the function of the population distribution of other ions).

Although we know that the total electrostatic potential at the reference ion can be obtained by knowing the distance of each surrounding ion from the reference ion, the number of ions is extremely large making it almost impossible to follow the conventional route to do so. So, instead of dealing with individual ions, Debye and Huckel proposed the revolutionary idea of considering all the surrounding ions as a continuous ionic cloud that surrounds the reference ions. They assigned the discrete charge only to the reference ion which is situated in the solvent's continuum of dielectric constant $\varepsilon$, i.e., water in this case.

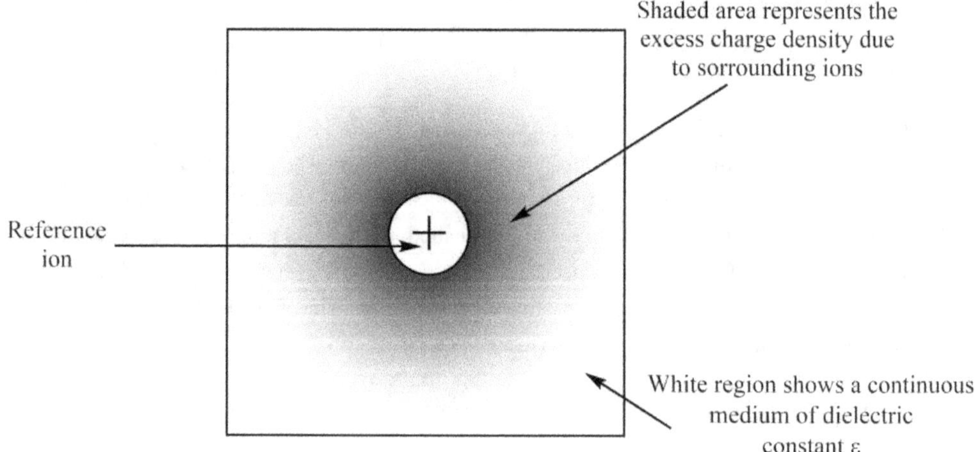


Figure 2. The pictorial representation of reference ion in its ionic cloud.

Hence, the Debye-Huckel's idea transformed the complex problem of computing the population density around the reference to the determination of excess charge density ($\rho$). At this point, one might think why the excess charge density is not zero everywhere because the solution as a whole is actually electroneutral. The reason for this behavior lies in the fact that positive centers attract negative density while a negative center tries to accumulate a positive cloud around it. Therefore, in a volume element near positive ion, there should be a more negative charge, i.e., excess negative charge density and vice-versa. This excess negative charge density near a positive center is counterbalanced by excess positive charge density near a negative center, keeping the overall electroneutrality maintained. It should also be mentioned here that since the thermal forces in the solution are always trying to randomize the ionic distribution, the excess charge density diminishes in moving away from the center.

## ➢ *Mathematical Development*

So far we have discussed the fundamentals aspects of how we can determine the partial molar free energy change generated by the ionic interactions of *i*th species. All we need is the electrostatic potential ($\psi$) of the reference ion due to the influence on it by the Coulombic forces of the surrounding. Since $\psi$, at a distance $r$ from the reference ion, depends upon the excess charge density at $r$, the first thing we need to develop it mathematically is the correlation between these two parameters, i.e., $\psi_r$ and $\rho_r$. One such relation for spherically symmetric charge distribution is the Poisson's equation in electrostatics that can be given as:

$$\frac{1}{r^2}\frac{d}{dr}\left(r^2\frac{d\psi_r}{dr}\right) = -\frac{4\pi}{\varepsilon}\rho_r \tag{4}$$

Where $\psi_r$ and $\rho_r$ are electrostatic potential and excess charge density in a very small volume element $dV$ situated at distance $r$ from the reference ion.

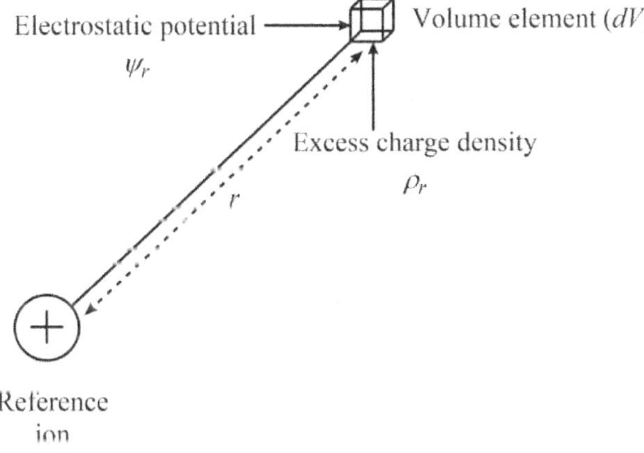


Figure 3. The depiction of electrostatic potential and excess charge density in $dV$ volume element at a distance $r$ from the reference ion.

Rearranging the Poisson's equation for excess charge density, we get

$$\rho_r = -\frac{\varepsilon}{4\pi}\left[\frac{1}{r^2}\frac{d}{dr}\left(r^2\frac{d\psi_r}{dr}\right)\right] \tag{5}$$

Also, the "linearized Boltzmann distribution" can be used to prove that

$$\rho_r = -\sum_i \frac{n_i^0 Z_i^2 e_0^2 \psi_r}{kT} \tag{6}$$

Where $n_i^0$ is the bulk concentration of the $i$th species and $k$ is simply the Boltzmann constant. Equating the Poisson's expression with the Boltzmann Formula, i.e., from equation (5) and (6), we get the linearized Poisson Boltzmann equation as:

$$-\frac{\varepsilon}{4\pi}\left[\frac{1}{r^2}\frac{d}{dr}\left(r^2\frac{d\psi_r}{dr}\right)\right] = -\sum_i \frac{n_i^0 Z_i^2 e_0^2 \psi_r}{kT} \tag{7}$$

or

$$\frac{1}{r^2}\frac{d}{dr}\left(r^2\frac{d\psi_r}{dr}\right) = \left(\frac{4\pi}{\varepsilon kT}\sum_i n_i^0 Z_i^2 e_0^2\right)\psi_r \tag{8}$$

Now assume a constant $\kappa^2$ with value

$$\kappa^2 = \frac{4\pi}{\varepsilon kT}\sum_i n_i^0 Z_i^2 e_0^2 \tag{9}$$

Using the value of equation (9) in equation (8), we get

$$\frac{1}{r^2}\frac{d}{dr}\left(r^2\frac{d\psi_r}{dr}\right) = \kappa^2 \psi_r \tag{10}$$

The above differential equation is the popular form of linearized Poisson-Boltzmann expression and its solution gives

$$\psi_r = \frac{Z_i e_0}{\varepsilon}\frac{e^{-\kappa r}}{r} \tag{11}$$

Now comparing equation (4) and equation (10), we get

$$\kappa^2 \psi_r = -\frac{4\pi}{\varepsilon}\rho_r \tag{12}$$

or

$$\rho_r = -\frac{\varepsilon\kappa^2\psi_r}{4\pi} \tag{13}$$

Now putting the value of electrostatic potential from equation (11) in equation (13), we get the expression for excess charge density as a function of $r$.

$$\rho_r = -\frac{\varepsilon\kappa^2}{4\pi} \times \frac{Z_i e_0}{\varepsilon} \frac{e^{-\kappa r}}{r} \tag{14}$$

$$\rho_r = -\frac{Z_i e_0 \kappa^2}{4\pi r} e^{-\kappa r} \tag{15}$$

Now because the magnitude of $\rho_r$ is a consequence of the unequal distribution of anions and cations, the above expression also defines the ionic-population-distribution around the reference ion.

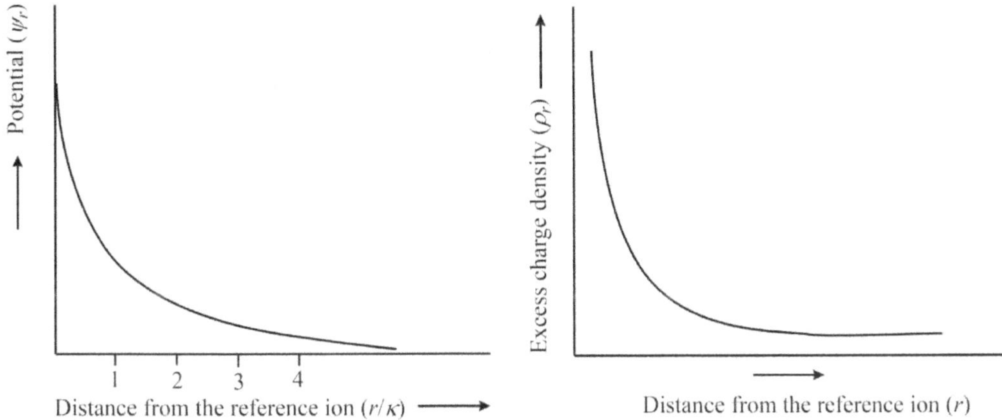



Figure 4. The variation of electrostatic potential and excess charge density as a function of the distance $r$ from the reference ion.

After knowing that the excess charge density decreases exponentially with $r$, we can determine the total excess charge ($dq$) at the same distance just by the following relation.

$$dq = \rho_r 4\pi r^2 dr \tag{16}$$

Where $4\pi r^2 dr$ is the volume element of a hollow spherical shell of thickness $dr$ with the inner and outer radius as $r$ and $r+dr$, respectively. After putting the value of $\rho_r$ from equation (15) in equation (16), we have

$$dq = \left( -\frac{Z_i e_0 \kappa^2}{4\pi r} e^{-\kappa r} \right) 4\pi r^2 dr \tag{17}$$

or

$$dq = -Z_i e_0 e^{-\kappa r} \kappa^2 r\, dr \tag{18}$$

Thus, the exponential part decreases with increasing $r$ while the non-exponential part shows a continuous increase as we move away from the reference ion.

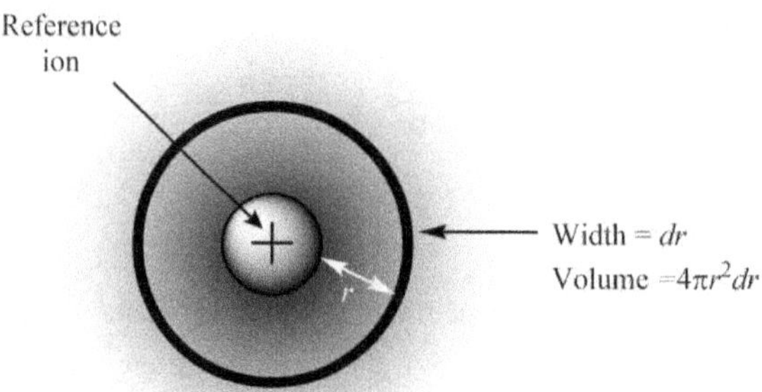


Figure 5. The depiction of excess charge density in a volume element of thickness '$dr$' as a function of the distance $r$ from the reference ion.

In order to find the distance of maximum "excess charge", we need to differentiate the equation (18) with respect to $r$, and then derivative needs to be put equal to zero i.e.

$$\frac{dq}{dr} = -Z_i e_0 \kappa^2 (e^{-\kappa r} - \kappa r e^{-\kappa r}) = 0 \tag{19}$$

Since $-Z_i e_0 \kappa^2$ is non zero for sure, the above equation holds true only if

$$e^{-\kappa r} - \kappa r e^{-\kappa r} = 0 \tag{20}$$

Which implies

$$r = \kappa^{-1} \tag{21}$$

Also from equation (9), we have

$$\kappa = \left( \frac{4\pi}{\varepsilon kT} \sum_i n_i^0 Z_i^2 e_0^2 \right)^{1/2} \tag{22}$$

Hence, the radius of maximum excess charge from the center of the reference ion can be obtained by putting the value of $\kappa$ from equation (22) in equation (21), we get $r_{max}$ or the "Debye-Huckel length" as

$$r_{max} = \left( \frac{\varepsilon kT}{4\pi} \frac{1}{\sum_i n_i^0 Z_i^2 e_0^2} \right)^{1/2} \tag{23}$$

At this point, we have understood most of the ideas of Debye-Huckel's theory of ion-ion interaction, and the question we raised initially is near to its end, i.e., to calculate the molar chemical potential change of ion-ion interaction. Now, although we have calculated the electrostatic potential, we cannot use the same in equation (3) because what we have obtained contains the contributions from the ionic cloud as well as from the reference ion itself. Therefore, before we calculate the partial molar free energy of ion-ion interaction, we need to separate potential from the ionic cloud and from the central ion first. The total electrostatic potential at a distance $r$ can be fragmented as given below.

$$\psi_r = \psi_{ion} + \psi_{cloud} \tag{24}$$

$$\psi_{cloud} = \psi_r - \psi_{ion} \tag{25}$$

From the formulation of potential due to a single charge at a distance $r$, we know that

$$\psi_{ion} = \frac{Z_i e_0}{\varepsilon r} \tag{26}$$

Now using the value of $\psi_r$ from equation (11) and of $\psi_{ion}$ from equation (26) into equation (25), we get

$$\psi_{cloud} = \frac{Z_i e_0}{\varepsilon} \frac{e^{-\kappa r}}{r} - \frac{Z_i e_0}{\varepsilon r} \tag{27}$$

$$\psi_{cloud} = \frac{Z_i e_0}{\varepsilon r} (e^{-\kappa r} - 1) \tag{28}$$

Now because the value of $\kappa$ is proportional to $\sum_i n_i^0 Z_i^2 e_0^2$, at very large dilution $\kappa$ becomes very small and $e^{-\kappa r}$ and can be expended as $1 - \kappa r$ ($e^x = 1 + x$). Therefore, equation (28) takes the form

$$\psi_{cloud} = \frac{Z_i e_0}{\varepsilon r} (1 - \kappa r - 1) \tag{29}$$

$$\psi_{cloud} = -\frac{Z_i e_0 \kappa}{\varepsilon} \tag{30}$$

Hence, potential due to ionic cloud is independent of the distance $r$ in this case. Now putting the value of $\psi_{cloud}$ from equation (30) into equation (3), we get

$$\Delta \mu_{i-I} = N_A w = \frac{N_A}{2} \left[ Z_i e_0 \times \left( -\frac{Z_i e_0 \kappa}{\varepsilon} \right) \right] \tag{31}$$

$$\Delta \mu_{i-I} = -\frac{N_A Z_i^2 e_0^2 \kappa}{2\varepsilon} \tag{32}$$

Hence, the Debye-Huckel's model gives the theoretical value of $\Delta \mu_{i-I}$ arising from ion-ion interactions. Now before we check the validity of the Debye-Huckel model, we will discuss the many aspects of mathematical treatment Debye-Huckel model in more detail in forthcoming sections.

### ❖ Potential and Excess Charge Density as a Function of Distance from the Central Ion

In order to determine the potential and excess charge density as a function of radial distance $r$, consider a very small volume element $dV$, situated at distance $r$ from the reference ion, in which $\psi_r$ and $\rho_r$ are electrostatic potential and excess charge density, respectively.

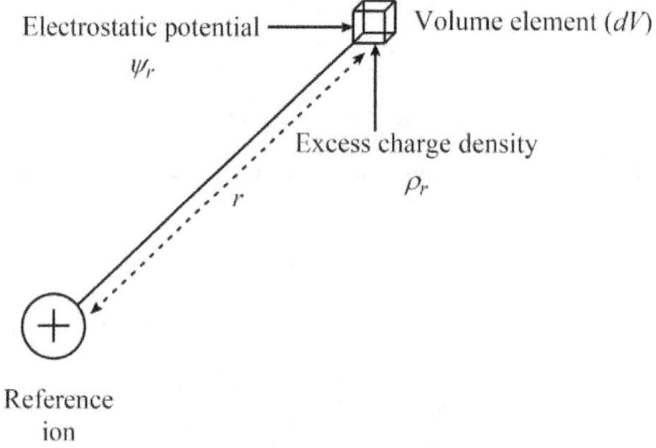


Figure 6. The depiction of electrostatic potential and excess charge density in $dV$ volume element at distance $r$ from the reference ion.

Since we want to study how electrostatic potential and excess charge density depend upon the distance from the reference ion, the first thing we need to develop it mathematically is a correlation of these two parameters i.e. $\psi_r$ and $\rho_r$. One such relation for spherically symmetric charge distribution is the Poisson's equation in electrostatics which can be given as:

$$\frac{1}{r^2}\frac{d}{dr}\left(r^2\frac{d\psi_r}{dr}\right) = -\frac{4\pi}{\varepsilon}\rho_r \tag{33}$$

Now the excess charge density in the volume element $dV$ can be obtained by multiplying the total number of ions per unit volume with their corresponding charges i.e.

$$\rho_r = n_1 Z_1 e_0 + n_2 Z_2 e_0 + n_3 Z_3 e_0 \dots n_i Z_i e_0 \tag{34}$$

or

$$\rho_r = \sum n_i Z_i e_0 \tag{35}$$

Where $n_1$ is the number of first kind ions with $Z_1 e_0$ charge, $n_2$ is the number of second kind ions with $Z_2 e_0$ charge and so on up to $i$th type.

Now, from the Boltzmann distribution law of classical statistical mechanics, we know that

$$n_i = n_i^0 e^{-U/kT} \tag{36}$$

Where $U$ represents the total change in the potential energy of the $i$th particle in going from the bulk concentration $n_i^0$ to the actual concentration of the $i$th particle i.e. $n_i$. At this stage, three cases arise, the magnitude of $U$ can be zero, positive or negative. If $U = 0$ i.e. there are no ion-ion interaction, $n_i = n_i^0$, which implies that concentration near the reference ion will be equal to the bulk concentration. If $U = -$, i.e., there are attractive ion-ion interaction, $n_i > n_i^0$, which implies that concentration near the reference ion will be higher to the bulk concentration. In the third scenario, if $U = +$, i.e., there are repulsive ion-ion interaction, $n_i < n_i^0$, which implies that concentration near the reference ion will be less to the bulk concentration. Now according to the Debye-Huckel model, only simple Coulombic forces need to be considered for very dilute solutions. Therefore, excluding all other short-range interactions like dispersion ones, the potential of average force U simply can be written as given below.

$$U = Z_i e_0 \psi_r \tag{37}$$

After using the value of $U$ from equation (37) in equation (36), we get

$$n_i = n_i^0 e^{-Z_i e_0 \psi_r / kT} \tag{38}$$

Putting the value of $n_i$ from equation (38) in equation (35), we have

$$\rho_r = \sum n_i^0 Z_i e_0 \, e^{-Z_i e_0 \psi_r / kT} \tag{39}$$

Now because the Debye-Huckel model considers the solutions in which $\psi_r$ is much less than $kT$, we can conclude that $Z_i e_0 \psi_r \ll kT$. Therefore, $e^{-Z_i e_0 \psi_r / kT}$ can be expended as

$$\rho_r = \sum n_i^0 Z_i e_0 \left( 1 - \frac{Z_i e_0 \psi_r}{kT} \right) \tag{40}$$

or

$$\rho_r = \sum n_i^0 Z_i e_0 - \sum \frac{n_i^0 Z_i^2 e_0^2 \psi_r}{kT} \tag{41}$$

The first term of the above equation gives the net charge on the whole of the solution, and it must be zero since the overall electrical neutrality is maintained. Therefore, the above equation takes the form

$$\rho_r = -\sum \frac{n_i^0 Z_i^2 e_0^2 \psi_r}{kT} \tag{42}$$

The above equation is the "linearized Boltzmann distribution".

Rearranging the Poisson's equation for excess charge density i.e. equation (33), we get

$$\rho_r = -\frac{\varepsilon}{4\pi}\left[\frac{1}{r^2}\frac{d}{dr}\left(r^2\frac{d\psi_r}{dr}\right)\right] \tag{43}$$

Where $n_i^0$ is the bulk concentration of the $i$th species and $k$ is simply the Boltzmann constant. Equating the Poisson's expression with the Boltzmann Formula i.e. from equation (43) and (42), we get the linearized Poisson Boltzmann equation as:

$$-\frac{\varepsilon}{4\pi}\left[\frac{1}{r^2}\frac{d}{dr}\left(r^2\frac{d\psi_r}{dr}\right)\right] = -\sum_i \frac{n_i^0 Z_i^2 e_0^2 \psi_r}{kT} \tag{44}$$

or

$$\frac{1}{r^2}\frac{d}{dr}\left(r^2\frac{d\psi_r}{dr}\right) = \left(\frac{4\pi}{\varepsilon kT}\sum_i n_i^0 Z_i^2 e_0^2\right)\psi_r \tag{45}$$

Now assume a constant $\kappa^2$ with value

$$\kappa^2 = \frac{4\pi}{\varepsilon kT}\sum_i n_i^0 Z_i^2 e_0^2 \tag{46}$$

Using the value of equation (46) in equation (45), we get

$$\frac{1}{r^2}\frac{d}{dr}\left(r^2\frac{d\psi_r}{dr}\right) = \kappa^2\psi_r \tag{47}$$

To solve the above differential equation, assume that $\psi_r$ is a function of a new variable, called $\mu$, as

$$\psi_r = \frac{\mu}{r} \tag{48}$$

Differentiating equation (48), we get

$$\frac{d\psi_r}{dr} = -\frac{\mu}{r^2} + \frac{1}{r}\frac{d\mu}{dr} \tag{49}$$

Multiplying both sides by $r^2$

$$r^2\frac{d\psi_r}{dr} = -\mu + r\frac{d\mu}{dr} \tag{50}$$

Now first multiplying both sides by $d/dr$ and then by $1/r^2$, we get

$$\frac{1}{r^2}\frac{d}{dr}\left(r^2\frac{d\psi_r}{dr}\right) = \frac{1}{r^2}\frac{d}{dr}\left(-\mu + r\frac{d\mu}{dr}\right) \tag{51}$$

or

$$\frac{1}{r^2}\frac{d}{dr}\left(r^2\frac{d\psi_r}{dr}\right) = \frac{1}{r^2}\left(-\frac{d\mu}{dr} + r\frac{d^2\mu}{dr^2} + \frac{d\mu}{dr}\right) \tag{52}$$

$$\frac{1}{r^2}\frac{d}{dr}\left(r^2\frac{d\psi_r}{dr}\right) = \frac{1}{r}\frac{d^2\mu}{dr^2} \tag{53}$$

Using the value of equation (48) and (53) into equation (47), we have

$$\frac{1}{r}\frac{d^2\mu}{dr^2} = \kappa^2\frac{\mu}{r} \tag{54}$$

or

$$\frac{d^2\mu}{dr^2} = \kappa^2\mu \tag{55}$$

The general solution of such an equation may be written as

$$\mu = Ae^{-\kappa r} + Be^{+\kappa r} \tag{56}$$

Where $A$ and $B$ are two unknown constants. Now using the value of $\mu$ from equation (56) into equation (48), we get

$$\psi_r = A\frac{e^{-\kappa r}}{r} + B\frac{e^{+\kappa r}}{r} \tag{57}$$

Since the potential at $r = \infty$ must vanish, this boundary condition is satisfied only if $B = 0$. Therefore, the acceptable form of the equation (57) should be like this

$$\psi_r = A\frac{e^{-\kappa r}}{r} \tag{58}$$

To evaluate the value of constant, imagine a situation in which ions are so apart from each other that there are no ion-ion interactions. Such a situation can be created by diluting the solution to a very large extent. In this state, potential around the reference ion will simply be due to the reference ion itself i.e.

$$\psi_r = \frac{z_i e_0}{\varepsilon r} \tag{59}$$

Furthermore, at such large dilution, the bulk concentration will almost be zero ($n_i^0 = 0$) which in turn would make $\kappa = 0$. Thus, the equation (58) in such a scenario will be

$$\psi_r = \frac{A}{r} \tag{60}$$

Equating the results of equation (59) and (60), we have

$$\frac{Z_i e_0}{\varepsilon r} = \frac{A}{r} \tag{61}$$

or

$$A = \frac{Z_i e_0 r}{\varepsilon r} \tag{62}$$

After putting the value of $A$ from equation (62) in equation (58), the final result for electrostatic potential is

$$\psi_r = \frac{Z_i e_0}{\varepsilon} \frac{e^{-\kappa r}}{r} \tag{63}$$

The expression is the solution of the linearized Poisson-Boltzmann equation.

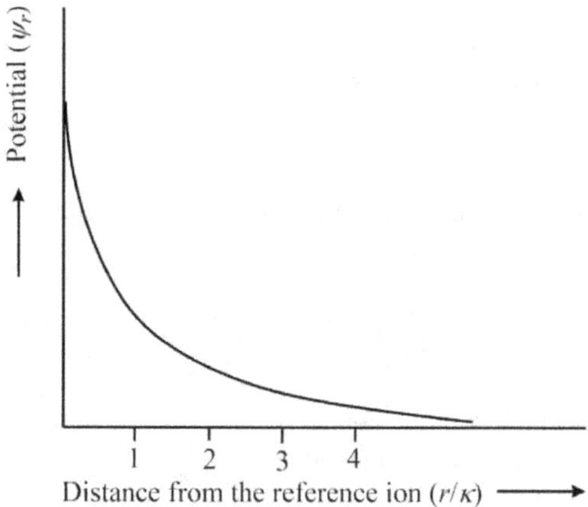


Figure 7. The variation of electrostatic potential as a function of the distance $r$ from the reference ion.

Now, in order to evaluate the excess charge density as a function of distance from the reference ion, compare equation (33) and equation (47) i.e.

$$\kappa^2 \psi_r = -\frac{4\pi}{\varepsilon} \rho_r \tag{64}$$

or

$$\rho_r = -\frac{\varepsilon \kappa^2 \psi_r}{4\pi} \tag{65}$$

Now putting the value of electrostatic potential from equation (63) in equation (65), we get the expression for excess charge density as a function of $r$.

$$\rho_r = -\frac{\varepsilon \kappa^2}{4\pi} \times \frac{Z_i e_0}{\varepsilon} \frac{e^{-\kappa r}}{r} \tag{66}$$

$$\rho_r = -\frac{Z_i e_0 \kappa^2}{4\pi r} e^{-\kappa r} \tag{67}$$

Now because the magnitude of $\rho_r$ is a consequence of the unequal distribution of anions and cations, the above also defines the ionic population distribution around the reference ion.

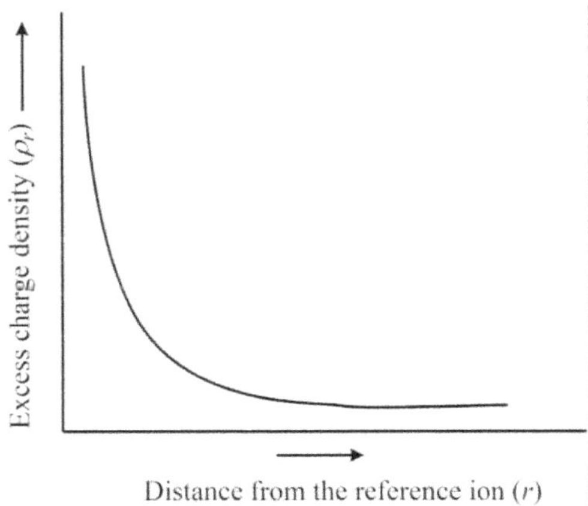


Figure 8. The variation of excess charge density as a function of the distance $r$ from the reference ion.

Hence, the magnitude of excess charge density declines exponentially as the distance from the reference increases. It is also worthy to note that the sign of excess charge around the reference ion is always opposite to the reference ion. A negatively charged reference ion has a positively charged surrounding atmosphere and vice-versa.

### ❖ Debye-Huckel Reciprocal Length

The concept of Debye-Huckel length can be understood in a better way only after knowing how much charge actually surrounds the reference ion. For this, assume that the total space around the reference ion is divided into the infinite number of hollow spherical shells of thickness $dr$ with the inner and outer radii as $r$ and $r+dr$, respectively. The volume of each such shell will be $4\pi r^2 dr$, and the total excess charge ($dq$) in the same can be obtained by multiplying this volume element with the excess charge density, i.e.,

$$dq = \rho_r 4\pi r^2 dr \tag{68}$$

Now, as we know that the excess charge density at distance $r$ from the reference ion is

$$\rho_r = -\frac{Z_i e_0 \kappa^2}{4\pi r} e^{-\kappa r} \tag{69}$$

Putting the value of equation (69) in equation (68), we get

$$dq = \left(-\frac{Z_i e_0 \kappa^2}{4\pi r} e^{-\kappa r}\right) 4\pi r^2 dr \tag{70}$$

or

$$dq = -Z_i e_0 e^{-\kappa r} \kappa^2 r \, dr \tag{71}$$

Since the exponential part becomes zero only at $r = \infty$, the total charge around the reference ion can be obtained by integrating the equation (71) from $r = 0$ to $r = \infty$.

$$q_{cloud} = \int_{r=0}^{r=\infty} dq = \int_{r=0}^{r=\infty} -Z_i e_0 e^{-\kappa r} \kappa^2 r \, dr \tag{72}$$

or

$$q_{cloud} = -Z_i e_0 \tag{73}$$

This is an important result which proves that a reference ion, with the charge $+Z_i e_0$, is surrounded by an exactly equal but oppositely charged cloud i.e. $-Z_i e_0$.

Now, since the exponential part decreases with increasing $r$ while the non-exponential part shows a continuous increase as we move away from the reference ion, there should be a distance of maximum charge.

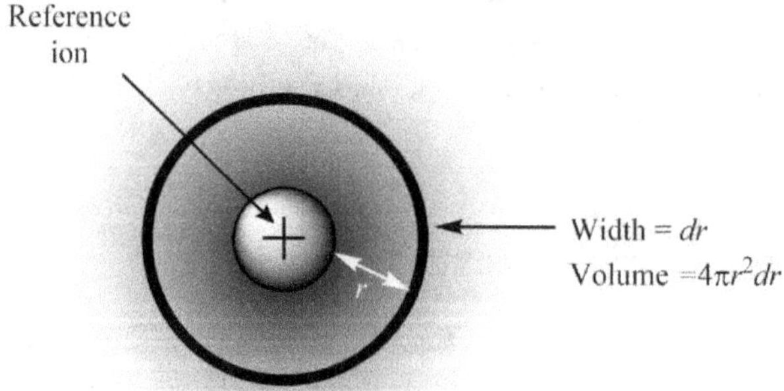


Figure 9. The depiction of excess charge density in a volume element of thickness '$dr$' as a function of the distance $r$ from the reference ion.

In order to find the distance of maximum "excess charge", we need to differentiate the equation (71) with respect to $r$, and then derivative needs to be put equal to zero i.e.

$$\frac{dq}{dr} = -Z_i e_0 \kappa^2 (e^{-\kappa r} - \kappa r e^{-\kappa r}) = 0 \tag{74}$$

Since $-Z_i e_0 \kappa^2$ is non zero for sure, the above equation holds true only if

$$e^{-\kappa r} - \kappa r e^{-\kappa r} = 0 \tag{75}$$

Which implies

$$r = \kappa^{-1} \tag{76}$$

Also as we know that

$$\kappa = \left( \frac{4\pi}{\varepsilon kT} \sum_i n_i^0 Z_i^2 e_0^2 \right)^{1/2} \tag{77}$$

Hence, the radius of maximum excess charge from the center of the reference ion can be obtained by putting the value of $\kappa$ from equation (77) in equation (76), we get $r_{max}$ or the "Debye-Huckel length" as

$$r_{max} = \left( \frac{\varepsilon kT}{4\pi} \frac{1}{\sum_i n_i^0 Z_i^2 e_0^2} \right)^{1/2} \tag{78}$$

It is also worthy to note that the cloud will tend to fade out ever more with the decreasing concentration.

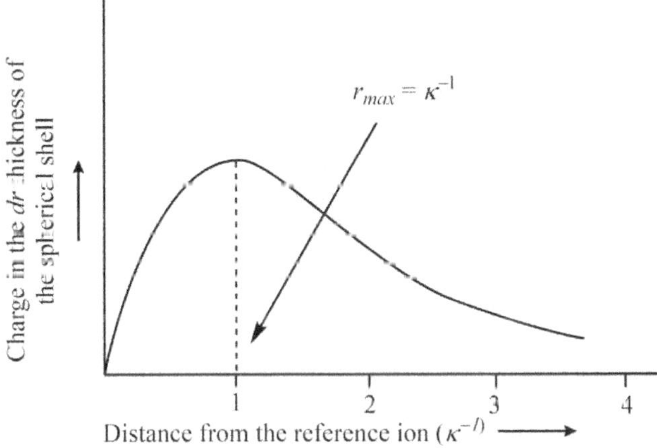


Figure 10. The variation of total charge enclosed in the $dr$ thickness of the spherical shell as a function of the distance $r$ from the reference ion.

## ❖ Ionic Cloud and Its Contribution to the Total Potential

According to the Debye-Huckel theory of ion-ion interaction, the molar chemical potential change ($\Delta\mu_{i-I}$) of ion-ion interaction can be calculated from the following relation.

$$\Delta\mu_{i-I} = \frac{N_A}{2}\,[Z_i e_0 \times \psi] \tag{79}$$

Where $N_A$ is the Avogadro number and $\psi$ the electrostatic potential at distance $r$ due to ionic cloud. The symbol $Z_i e_0$ is the charge on the $i$th species. The value of $\psi_r$ is

$$\psi_r = \frac{Z_i e_0}{\varepsilon}\,\frac{e^{-\kappa r}}{r} \tag{80}$$

Where $\varepsilon$ is the dielectric constant of the surrounding medium and $\kappa$ is a constant defined by

$$\kappa = \left(\frac{4\pi}{\varepsilon kT}\sum_i n_i^0 Z_i^2 e_0^2\right)^{1/2} \tag{81}$$

The value of electrostatic potential given by equation (80) cannot be used in equation (79) because contains the contribution form the ionic cloud as well as from the reference ion itself. Therefore, before we calculate the partial molar free energy of ion-ion interaction, we need to separate potential from the ionic cloud and from the central ion first.

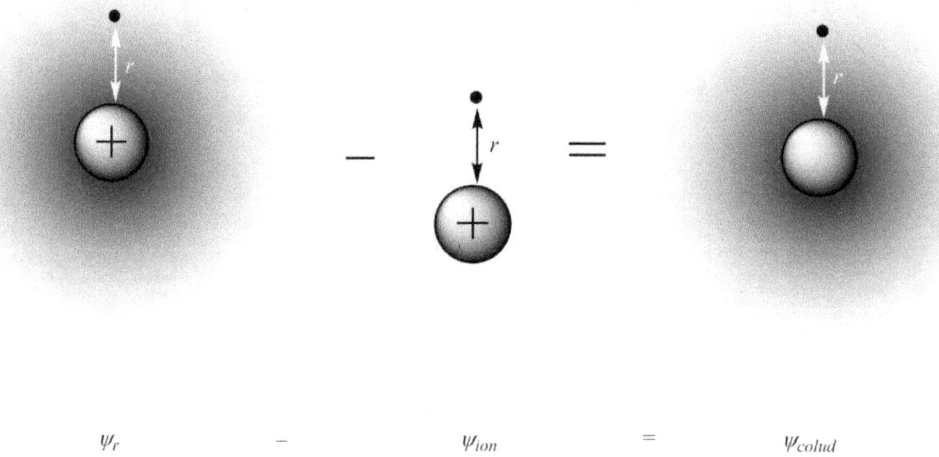


Figure 11. The pictorial representation of the superposition of the potentials at distance $r$ from the reference ion due to the ionic cloud ion and the potential due to the central ion itself.

The total electrostatic potential at a distance $r$ can be fragmented as given below.

$$\psi_r = \psi_{ion} + \psi_{cloud} \tag{82}$$

or

$$\psi_{cloud} = \psi_r - \psi_{ion} \tag{83}$$

From the formulation of potential due to a single charge at a distance $r$, we know that

$$\psi_{ion} = \frac{Z_i e_0}{\varepsilon r} \tag{84}$$

Now using the value of $\psi_r$ from equation (80) and $\psi_{ion}$ from equation (84) into equation (83), we get

$$\psi_{cloud} = \frac{Z_i e_0}{\varepsilon} \frac{e^{-\kappa r}}{r} - \frac{Z_i e_0}{\varepsilon r} \tag{85}$$

or

$$\psi_{cloud} = \frac{Z_i e_0}{\varepsilon r} (e^{-\kappa r} - 1) \tag{86}$$

Now because the value of $\kappa$ is proportional to $\sum_i n_i^0 Z_i^2 e_0^2$, at very large dilution $\kappa$ becomes very small and $e^{-\kappa r}$ and can be expended as $1 - \kappa r$ ($e^x = 1 + x$). Therefore, equation (86) takes the form

$$\psi_{cloud} = \frac{Z_i e_0}{\varepsilon r} (1 - \kappa r - 1) \tag{87}$$

or

$$\psi_{cloud} = -\frac{Z_i e_0 \kappa}{\varepsilon} \tag{88}$$

Hence, potential due to ionic cloud is independent of the distance $r$ in this case. Hence, potential due to ionic cloud is independent of the distance $r$ in this case. Now putting the value of $\psi_{cloud}$ from equation (88) into equation (79), we get

$$\Delta \mu_{i-I} = N_A w = \frac{N_A}{2} \left[ Z_i e_0 \times \left( -\frac{Z_i e_0 \kappa}{\varepsilon} \right) \right] \tag{89}$$

or

$$\Delta \mu_{i-I} = -\frac{N_A Z_i^2 e_0^2 \kappa}{2\varepsilon} \tag{90}$$

Hence, Debye-Hückel model gives the theoretical value of $\Delta \mu_{i-I}$ arising from ion-ion interactions.

## ❖ Debye-Huckel Limiting Law of Activity Coefficients and Its Limitations

So far in this chapter, the idea of ions in electrolytic solutions and their mutual interactions was taken for granted. However, the conceptual understanding of these electrolytic solutions thrived via a different route. Initially, the scientific community thought that the electrolytic solutions are the same as the non-electrolytic ones. In other words, they did not pay any special attention to the charged species, and therefore, treated them just like the non-electrolytes.

### ➢ *Derivation of Debye-Huckel Limiting Law*

Classically, the partial molar free energy ($\mu_i$) of the $i$th species in a non-electrolyte is given by the following relation.

$$\mu_i = \mu_i^0 + RT \ln x_i \tag{91}$$

Where $\mu_i^0$ is the partial molar free energy when the concentration ($x_i$) is unity (standard state). Mathematically, we can say that is if $x_i = 1$, $\mu_i = \mu_i^0$.

At this point, it worthy to point out that the Coulombic interactions (long-range) can be neglected in the case of non-electrolytic solutions, which is obviously due to the uncharged or neutral nature of the solute particles. However, some short-range electrostatic interactions like the London dispersion forces or the dipole-dipole may play a significant role if the average inter-particle distance is small enough. This means that if the dilution is large enough, the ion-ion interactions (long-range Coulombic effects) can simply be ignored in a non-electrolytic solution. Thus, the equation (91) is valid only for those solutions in which no long-range electrostatic interactions occur, and such solutions are called as ideal solutions. However, when the equation (91) was applied to the electrolytic solutions, it was found that

$$\mu_i - \mu_i^0 \neq RT \ln x_i \tag{92}$$

Such solutions which show deviation from equation (91) are called as non-ideal solutions. This means that unlike non-electrolytes, the ion-ion interactions must be considered before any theoretical treatment for the chemical potential of ionic solutions is carried out.

Therefore, to rationalize this deviation, the scientific community developed a unique approach even before the Debye-Huckel's ionic-cloud theory. They tried to quantify the deviation from "idealistic behavior" by incorporating an empirical parameter $f_i$ in the equation (92) as

$$\mu_i - \mu_i^0 = RT \ln x_i f_i \tag{93}$$

Where $\mu_i^0$ is the partial molar free energy when the concentration ($x_i$) and correction factor ($f_i$) both are unity. Mathematically, we can say that if $x_i = 1$ and $f_i = 1$, $\mu_i = \mu_i^0$; which is obviously a hypothetical situation in which standard state behaves ideally. Moreover, it should also be noted that term $x_i$ represents the actual concentration of the ions of $i$th type which may or may not be equal to expected concentration because weak electrolytes do not dissociate completely like the strong one. After that, it was assumed that it is not the actual

concentration $(x_i)$ but the effective concentration $(x_i f_i)$ that dictates the chemical potential change. For simplicity, the effective concentration if $i$th species was simply labeled as the activity $(a_i)$. Mathematically, we can say that

$$a_i = x_i f_i \tag{94}$$

The correction factor $f_i$ is also called as the "activity coefficient" and has a value of unity for ideal solutions i.e. when $f_i = 1$, $a_i = x_i$.

Consequently, the partial molar free energy change in going from the ideal state to the real state in ionic solutions can be written as

$$\mu_i - \mu_i^0 = RT \ln x_i + RT \ln f_i \tag{95}$$

The above equation is an empirical formulation of the behavior of ionic solutions and cannot give any theoretical result for $f_i$. Therefore, in order to find out the physical significance of the activity coefficient, we need to assume an ionic solution that can be switched from ideal (no ion-ion interaction) to the real situation (with ion-ion interaction). For ideal solutions, we have

$$\mu_i(\text{ideal}) = \mu_i^0 + RT \ln x_i \tag{96}$$

For real solutions,

$$\mu_i(\text{real}) = \mu_i^0 + RT \ln x_i + RT \ln f_i \tag{97}$$

The chemical potential change from ion-ion interaction can be obtained by subtracting equation (96) from equation (97) i.e.

$$\mu_i(\text{real}) - \mu_i(\text{ideal}) = \Delta\mu_{i-I} = \mu_i^0 + RT \ln x_i + RT \ln f_i - \mu_i^0 - RT \ln x_i \tag{98}$$

$$\Delta\mu_{i-I} = RT \ln f_i \tag{99}$$

Hence, the activity coefficient is a measure of chemical potential change due to the interaction of $i$th types with the rest of the ionic species. Now, according to the Debye-Huckel theory of ion-ion interaction, the partial molar free energy change due to ion-ion interaction is

$$\Delta\mu_{i-I} = -\frac{N_A Z_i^2 e_0^2 \kappa}{2\varepsilon} \tag{100}$$

Where $Z_i$ is the charge number on the $i$th type of ion while $e_0$ is the electronic charge. $N_A$ is the Avogadro number and $\varepsilon$ is the dielectric constant of the surrounding medium, i.e., solvent. The symbol $\kappa$ with $n_i^0$ as the bulk concentration is

$$\kappa = \left( \frac{4\pi}{\varepsilon kT} \sum_i n_i^0 Z_i^2 e_0^2 \right)^{1/2} \tag{101}$$

Now, from equation (99) and equation (100), we have

$$RT \ln f_i = -\frac{N_A Z_i^2 e_0^2 \kappa}{2\varepsilon} \tag{102}$$

The above result implies that the Debye-Huckel's ionic cloud theory enables us to determine the activity coefficient in a theoretical framework.

Furthermore, if the actual concentration of the $i$th type of ion is represented in terms of molarity ($c_i$) or the molality ($m_i$), the equation (96) takes the form

$$\mu_i = \mu_i^0(c) + RT \ln c_i \tag{103}$$

and

$$\mu_i = \mu_i^0(m) + RT \ln m_i \tag{104}$$

Similarly, for real solutions, equation (97) takes the forms

$$\mu_i = \mu_i^0(c) + RT \ln c_i + RT \ln f_c \tag{105}$$

and

$$\mu_i = \mu_i^0(m) + RT \ln m_i + RT \ln f_m \tag{106}$$

Where $\mu_i^0(c)$ and $\mu_i^0(m)$ are corresponding standard chemical potential at a molarity $c_i$ and molality $m_i$, respectively.

At this point, some experimental limitations must be discussed before any comparative analysis of the activity coefficient is carried out. It is a quite well-known fact that we cannot measure the hydration energy, i.e., ion-solvent interaction of individual ionic species because the addition of only cations or the anions is not possible practically. Even if it was possible, it would result in a negatively or positively charged solution depending upon the nature of the ions added, which eventually, would cause undesired interactions. Owing to similar arguments, it is also practically impossible to measure the activity coefficient $f_i$ which is also a function chemical potential change arising from ion-ion interaction. The only way to avoid the situation is to add the electroneutral electrolytes to the solvent which would eventually produce the positive and negative ions simultaneously irrespective of whether it is strong or weak. Therefore, one can determine the activity of a net electrolyte consisted of minimum two ionic species, and we need something that can connect activity coefficient of electrolyte with the individual ionic species. In other words, the activity coefficient of a single ionic species can be determined only via a theoretical route. All this resulted in the idea of "mean ionic activity coefficient". To illustrate mathematically, consider a NaCl-type univalent electrolyte MA. The chemical potential of cations ($M^+$) and anions ($A^-$) can be written as

$$\mu_{M^+} = \mu_{M^+}^0 + RT \ln x_{M^+} + RT \ln f_{M^+} \tag{107}$$

and

$$\mu_{A^-} = \mu_{A^-}^0 + RT \ln x_{A^-} + RT \ln f_{A^-} \tag{108}$$

Combining the equation (107) with equation (108), we get

$$\mu_{M^+} + \mu_{A^-} = (\mu_{M^+}^0 + \mu_{A^-}^0) + RT \ln (x_{M^+}x_{A^-}) + RT \ln (f_{M^+}f_{A^-}) \tag{109}$$

The above equation gives the free energy of the system due to two moles of ions i.e. one mole of $M^+$ and mole of $A^-$; or due to one mole of electroneutral electrolyte. Since we are interested in the average input to the total free energy due to one mole of ions only, therefore, we must divide the equation (109) by 2, i.e.,

$$\frac{\mu_{M^+} + \mu_{A^-}}{2} = \frac{\mu_{M^+}^0 + \mu_{A^-}^0}{2} + RT \ln (x_{M^+}x_{A^-})^{1/2} + RT \ln (f_{M^+}f_{A^-})^{1/2} \tag{110}$$

If we consider

$$\mu_{\pm} = \frac{\mu_{M^+} + \mu_{A^-}}{2} \tag{111}$$

$$\mu_{\pm}^0 = \frac{\mu_{M^+}^0 + \mu_{A^-}^0}{2} \tag{112}$$

and

$$x_{\pm} = (x_{M^+}x_{A^-})^{1/2} \tag{113}$$

$$f_{\pm} = (f_{M^+}f_{A^-})^{1/2} \tag{114}$$

Where $\mu_{\pm}$ and $\mu_{\pm}^0$ are the mean chemical potential and standard mean chemical potential, respectively. The symbol $x_{\pm}$ and $f_{\pm}$ represent the mean ionic mole fraction and mean ionic activity coefficient, respectively. Here, it is also worthy to note that $\mu_{\pm}$ and $\mu_{\pm}^0$ are simply the arithmetic means whereas $x_{\pm}$ and $f_{\pm}$ are the geometric mean quantities. Now using values from equations (111−114) into equation (110), we get

$$\mu_{\pm} = \mu_{\pm}^0 + RT \ln x_{\pm} + RT \ln f_{\pm} \tag{115}$$

Since it is for one mole instead two, we can write

$$\frac{1}{2}\mu_{MA} = \mu_{\pm} = \mu_{\pm}^0 + RT \ln x_{\pm} + RT \ln f_{\pm} \tag{116}$$

Hence, the experimental value of $f_{\pm}$ can be obtained just by knowing the free energy of one mole of electrolytic solution at a particular concentration. Once the value of $f_{\pm}$ is known, the product of individual activity coefficients can be obtained using equation (114). The individual values of activity coefficients obtained from equation (102) can be put into equation (114) to compare with experimentally observed value so that the Debye-Huckel model can be tested.

Furthermore, if one mole of electroneutral electrolyte generates $v_+$ and $v_-$ moles of cations and anions, then equation (107) and equation (108) will become

$$v_+\mu_+ = v_+\mu_+^0 + v_+RT \ln x_+ + v_+RT \ln f_+ \tag{117}$$

and

$$v_-\mu_- = v_-\mu_-^0 + v_-RT \ln x_- + v_-RT \ln f_- \tag{118}$$

To find the free energy change due to per mole of cation and anion, add equation (117) to equation (118) and then divide by $v = v_+ + v_-$.

$$\frac{v_+\mu_+ + v_-\mu_-}{v} = \frac{v_+\mu_+^0 + v_-\mu_-^0}{v} + RT \ln \left(x_+^{v_+}x_-^{v_-}\right)^{1/v} + RT \ln \left(f_+^{v_+}f_-^{v_-}\right)^{1/v} \tag{119}$$

If we consider

$$\mu_\pm = \frac{v_+\mu_+ + v_-\mu_-}{v} \tag{120}$$

$$\mu_\pm^0 = \frac{v_+\mu_+^0 + v_-\mu_-^0}{v} \tag{121}$$

and

$$x_\pm = \left(x_+^{v_+}x_-^{v_-}\right)^{1/v} \tag{122}$$

$$f_\pm = \left(f_+^{v_+}f_-^{v_-}\right)^{1/v} \tag{123}$$

Now using values from equations (120–123) into equation (119), we get

$$\mu_\pm = \mu_\pm^0 + RT \ln x_\pm + RT \ln f_\pm \tag{124}$$

Taking logarithm both side of equation (123), we have

$$\ln f_\pm = \frac{1}{v}(v_+ \ln f_+ + v_- \ln f_-) \tag{125}$$

Now putting the value of $\ln f_+$ and $\ln f_-$ using equation (102), we get

$$\ln f_\pm = -\frac{1}{v}\left[\frac{N_A e_0^2 \kappa}{2\varepsilon RT}(v_+Z_+^2 + v_-Z_-^2)\right] \tag{126}$$

Now owing to the electroneutrality of the solution, $v_+Z_+$ must be equal to $v_-Z_-$, i.e.,

$$v_+Z_+^2 + v_-Z_-^2 = v_+Z_+Z_- + v_-Z_-Z_+ \tag{127}$$

$$v_+Z_+^2 + v_-Z_-^2 = Z_+Z_-(v_+ + v_-) \tag{128}$$

$$v_+Z_+^2 + v_-Z_-^2 = Z_+Z_-v \tag{129}$$

After putting the value of equation (129) into equation (126), we have

$$\ln f_\pm = -\frac{N_A e_0^2 \kappa}{2\varepsilon RT} Z_+Z_- \tag{130}$$

Furthermore, using the value of $\kappa$ using equation (101), the above equation takes the form

$$\ln f_\pm = -\frac{N_A e_0^2 Z_+Z_-}{2\varepsilon RT}\left(\frac{4\pi}{\varepsilon kT}\sum_i n_i^0 Z_i^2 e_0^2\right)^{1/2} \tag{131}$$

Since $n_i^0 = c_i N_A/1000$, the equation (131) becomes

$$\ln f_\pm = -\frac{N_A e_0^2 Z_+Z_-}{2\varepsilon RT}\left(\frac{4\pi}{\varepsilon kT}\sum_l \frac{c_i N_A Z_i^2 e_0^2}{1000}\right)^{1/2} \tag{132}$$

Multiply and divide the equation (132) by 2, and then put $c_i Z_i^2/2 = I$, i.e., ionic strength, we get

$$\ln f_\pm = -\frac{N_A e_0^2 Z_+Z_-}{2\varepsilon RT}\left(\frac{8\pi N_A e_0^2}{1000\varepsilon kT}\right)^{1/2}\sqrt{I} \tag{133}$$

$$\ln f_\pm = -\frac{N_A e_0^2 Z_+Z_-}{2\varepsilon RT}B\sqrt{I} \tag{134}$$

Where the constant $B$ is defined as

$$B = \left(\frac{8\pi N_A e_0^2}{1000\varepsilon kT}\right)^{1/2} \tag{135}$$

Converting the natural logarithm to the common logarithm, the equation (134) becomes

$$\log f_\pm = -\frac{1}{2.303}\frac{N_A e_0^2 Z_+Z_-}{2\varepsilon RT}B\sqrt{I} \tag{136}$$

After a new constant $A$ as

$$A = \frac{1}{2.303}\frac{N_A e_0^2}{2cRT}B \tag{137}$$

The equation (136) can be further simplified as given below.

$$\log f_\pm = -A(Z_+Z_-)\sqrt{I} \tag{138}$$

Which is the Debye-Huckel limiting law of activity coefficients.

> ### *Limitations of Debye-Huckel Limiting Law*

The negative sign in the Debye-Huckel limiting law implies the fact that the activity coefficient is always less than unity. In order to discuss the limitations of this law, recall the popular relationship, i.e.,

$$\log f_\pm = -A(Z_+Z_-)\sqrt{I} \tag{139}$$

Since $A$ is constant and the product $Z_+Z_-$ is also constant for a particular electrolyte, the logarithm of activity coefficient must decrease linearly with the square root of ionic strength. In other words, the slope $\log f_\pm$ and $\sqrt{I}$ can be determined simply by knowing the valences of the ions involved and by the knowledge of some physical constants. Furthermore, it should also be noted that the slope is independent of the very nature of the electrolyte and is a function of valences of cations and anions only. For instance, since the value of $A$ for the water as solvent is 0.509, the slope of for NaCl as well as for KCl must be equal to 1×0.509 only. For uni-bivalent and the bi-bivalent electrolyte is should be 2×0.509 and 4×0.509. The logarithmic variation of mean ionic activity coefficient ($f_\pm$) with the square root of the ionic strength ($\sqrt{I}$) for electrolytes of different valences follows the equation (139) pretty strictly.

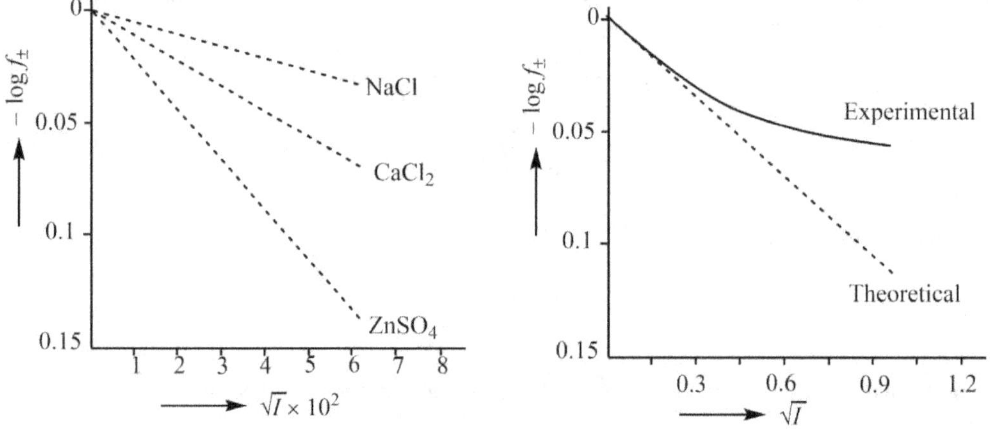


Figure 12. The logarithmic variation of $f_\pm$ with the square root of the ionic strength ($\sqrt{I}$) for electrolytes of different valences in very dilute (left) and in solutions up to large concentration (right).

It is obvious from both the graphs that $\log f_\pm$ becomes zero when the dilution is very large indicating that the activity coefficient is unity, which is according to the Debye-Huckel limiting law. However, it should also be remembered that any theory is always a simplification of the real problem, and therefore, some deviations are expected. The same has been observed when the ionic strength is increased. It can be clearly seen that the deviation of the experimental result increases with the rise in the square root of the ionic strength.

## ❖ Ion-Size Effect on Potential

In the Debye-Huckel theory of ion-ion interaction, the central ion was simply treated as a point charge instead of its actual size. The scientific community realized that it might be one of the reasons behind the deviations observed from Debye-Huckel limiting law of activity coefficients. In order to understand the effect of ion size on the potential, recall the expression for the Debye-Huckel length ($r_{max}$) or $\kappa^{-1}$, i.e.,

$$r_{max} = \left( \frac{\varepsilon kT}{4\pi} \frac{1}{\sum_i n_i^0 Z_i^2 e_0^2} \right)^{1/2} \tag{140}$$

Where $n_i^0$ is the bulk concentration of the $i$th species and $k$ is simply the Boltzmann constant. The symbol $\varepsilon$ represents the dielectric constant of the surrounding medium. The symbol $Z_i$ shows the charge number of the ion whereas $e_0$ represents the electronic charge. It is obvious from the equation (140) that the mean thickness of the ionic cloud ($\kappa^{-1}$) is actually inversely proportional to the concentration. This means that at higher concentrations, the mean thickness of the ionic cloud will be small and cannot outrank the size of the central ion anymore. In other words, at higher concentration, the size of the reference ion cannot be neglected at all, and therefore, the point charge approximation is no longer valid. For instance, At a concentration of $0.1N$, the mean thickness of the ionic cloud is only ten times the radius of the reference ion.

Consider $4\pi r^2 dr$ as the volume element of a hollow spherical shell of thickness $dr$ with the inner and outer radius as $r$ and $r+dr$, respectively, with the size of the ion represented by the parameter $a$.

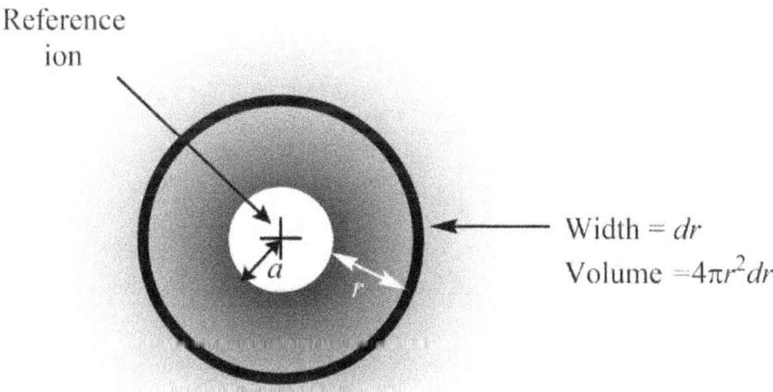


Figure 13  The depiction of excess charge density as a function of the distance $r$ from the reference ion of finite size $a$.

Therefore, the ion-size must be considered for a more realistic picture of the actual situation. To do so, consider the linearized Poisson-Boltzmann equation which is free from both approximations, i.e., ion-size matters and the point charge consideration.

$$\frac{1}{r^2}\frac{d}{dr}\left(r^2\frac{d\psi_r}{dr}\right) = \kappa^2\psi_r \tag{141}$$

Where $\psi_r$ is the potential at a distance $r$ from the reference ion. The symbol $\kappa$ is defined as

$$\kappa = \left(\frac{4\pi}{\varepsilon kT}\sum_i n_i^0 Z_i^2 e_0^2\right)^{1/2} \tag{142}$$

The general solution of the above equation is

$$\psi_r = A\frac{e^{-\kappa r}}{r} + B\frac{e^{+\kappa r}}{r} \tag{143}$$

Since the potential at $r = \infty$ must vanish, this boundary condition is satisfied only if $B = 0$. Therefore, the acceptable form of the equation (143) should be like this

$$\psi_r = A\frac{e^{-\kappa r}}{r} \tag{144}$$

To evaluate the value of constant A in this scenario, a slightly different route is followed. As we know that the total charge in the $dr$ thickness of a spherical shell at distance $r$ is

$$dq = \rho_r 4\pi r^2 dr \tag{145}$$

Now, the excess charge density at distance $r$ from the reference ion will be

$$\rho_r = -\frac{\varepsilon}{4\pi}\left[\frac{1}{r^2}\frac{d}{dr}\left(r^2\frac{d\psi_r}{dr}\right)\right] = -\frac{\varepsilon\kappa^2\psi_r}{4\pi} \tag{146}$$

Using the value of $\psi_r$ from equation (144) into equation (146), we get

$$\rho_r = -\frac{\varepsilon\kappa^2}{4\pi}A\frac{e^{-\kappa r}}{r} \tag{147}$$

After putting the value of equation (147) into equation (145), we have

$$dq = -\frac{\varepsilon\kappa^2}{4\pi}A\frac{e^{-\kappa r}}{r}4\pi r^2 dr \tag{148}$$

or

$$dq = -\varepsilon\kappa^2 Ae^{-\kappa r}rdr \tag{149}$$

Since the exponential part becomes zero only at $r = \infty$, the total charge around the reference ion can be obtained by integrating the equation (149) from an unknown parameter $r = a$ to $r = \infty$.

$$q_{cloud} = \int_{r=a}^{r=\infty} dq \, dr = \int_{r=a}^{r=\infty} -\varepsilon\kappa^2 A e^{-\kappa r} r \, dr \tag{150}$$

or

$$q_{cloud} = \int_{r=a}^{r=\infty} dq \, dr = -A\varepsilon e^{-\kappa a}(1 + \kappa a) \tag{151}$$

Also as we know that the total charge on the ionic cloud must be equal and opposite to the charge on the reference ion i.e.

$$q_{cloud} = -Z_i e_0 \tag{152}$$

Equating the results of equation (151) and equation (152), we have

$$-A\varepsilon e^{-\kappa a}(1 + \kappa a) = -Z_i e_0 \tag{153}$$

or

$$A = \frac{Z_i e_0}{\varepsilon} \frac{e^{\kappa a}}{(1 + \kappa a)} \tag{154}$$

Using the value of $A$ from equation (154) in equation (144), we get

$$\psi_r = \frac{Z_i e_0}{\varepsilon} \frac{e^{\kappa a}}{(1 + \kappa a)} \frac{e^{-\kappa r}}{r} \tag{155}$$

### ❖ Ion-Size Parameter and the Theoretical Mean - Activity Coefficient in the Case of Ionic Clouds with Finite-Sized Ions

As we know that the total electrostatic potential at a distance $r$ can simply be fragmented as given below.

$$\psi_r = \psi_{ion} + \psi_{cloud} \tag{156}$$

or

$$\psi_{cloud} = \psi_r - \psi_{ion} \tag{157}$$

The electrostatic potential at a distance $r$ from an ion of finite size is

$$\psi_r = \frac{Z_i e_0}{\varepsilon} \frac{e^{\kappa a}}{(1 + \kappa a)} \frac{e^{-\kappa r}}{r} \tag{158}$$

Also, from the formulation of potential due to a single charge at a distance $r$, we know that

$$\psi_{ion} = \frac{Z_i e_0}{\varepsilon r} \tag{159}$$

After putting the values of $\psi_r$ and $\psi_{ion}$ from equation (158) and equation (159) into equation (157), we have

$$\psi_{cloud} = \frac{Z_i e_0}{\varepsilon} \frac{e^{\kappa a}}{(1 + \kappa a)} \frac{e^{-\kappa r}}{r} - \frac{Z_i e_0}{\varepsilon r} \tag{160}$$

or

$$\psi_{cloud} = \frac{Z_i e_0}{\varepsilon r} \left[ \frac{e^{\kappa(a-r)}}{(1 + \kappa a)} - 1 \right] \tag{161}$$

At this point, we must recall the relationship of mean activity coefficient and chemical potential change of the ion-ion interaction, i.e.,

$$RT \ln f_i = \Delta\mu_{i-I} \tag{162}$$

also

$$\Delta\mu_{i-I} = \frac{N_A Z_i e_0}{2} \psi \tag{163}$$

From equation (162) and (163), we have

$$RT \ln f_i = \frac{N_A Z_i e_0}{2} \psi \tag{164}$$

$$\ln f_i = \frac{N_A Z_i e_0}{2RT} \psi \tag{165}$$

Moreover, as the potential at the surface ($r = a$) of the ion must be

$$\psi = \psi_{cloud} \tag{166}$$

Which implies that $\psi_{cloud}$ in equation (161) at $r = a$ should become

$$\psi_{cloud} = \frac{Z_i e_0}{\varepsilon a} \left[ \frac{1}{(1 + \kappa a)} - 1 \right] \tag{167}$$

or

$$\psi_{cloud} = \frac{Z_i e_0}{\varepsilon a} \left[ \frac{1 - 1 - \kappa a}{(1 + \kappa a)} \right] \tag{168}$$

$$\psi_{cloud} = - \frac{Z_i e_0}{\varepsilon \kappa^-} \frac{1}{(1 + \kappa a)} \tag{169}$$

After using the value of equation (169) in equation (165), we get

$$\ln f_i = -\frac{N_A Z_i^2 e_0^2}{2RT\varepsilon\kappa^{-1}}\frac{1}{(1+\kappa a)} \tag{170}$$

The mean ionic activity coefficient, in this case, will be

$$\log f_\pm = -\frac{AZ_+ Z_-}{(1+\kappa a)}I^{1/2} \tag{171}$$

Where $I$ is the ionic strength and $A$ is Debye-Huckel constant. Now since the thickness of the ionic cloud is defined as

$$\kappa = BI^{1/2} \tag{172}$$

After putting the value of $\kappa$ from equation (172) into equation (171), we have

$$\log f_\pm = -\frac{AZ_+ Z_- I^{1/2}}{1+a\,BI^{1/2}} \tag{173}$$

Which is the equation for the mean ionic activity coefficient when the central ion has a finite size. Now recall the equation for mean ionic activity coefficient when the central ion was considered as point charge i.e.

$$\log f_\pm = -AZ_+ Z_- I^{1/2} \tag{174}$$

Hence, the two equations differ only in respect of denominator $1+\kappa a$. The rearranged form of the equation (171) is

$$\log f_\pm = -\frac{AZ_+ Z_-}{(1+a/\kappa^{-1})}I^{1/2} \tag{175}$$

At large dilution, $a \ll \kappa^{-1}$, and the denominator tends to approach unity. Therefore, at very large dilution, the equation (175) becomes equal to equation (174), proving the correspondence principle.

The most obvious approach to estimate the ion size parameter is the sum of the crystallographic radii of the cation and anions present in the electrolytic-solution. This is because the two ions cannot come closer than the sum of their individual ionic radii. However, since ions are hydrated in aqueous solutions, one might think the ion size parameter as the sum of the hydrated radii instead. Nevertheless, the hydrated radii would face some compression during the course of the collision of ion. All this suggests that the magnitude of '$a$' must be greater than the sum of the individual crystallographic radii and should be less than the sum of their individual hydrated radii. Therefore, the "mean distance of closest approach" should be more appropriate for this situation. The best way to estimate the ion-size parameter is to calibrated equation (175) to match the experimental value of mean ionic activity coefficient. After knowing the ion size parameter at one concentration, the value of mean ionic activity coefficient can easily be determined at other concentrations.

## ❖ Debye-Huckel-Onsager Treatment for Aqueous Solutions and Its Limitations

It is a well-known fact that the conductance of weak electrolytic solutions increases with the increase in dilution. This can be easily explained on the basis of Arrhenius's theory of electrolytic dissociation which says that the magnitude of dissociated electrolyte, and hence the number of charge carriers, increases with the increase in dilution. However, the problem arises when the strong or true electrolytes show the same trend but at a much lower scale. We used the word "problem" because even at the higher concentration, the electrolyte dissociates completely inferring that there is no possibility of further dissociation with dilution. This means that there should be no increase in the conductance of strong electrolytes with the addition of water.

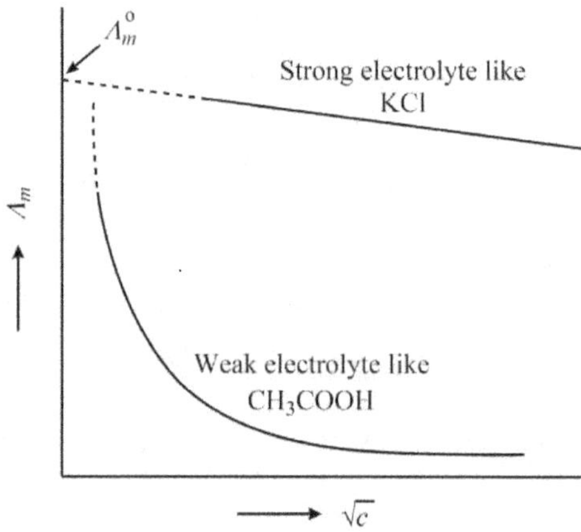


Figure 14. The typical variation of molar conductance ($\Lambda_m$) with the square root of the concentration ($\sqrt{c}$) for strong and weak electrolytes.

The primary reason behind this weird behavior of strong electrolyte is that the conductance of any electrolytic solution depends not only upon the number of charge carriers but also upon the speed of these charge carriers. Therefore, if the dilution does not affect the number of charge carriers in strong electrolytes, it must be affecting the speed of ions to change its conductance. The main factor that is responsible for governing the ionic mobility is ion-ion interactions. Now since these ion-ion interactions are dependent upon the interionic distances, they eventually vary with the population density of charge carriers. Higher population density means smaller interionic distances and therefore stronger ion-ion interactions. On the other hand, the lesser population density of ions would result in larger interionic separations and hence weaker ion-ion interactions.

In the case of weak electrolytes, the degree of dissociation is very small at high concentrations yielding a very low population density of charge carriers. This would result in almost zero ion-ion interactions

at high concentrations. Now although the degree of dissociation increases with dilution which in turn also increases the total number of charge carriers, the population density remains almost unchanged since extra water has been added for these extra ions. Thus, we can conclude that there are no ion-ion interactions in weak electrolytes neither at high nor at the low concentration; and hence the rise in conductance with dilution almost a function dissociation only.

In the case of strong electrolytes, the degree of dissociation is a hundred percent even at high concentrations yielding a very high population density of charge carriers. This would result in very strong ion-ion interactions at high concentrations, hindering the speed of various charge carriers. Now when more and more solvent is added, the total number of charge carriers remains the same but the population density decreases continuously creating large interionic separations. This would result in a decrease in ion-ion interaction with increasing dilution, and therefore, the charge carriers would be freer to move in the solution. Thus, we can conclude that though there is no rise in the number of charge carriers with dilution, the declining magnitude of ion-ion interaction creates faster ions and larger conductance.

> ➢ *Factor Affecting the Conductance of Strong Electrolytic Solutions*

In 1923, Peter Debye and Erich Huckel proposed an extremely important idea to quantify the conductance of strong electrolytes in terms of these interionic interactions. In this model, a reference ion is thought to be suspended in solvent-continuum of dielectric constant $\varepsilon$ and is surrounded by oppositely charged ions. Besides the ion-ion interaction, the viscosity of the solvent also affects the overall speed of the moving ion. The primary effects which are responsible for controlling the ionic mobility are discussed below.

**1. Asymmetry effect or the relaxation effect:** In the absence of applied electric field, the ionic atmosphere of the reference ion remains spherically symmetrical. This means that the electrostatic force of attraction on the reference ion from all the directions would be the same. However, when the electric field is applied, the ion starts to move towards the oppositely charged electrode. This, in turn, would destroy the spherical symmetry of the cloud, and more ions would be left behind creating a net backward pull to the reference ion. This effect, therefore, would slow down the moving ions and the conductance would be decreased.

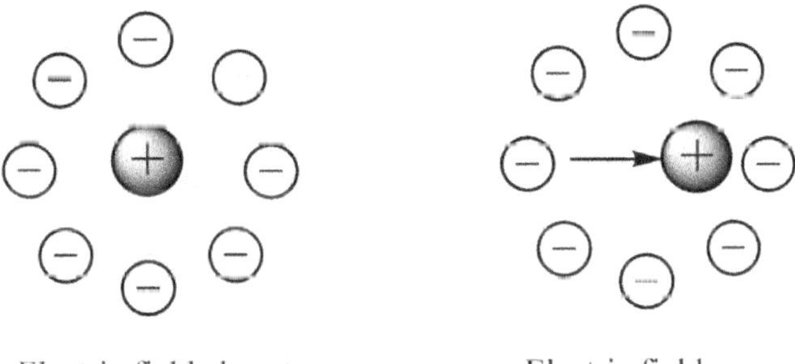


Figure 15. The asymmetric or relaxation effect in the conductance of strong electrolytes.

Alternatively, this can also be visualized in terms of cloud-destruction and cloud-building around the reference ion. In other words, during the movement of ion, the ionic cloud around the reference ion must rebuild itself to keep things natural. Since this rebuilding is not instantaneous and takes some time called as relaxation time, the old cloud exerts a backward pull on the reference ion opposing oppressing its speed. All this results in a diminished magnitude of the conductance.

**2. Electrophoretic effect:** After the application of the external electric field, the reference ion and ionic could move in opposite directions. During the course of this movement, the solvent associated with the surrounding ions also moves in a direction opposite to the central ion. In other words, we can say that the reference ion has to move against a solvent stream, which makes it somewhat slower than usual. This phenomenon is called as the electrophoretic effect.

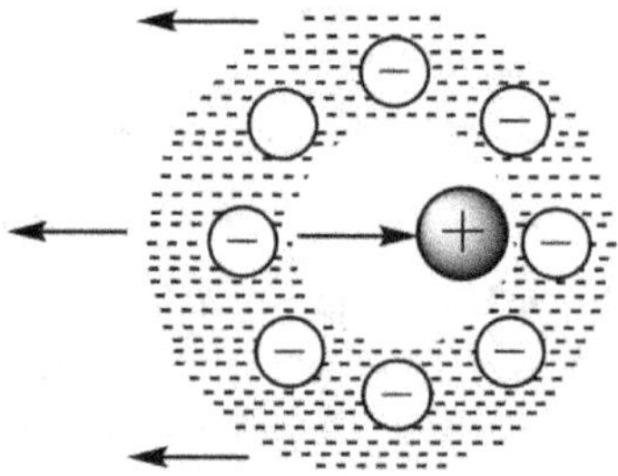

Figure 16. The asymmetric or relaxation effect in the conductance of strong electrolytes.

Since the electrophoretic effect reduces the speed of the ion, the conductance of the electrolytic solution is also affected considerably.

**3. Viscous Effect:** In addition to the asymmetric and electrophoretic effects, another type of resistance also exists which affects the conductance of electrolytic solutions, the "viscous effect". This is simply the frictional resistance created by the viscosity of the solvent used. For an ion of given charge and size, the ionic mobility as well the conductance decrease with the increase in the magnitude of the viscosity of the solvent used. In other words, less viscous solvents yield higher conductance and vice-versa.

> ➤ *Mathematical Development of Debye-Huckel-Onsager theory of Strong Electrolytes*

It is a well-known fact that the equivalent conductivity ($\Lambda$) of an electrolytic solution is correlated to the ionic mobilities ($u$) ions involved as

$$\Lambda = F(u_+ + u_-) \tag{176}$$

Where $F$ is the Faraday constant. Now, recall the ionic mobilities of the cation and anion i.e.

$$u_+ = u_+^0 - \kappa \left( \frac{Z_+ e_0}{6\pi\eta} + \frac{e_0^2 \omega}{6\varepsilon kT} u_+^0 \right) \tag{177}$$

$$u_- = u_-^0 - \kappa \left( \frac{Z_- e_0}{6\pi\eta} + \frac{e_0^2 \omega}{6\varepsilon kT} u_-^0 \right) \tag{178}$$

Where $u_+^0$ and $u_-^0$ are the ionic mobilities of the cation and anions at infinite dilution, respectively. The symbol $\varepsilon$ represents the dielectric constant of the medium whereas $\eta$ is the coefficient of viscosity. $Z_+$ and $Z_-$ are charge numbers of the cation and anion, respectively. The symbol $e_0$ simply shows the electronic charge. The symbol $\kappa$ represents ($n_i^0$ is the bulk concentration)

$$\kappa = \left( \frac{4\pi}{\varepsilon kT} \sum_i n_i^0 Z_i^2 e_0^2 \right)^{1/2} = \left( \frac{4\pi Z^2 e_0^2 c}{\varepsilon kT} \right)^{\frac{1}{2}} \left( \frac{N_A}{1000} \right)^{1/2} \tag{179}$$

The quantity $\omega$ is defined as

$$\omega = \frac{Z_+ Z_- 2q}{1 + \sqrt{q}} \qquad where \qquad q = \frac{Z_+ Z_-}{Z_+ + Z_-} \frac{\lambda_+ + \lambda_-}{Z_+ \lambda_+ + Z_- \lambda_-} \tag{180}$$

After putting the values of $u_+$ and $u_-$ from equation (177) and (178) in equation (176), we get

$$\Lambda = F\left[ u_+^0 - \kappa \left( \frac{Z_+ e_0}{6\pi\eta} + \frac{e_0^2 \omega}{6\varepsilon kT} u_+^0 \right) \right] + F\left[ u_-^0 - \kappa \left( \frac{Z_- e_0}{6\pi\eta} + \frac{e_0^2 \omega}{6\varepsilon kT} u_-^0 \right) \right] \tag{181}$$

In the case of symmetrical electrolytes, we can put $Z_+ = Z_- = Z$, and therefore $Z_+ + Z_- = 2Z$. Thus, the above equation for such cases takes the form

$$\Lambda = F(u_+^0 + u_-^0) - \left[ \frac{FZ\kappa e_0}{3\pi\eta} + \frac{e_0^2 \omega k}{6\varepsilon kT} F(u_+^0 + u_-^0) \right] \tag{182}$$

Since $F(u_+^0 + u_-^0) = \Lambda^0$, the equation (182) becomes

$$\Lambda = \Lambda^0 - \left[ \frac{FZ\kappa e_0}{3\pi\eta} + \frac{e_0^2 \omega k}{6\varepsilon kT} \Lambda^0 \right] \tag{183}$$

Now expending above equation further by putting the value of $\kappa$ from equation (179), we get

$$\Lambda = \Lambda^0 - \left[ \frac{FZ e_0}{3\pi\eta} \left( \frac{8\pi Z^2 e_0^2 N_A}{1000\varepsilon kT} \right)^{\frac{1}{2}} + \frac{e_0^2 \omega}{6\varepsilon kT} \left( \frac{8\pi Z^2 e_0^2 N_A}{1000 c kT} \right)^{\frac{1}{2}} \Lambda^0 \right] \sqrt{c} \tag{184}$$

Define two constant $A$ and $B$ as

$$A = \frac{FZ e_0}{3\pi\eta} \left( \frac{8\pi Z^2 e_0^2 N_A}{1000\varepsilon kT} \right)^{\frac{1}{2}} \qquad and \qquad B = \frac{e_0^2 \omega}{6\varepsilon kT} \left( \frac{8\pi Z^2 e_0^2 N_A}{1000\varepsilon kT} \right)^{\frac{1}{2}} \tag{185}$$

Therefore, the equation (184) can be simplified as

$$\Lambda = \Lambda^0 - (A + B\Lambda^0)\sqrt{c} \qquad\qquad (186)$$

or

$$\Lambda = \Lambda^0 - constant\sqrt{c} \qquad\qquad (187)$$

The equation (184) and equation (186) are the popular forms famous Debye-Huckel-Onsager equation for electrolyte solutions. The constants $A$ and $B$ can easily be determined from the knowledge of temperature T, valence type of the electrolyte z, the viscosity of the medium, the dielectric constant and other universal constants like Avogadro number. From Equation (187), it is obviously a straight line equation which means the for symmetrical electrolytic solutions, we can plot the conductance vs square root of the concentration for which the slope will be negative. The intercept after extrapolation gives the value of conductance of such solutions at infinite dilution.

> ➢ *Limitations of Debye-Huckel-Onsager Equation*

Since the plot of conductance vs square root of the concentration is linear with negative slope and positive intercept, it seems quite straightforward to study the strong electrolytes. However, it has been observed that the equation (187) is followed only up low and moderate concentrations.

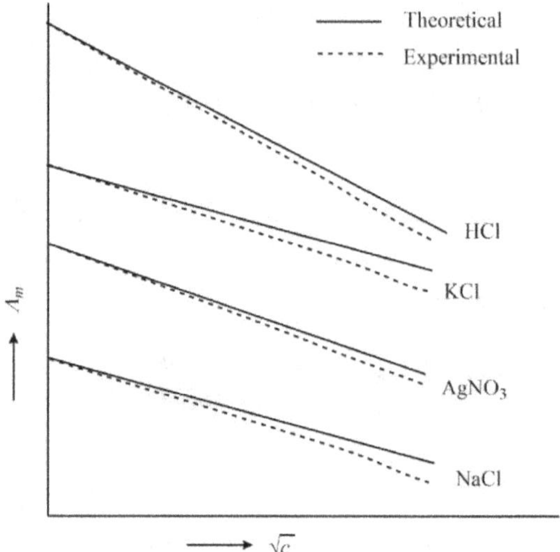


Figure 17. The comparison of theoretical and experimental conductance as a function of concentration for some symmetric electrolytes.

It can be clearly seen that the theory and experiment move apart as the concentration increases. This is simply because some approximation used to derive are Debye-Huckel-Onsager equation are not valid.

## ❖ Debye-Huckel-Onsager Theory for Non-Aqueous Solutions

Before we discuss the Debye-Huckel-Onsager theory for non-aqueous solutions, recall the same for aqueous solutions i.e.

$$\Lambda = \Lambda^0 - constant\sqrt{c} \tag{187}$$

or

$$\Lambda = \Lambda^0 - (A + B\Lambda^0)\sqrt{c} \tag{188}$$

Where the two constants, $A$ and $B$, are defined as

$$A = \frac{FZe_0}{3\pi\eta} \left(\frac{8\pi Z^2 e_0^2 N_A}{1000\varepsilon kT}\right)^{\frac{1}{2}} \tag{189}$$

and

$$B = \frac{e_0^2 \omega}{6\varepsilon kT} \left(\frac{8\pi Z^2 e_0^2 N_A}{1000\varepsilon kT}\right)^{\frac{1}{2}} \tag{190}$$

Where $F$ is the Faraday constant and $N_A$ is the Avogadro number. The symbol $\varepsilon$ represents the dielectric constant of the medium whereas $\eta$ is the coefficient of viscosity. $Z$ is charge numbers of the cation and anion. The symbol $e_0$ simply shows the electronic charge. The quantity $\omega$ is defined as

$$\omega = \frac{Z_+ Z_- 2q}{1 + \sqrt{q}}$$

Where $q$ is defined as

$$q = \frac{Z_+ Z_-}{Z_+ + Z_-} \frac{\lambda_+ + \lambda_-}{Z_+ \lambda_+ + Z_- \lambda_-}$$

It is obvious from the Debye-Huckel-Onsager equation that the plot of conductance vs square root of the concentration will be a straight line with a negative slope and positive intercept. The intercept after extrapolation gives the value of conductance of such solutions at infinite dilution.

Now, it has been observed that the Debye-Huckel-Onsager equation can also be applied to non-aqueous solutions up to the fairly good agreement. For instance, consider the variation of equivalent conductivity as a function of the square root of the concentration for different alkali sulfocyanates in methanol as the solvent. The theoretical predictions show that the results of the Debye-Hückel-Onsager equation are in good agreement with the experiment up to 0.002 mol dm$^{-3}$.

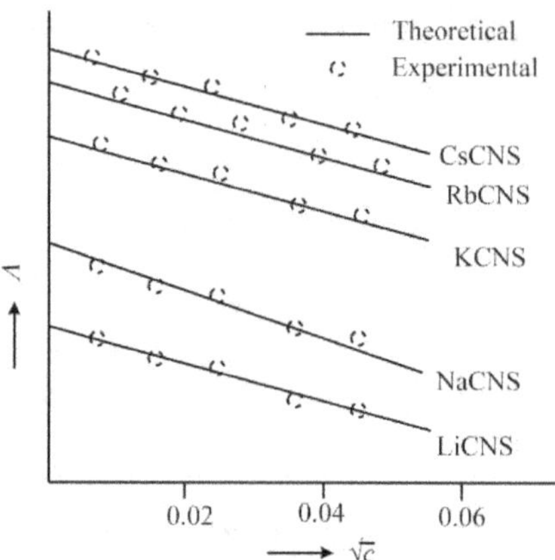


Figure 18. The variation of equivalent conductivity of alkali sulfocyanates vs $c^{1/2}$ in $CH_3OH$.

In going from water to nonaqueous solvent, a significant variation in the quantities like dielectric constant of the medium, the distance of the closest approach, or viscosity is observed. Now since the Debye-Hückel-Onsager equation does have these quantities, the slope and intercept of the $\Lambda$ vs $c^{1/2}$ may also vary drastically.

### ❖ The Solvent Effect on the Mobility at Infinite Dilution

As we know that the asymmetry and electrophoretic effects are not active at infinite dilution because of their dependence on the size of ionic-cloud, the mobility of ions in such cases can be formulated simply from the Stokes law, i.e.,

$$u^0_{con} = \frac{Ze_0}{6\pi\eta r} \tag{191}$$

Where $r$ represents the radius of the solvated ion and $\eta$ is the coefficient of viscosity. $Z$ is charge numbers of the cation and anion. The symbol $e_0$ simply shows the electronic charge. Now, if we imagine the same electrolyte in different electrolytic solutions, we can say

$$u^0_{con}\eta r = constant \tag{192}$$

Assuming further that the $r$ is also independent of solvent type, we have

$$u^0_{con}\eta = constant \tag{193}$$

Thus, it is obvious from the equation (193) that ionic mobility is inversely proportional to the coefficient of viscosity. More viscous solvents would result in a slow drift of ions and vice-versa. It has also been found that the equation (193) is more valid for solvents other than water.

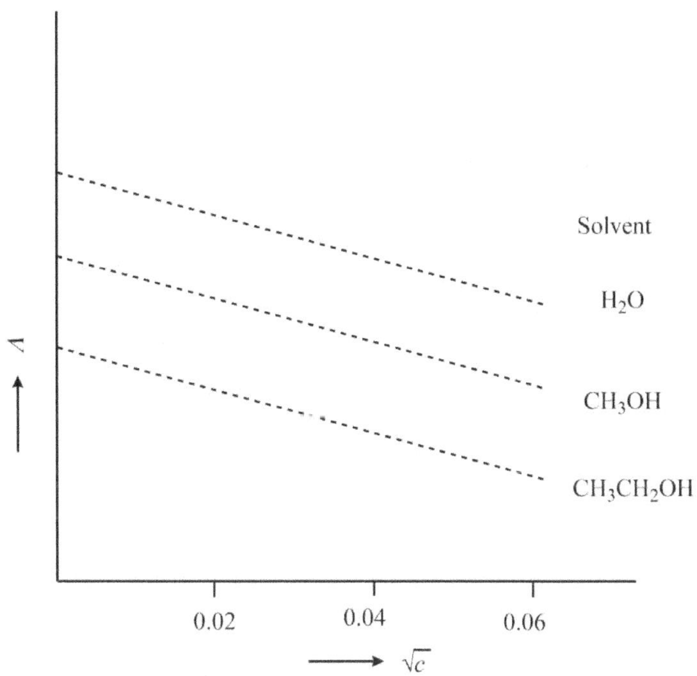


Figure 19. The variation of equivalent conductivity vs $c^{1/2}$ in different solvents.

Moreover, the radius of solvated ion may vary drastically in going from one solvent to another (in some cases it gets even double), the result of equation (193) does not find a very large application domain. This variation in the solvated radius is primarily due to the difference in the size of solvent molecules. For instance, the size of the solvated ion is larger in ethanol than in methanol, which in turn, is larger than what is observed in water. All this results in an opposite trend in the ionic mobilities, and therefore, in the equivalent conductivities as well. Recalling the simple Walden's rule i.e.

$$\Lambda\eta = constant \tag{194}$$

Where $\Lambda$ is the equivalent conductivity of the electrolytic solution at concentration $c$. This simply means that equivalent conductivity and the viscosity of the solvent are inversely proportional to each other. However, the effect of the solvated radius must be considered for more accurate results. Therefore, we must more acceptable form of Walden's rule, i.e.,

$$u^0\eta r = constant \tag{195}$$

The symbol $r$ represents the radius of the ionic species considered in the solvent under examination.

## ❖ Equivalent Conductivity ($\Lambda$) vs Concentration $C^{1/2}$ as a Function of the Solvent

In order to understand the variation of equivalent conductivity with the square root of concentration for different solvents, recall the generalized Walden's rule, i.e.,

$$u^0 \eta r = constant \tag{196}$$

The $u^0$ is the ionic mobility at infinite dilution and symbol $r$ represents the radius of the ionic species considered in the solvent under examination. The symbol $\eta$ represents the coefficient of viscosity of the solvent used. Also, as we know that the equivalent conductivity at infinite dilution can be obtained from the relation given below.

$$\Lambda^0 = F u^0 \tag{197}$$

Now, if the ionic mobility obtained using equation (196) is used in equation (197) for different nonaqueous solutions, it has been found that predicted values of equivalent conductivity are quite large and sometimes even outnumber the equivalent conductivity in water as the solvent. All this suggests that there should be no problem in using non-aqueous solutions in electrochemical systems. However, the quantity that is more important for practical applications is the specific conductivity ($\sigma$) at a finite concentration rather than the equivalent conductivity at infinite dilution. In other words, the conjugative relationship of electrode geometry and specific conductivity dictates the overall electrolytic resistance of the system under consideration. At this point, one might ask why the electrolytic resistance is important. The answer lies in the fact that the magnitude of useful power wasted as heat in the electrolytic solution is $I^2R$ ($I$ is the current passed); and therefore, it is the electrolytic resistance that $R$ must possess a lower value for feasibility. The lower value of $R$ implies a higher value of specific conductivity, and the relation of specific conductivity with equivalent conductivity at a certain concentration '$c$' is

$$\sigma = \Lambda Z c \tag{198}$$

Now since the equivalent conductance $\Lambda$ also depends on concentration, we cannot use equivalent conductance at infinite dilution in equation (198) to find out the specific conductivity. First of all, one must determine the equivalent conductivity at concentration '$c$', which is possible from the Debye-Huckel-Onsager equation if the values of constants $A$ and $B$ are known. Hence, to proceed further, we must recall the Debye-Huckel-Onsager equation for the non-aqueous solution first, i.e.,

$$\Lambda = \Lambda^0 - (A + B\Lambda^0)\sqrt{c} \tag{199}$$

Where the two constants, $A$ and $B$, are defined as

$$A = \frac{FZe_0}{3\pi\eta} \left( \frac{8\pi Z^2 e_0^2 N_A}{1000 \varepsilon k T} \right)^{\frac{1}{2}} \tag{200}$$

$$B = \frac{e_0^2\,\omega}{6\varepsilon kT}\left(\frac{8\pi Z^2 e_0^2 N_A}{1000\varepsilon kT}\right)^{\frac{1}{2}} \tag{201}$$

Where $F$ is the Faraday constant and $N_A$ is the Avogadro number. The symbol $\varepsilon$ represents the dielectric constant of the medium whereas $\eta$ is the coefficient of viscosity. $Z$ is charge numbers of the cation and anion. The symbol $e_0$ simply shows the electronic charge. $\omega$ is a parameter defined earlier in this chapter.

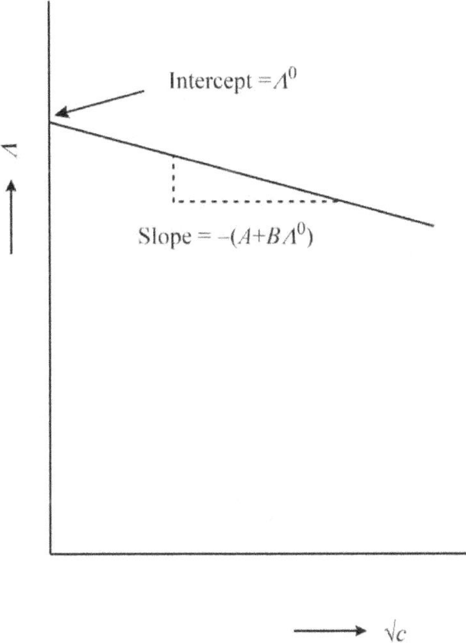


Figure 20. The variation of equivalent conductivity vs $c^{1/2}$ in any arbitrary solvent.

It can be clearly seen that both the constants in the slope have the dielectric constant $\varepsilon$ in the denominator, and therefore, the equivalent conductivity as well as specific conductivity show decrease with the increasing $\varepsilon$. In other words, we can say that the lower dielectric constant would result in stronger ion-ion interaction and vice-versa. Therefore, in the case of non-aqueous solvents with very low dielectric constants, the relative variation in the magnitude of equivalent conductivity with the square root of the concentration is very large. Consequently, all this result in a very low specific conductivity at any practical electrolytic concentration in non-aqueous solution than in water. Furthermore, in addition to the larger electrolytic resistance, non-aqueous solutions are hard to keep so because they can always absorb some moisture from the atmosphere.

### ❖ Effect of Ion Association Upon Conductivity (Debye-Huckel-Bjerrum Equation)

According to the Debye-Huckel theory of ion-ion interaction, the ions produced by the dissociation of electrolytes are randomly distributed in the solution with oppositely charged ionic-cloud surrounding them. Nevertheless, the possibility that some negative ions might get very close to the positively charged reference ion was neglected. In such a situation, the translational thermal energy would not be enough to make the ions to move independently, and an ionic-pair may be formed. Being a combined entity of two equal and opposite charges, these ion pairs are completely neutral and distributed randomly in the electrolytic solutions. These neutral ion-pairs are not influenced by the reference ion under consideration. Therefore, the fraction of ions participating in these ion pairs must be obtained for a more accurate picture of the electrochemical properties of the solutions.

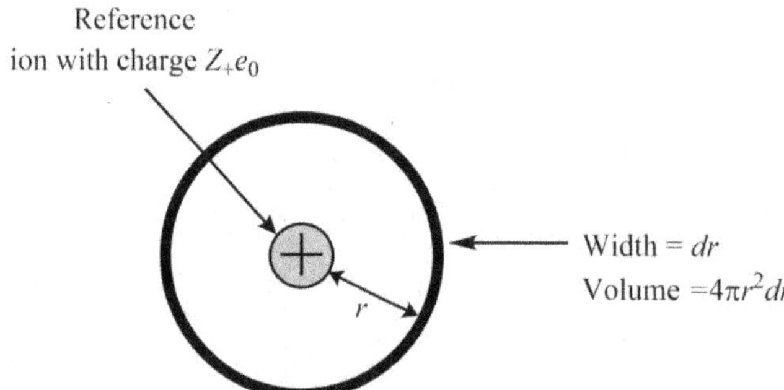


Figure 21. The probability of finding an oppositely charged ion i.e. $Z_-e_0$ at distance $r$ in $dr$ thickness around the reference ion.

In order to understand the concept, imagine a spherical shell at a distance $r$ from the reference ion with $dr$ thickness. The probability of finding the anion ($P_r$) in the spherical shell is proportional to three main factors. The first one is the ratio $4\pi r^2 dr$ volume element to the total volume $V$, the second one is the total number of anions i.e. $N_-$ and the third is the $e^{-U/kT}$ ($U$ represents the potential energy of anion situated at distance $r$). Mathematically, we can formulate this as

$$P_r = \left(\frac{4\pi r^2 dr}{V}\right)(N_-)\left(e^{-U/kT}\right) \tag{202}$$

Now because $N_-/V$ represents the anions' concentration $n_-^0$, the above equation takes the form

$$P_r = 4\pi r^2 dr\, n_-^0 e^{-U/kT} \tag{203}$$

Furthermore, after recalling the value of $U$, i.e.,

---

$$U = -\frac{Z_+Z_- e_0^2}{\varepsilon r} \tag{204}$$

Equation (203) becomes

$$P_r = 4\pi r^2\, n_-^0\, e^{Z_+Z_- e_0^2/\varepsilon r kT}\, dr \tag{205}$$

Where $\varepsilon$ is the dielectric constant of the medium and $e_0$ is the electronic charge. Now define a new parameter $\lambda$ as

$$\lambda = \frac{Z_+Z_- e_0^2}{\varepsilon kT} \tag{206}$$

Therefore, we can write

$$P_r = (4\pi\, n_-^0)e^{\lambda/r}r^2\, dr \tag{207}$$

Similarly, the probability of finding a positively charged ion at distance $r$ in $dr$ thickness around a negatively charged reference ion will be

$$P_r = (4\pi\, n_+^0)e^{\lambda/r}r^2\, dr \tag{208}$$

Therefore, the probability of finding $i$th type of ion at distance $r$ in $dr$ thickness around a $k$th type of reference ion can be formulated as

$$P_r = (4\pi\, n_i^0)e^{\lambda/r}r^2\, dr \tag{208}$$

where

$$\lambda = \frac{Z_i Z_k e_0^2}{\varepsilon kT} \tag{209}$$

The variation of probability of finding one type ion around other types shows a strange behavior with distance.

Table 1. The total number of oppositely charged ions as a function of distance.

| $r$ (pm) | Number of ions in shell $\times\ 10^{22}$ |
|---|---|
| 200 | 1.77 |
| 250 | 1.37 |
| 300 | 1.22 |
| 350 | 1.17 |
| 400 | 1.21 |

When $r$ varies from very small to moderate value, the magnitude of $P_r$ decreases because of the factor $e^{\lambda/r}$. However, after a certain value, the probability starts to rise due to the dominance of the directly correlated factor of $r^2$. Hence, we can say that the probability of finding the oppositely charged ion at distance $r$ from the reference ion first decreases and then increases as we move away from the central ion.

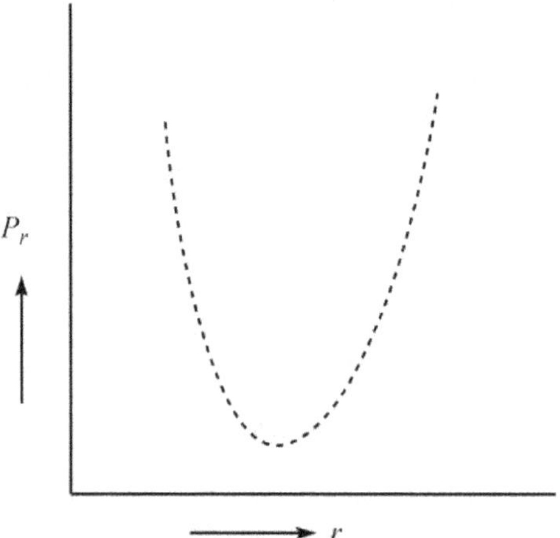


Figure 22. The variation of the overall probability of finding an ion around ion under consideration as a function of distance.

Since the probability of finding a negative ion within a certain distance around a positive ion is obtained by applying the lower and upper bounds, we need to find these limits first. Now, because the ion-pair can be formed only if the participating ions close enough so that the electrostatic attraction can outshine the thermal forces; let this distance be represented by $q$.

In other words, we can say that the formation of ion-pair takes place only if the separation between the cation and anion is less than $q$. Since the cation and anion cannot approach each other closer than the ionic-size parameter $(a)$, i.e., (the distance of closest approach), the ion-pair will be formed for an interionic separation of greater than $a$ but less than $q$. The probability of the ion-pair formation is the number of $i$th type of ion participating in the ion-association to the total number of the same type of ions. Therefore, the probability of ion-pair formation $(\theta)$ can be formulated as

$$\theta = \int_a^q P_r\,dr = \int_a^q (4\pi\,n_i^0)e^{\lambda/r}r^2\,dr \tag{210}$$

Owing to the divergent nature of the integral, the upper limit can be set to the overall probability minima $(q)$ as shown below.

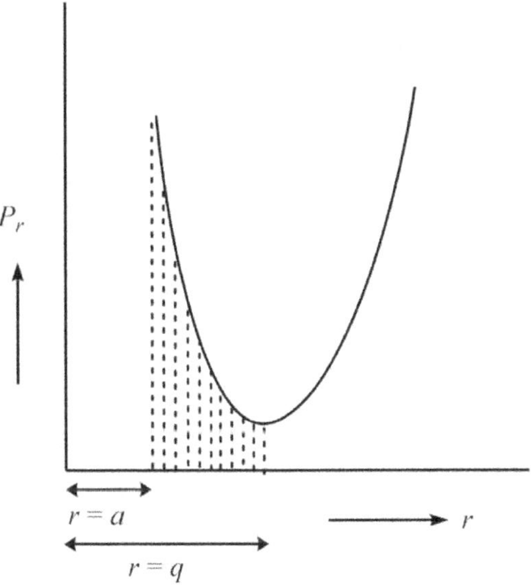



Figure 23. The depiction of the overall probability of finding an oppositely charged ion around reference ion from $a$ to $q$.

To find the minima, differentiate equation (208) with respect to $r$ and then put equal to zero, i.e.,

$$\frac{dP_r}{dr} = 4\pi\, n_i^0 e^{\frac{\lambda}{r}}\, 2r - 4\pi\, n_i^0 r^2 e^{\frac{\lambda}{r}}\, \frac{\lambda}{r^2} \tag{211}$$

or

$$2r_{min} - \lambda = 0 \tag{212}$$

or

$$q = r_{min} = \frac{\lambda}{2} \tag{213}$$

After putting the value of $\lambda$ from equation (206) into (213), we get

$$q = \frac{Z_1 Z\, e_0^2}{2\varepsilon kT} \tag{214}$$

Bjerrum claimed that only Coulombic interactions (short-range) lead to the formation of ion-pair; and if the separation of cation and anion larger than $q$, they should be treated as free ions. In other words, we can say that the formation of ion-pair is feasible if $a < q$ and infeasible if $a > q$.

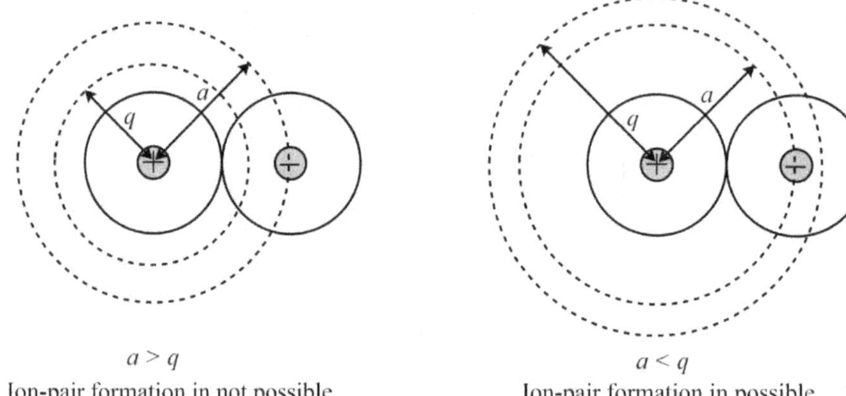



Figure 24. The pictorial representation of ion-pair in the electrolytic solution.

Using the upper limit as $q = \lambda/2$, the probability of ion-pair formation ($\theta$) becomes

$$\theta = 4\pi\, n_i^0 \int_a^{\lambda/2} e^{\lambda/r} r^2 dr \tag{215}$$

At this stage, after define a new variable for simplicity $y = \lambda/r = 2q/r$ and a new constant $b = \lambda/a$, the above equation takes the form

$$\theta = 4\pi\, n_i^0 \left(\frac{Z_+ Z_- e_0^2}{\varepsilon kT}\right)^3 \int_2^b e^y y^{-4} dy \tag{216}$$

Hence, using the above equation we can easily determine the fraction of the ions reserved in pairing and cannot move freely. Nevertheless, it would be more beneficial if we could use some other simple number to quantify the magnitude of ion-pair formation instead of $\theta$.

From the Arrhenius theory of electrolytic dissociation, we know that the $A^+B^-$ type electrolyte would give monovalent cations and anions in water i.e.

$$AB \rightleftharpoons A^+ + B_- \tag{217}$$

Using the law of mass action, the dissociation constant can be given as

$$K = \frac{a_A^+ a_B^-}{a_{AB}} \tag{218}$$

Similarly, if a cation $M^+$ and anion $A^-$ form an ion-pair i.e.

---

$$M^+ + A^- \rightleftharpoons IP \tag{219}$$

The law of mass action can also be used to give association constant as

$$K_A = \frac{a_{IP}}{a_M^+ a_A^-} \tag{220}$$

Where $a_M^+$, $a_A^-$ and $a_{IP}$ are the activities of the cation, anion and ion-pair, respectively. Moreover, it can also be seen that the association constant is just the reciprocal of dissociation constant. Now, $\theta$ is the fraction of ions forming ion-pair, whereas $\theta c$ and $(1 - \theta)c$ are the concentrations of ion-pair and free ions, respectively. Therefore, the association constant in terms of activity coefficients can be written as

$$K_A = \frac{\theta c f_{IP}}{(1 - \theta)c f_+ (1 - \theta)c f_-} \tag{221}$$

or

$$K_A = \frac{\theta}{(1 - \theta)^2} \frac{1}{c} \frac{f_{IP}}{f_+ f_-} \tag{222}$$

Where $f_+$, $f_-$ and $f_{IP}$ are the activity coefficients for cation, anion and ion-pair, respectively. In terms of mean ionic activity coefficient and analytical concentration $c_a$, the equation (222) can also be written as

$$K_A = \frac{\theta}{(1 - \theta)^2} \frac{1}{c_a} \frac{f_{IP}}{f_\pm^2} \tag{223}$$

Now because the ion pairs are neutral species, and therefore, do not participate in ion-ion interactions; the activity coefficients ion-pairs can be taken as unity. Therefore, equation (223) can be written as

$$K_A = \frac{\theta}{(1 - \theta)^2} \frac{1}{c_a} \frac{1}{f_\pm^2} \tag{224}$$

or

$$\theta = K_A (1 - \theta)^2 c_a f_\pm^2 \tag{225}$$

At this stage, the only thing that is needed for further treatment is the relation between $\theta$ and the conductivity of the electrolyte used. To do so, recall the correlation between the concentration of free ions and specific conductivity ($\sigma$) i.e.

$$\sigma = ZF(u_+ + u_-)c_{free\ ions} \tag{226}$$

Where $F$ is the Faraday constant.

$$\sigma = ZF(u_+ + u_-)\frac{c_{free\ ions}}{c_a} c_a \tag{227}$$

The term $c_{free\ ions}/c_a$ is the fraction of ions that are free (not associated), and therefore

$$\frac{c_{free\ ions}}{c_a} = 1 - \theta \tag{228}$$

Using the result of equation (228) in equation (227), we get

$$\sigma = ZF(u_+ + u_-)(1 - \theta)c_a \tag{229}$$

Furthermore, after converting the specific conductivity to equivalent conductivity ($\Lambda = \sigma/Zc_a$), the equation (229) takes the form

$$\Lambda = F(u_+ + u_-)(1 - \theta) \tag{230}$$

Imagine a situation if no ion-association occurs ($\theta = 0$), then the equation (230) would reduce to

$$\Lambda_{\theta=0} = F(u_+ + u_-) \tag{231}$$

Using equation (231) in equation (230), we get

$$\Lambda = \Lambda_{\theta=0}(1 - \theta) \tag{232}$$

or

$$\frac{\Lambda}{\Lambda_{\theta=0}} = 1 - \theta \tag{233}$$

or

$$\theta = 1 - \frac{\Lambda}{\Lambda_{\theta=0}} \tag{234}$$

Now putting the values of $\theta$ and $1-\theta$ from equation (234) and (233) in equation (225), we have

$$1 - \frac{\Lambda}{\Lambda_{\theta=0}} = K_A\left(\frac{\Lambda}{\Lambda_{\theta=0}}\right)^2 c_a f_\pm^2 \tag{235}$$

Rearranging, we get

$$\frac{1}{\Lambda} = \frac{1}{\Lambda_{\theta=0}} + \frac{K_A f_\pm^2}{\Lambda_{\theta=0}^2}\Lambda c_a \tag{236}$$

Rearranging equation (229) for $F(u_+ + u_-)$ and putting in equation (231), we get

$$\Lambda_{\theta=0} = F(u_+ + u_-) = \frac{\sigma}{Z(1 - \theta)c_a} \tag{237}$$

The above equation shows equivalent conductivity at zero ion-association at $(1 - \theta)c_a$ concentration. Therefore, we must use Debye-Huckel-Onsager equation to express the more appropriate conductivity at small concentrations i.e.

$$\Lambda_{\theta=0} = \Lambda^0 - (A + B\Lambda^0)\sqrt{1 - \theta}\sqrt{c_a} \tag{238}$$

Where the two constants, $A$ and $B$, are defined as

$$A = \frac{FZe_0}{3\pi\eta}\left(\frac{8\pi Z^2 e_0^2 N_A}{1000\varepsilon kT}\right)^{\frac{1}{2}} \quad and \quad B = \frac{e_0^2 \omega}{6\varepsilon kT}\left(\frac{8\pi Z^2 e_0^2 N_A}{1000\varepsilon kT}\right)^{\frac{1}{2}} \tag{239}$$

Where $F$ is the Faraday constant and $N_A$ is the Avogadro number. The symbol $\varepsilon$ represents the dielectric constant of the medium whereas $\eta$ is the coefficient of viscosity. $Z$ is charge numbers of the cation and anion. The symbol $e_0$ simply shows the electronic charge. $\omega$ is a parameter defined earlier in this chapter. The equation (238) can be simplified as

$$\Lambda_{\theta=0} = \Lambda^0 W \tag{240}$$

Where $W$ represents a continued fraction i.e.

$$W = 1 - w\left\{1 - w\left[1 - w(...)^{-1/2}\right]^{-1/2}\right\}^{-1/2} \tag{241}$$

Provided that

$$w = \frac{(A + B\Lambda^0)\sqrt{c_a}\sqrt{\Lambda}}{(\Lambda^0)^{3/2}} \tag{242}$$

Using the result of equation (240) into (236), we get

$$\frac{1}{\Lambda} = \frac{1}{\Lambda^0 W} + \frac{K_A f_\pm^2}{(\Lambda^0 W)^2}\Lambda c_a \tag{243}$$

or

$$\frac{W}{\Lambda} = \frac{W}{\Lambda^0} + \frac{K_A}{(\Lambda^0)^2}\frac{f_\pm^2 \Lambda c_a}{W} \tag{244}$$

After looking at the above equation, it is obvious that the formation of ion-pairs has changed the variation of equivalent conductivity with concentration drastically. When the formation of ion-pairs was ignored, the variation of equivalent conductivity with concentration was empirically explained by the Kohlrausch's law.

     Nevertheless, in the case of non-aqueous solvents with low dielectric constant, a significant amount of ion-pair formation occurs, and therefore, $W/\Lambda$ is plotted vs $f_\pm^2 \Lambda c_a/W$ with $W/\Lambda^0$ as intercept and $K_A/(\Lambda^0)^2$ as slope.

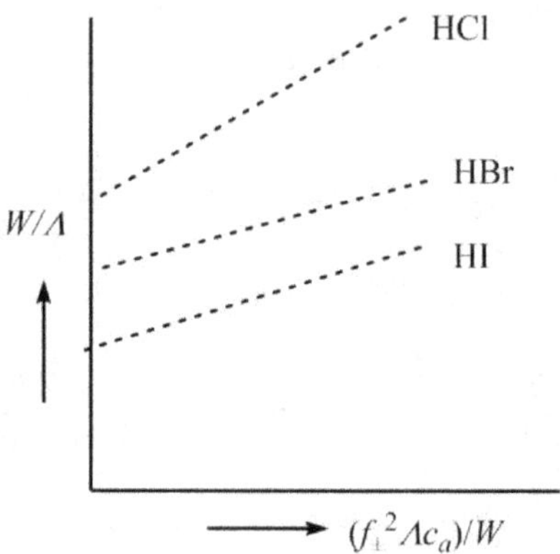



Figure 25. The variation of $W/\Lambda$ is plotted vs $f_\pm^2 \Lambda c_a/W$ for some typical electrolytic solutions.

## ❖ Problems

Q 1. Discuss the Debye-Huckel theory of ion-ion interaction in detail.

Q 2. What is excess charge density? How does it vary with the distance from the central ion?

Q 3. Define Debye-Huckel reciprocal length.

Q 4. Derive the expression for the contribution of the ionic cloud to the total potential at a particular distance from the reference ion in strong electrolytes.

Q 5. State and explain the Debye-Huckel limiting law of activity coefficient. Also, discuss its limitations.

Q 6. What is the ion-size parameter? How does it affect total potential around the central ion?

Q 7. Discuss the asymmetry effect in the conductance of strong electrolytes?

Q 8. Derive and discuss the Debye-Huckel-Onsager equation for aqueous solutions.

Q 9. Discuss the effect of the nature of the solvent on the ionic mobility at infinite dilution.

Q 10. How does the equivalent conductivity vary with the square root of the concentration if the dielectric constant of the solvent is very low?

Q 11. Define ion-association in strong electrolytic solutions. How does this affect overall conductivity?

### ❖ Bibliography


[1] J. Bockris, A. Reddy, *Modern Electrochemistry – Volume 1: Ionics*, Kluwer Academic Publishers, New York, USA 2002.

[2] B. R. Puri, L. R. Sharma, M. S. Pathania, *Principles of Physical Chemistry*, Vishal Publications, Jalandhar, India, 2008.

[3] P. Atkins, J. Paula, *Physical Chemistry*, Oxford University Press, Oxford, UK, 2010.

[4] E. Steiner, *The Chemistry Maths Book*, Oxford University Press, Oxford, UK, 2008.

[5] M. R. Wright, *An Introduction to Aqueous Electrolyte Solutions*, John Wiley & Sons Ltd, Sussex, UK, 2007.

[6] P. Debye, E. Hückel, *The Theory of Electrolytes. I. Lowering of Freezing Point and Related Phenomena*, Physikalische Zeitschrift., 24 (1923) 185-206.

[7] V. S. Bagotsky, *Fundamentals of Electrochemistry*, John Wiley & Sons, New Jersey, USA, 2006.

[8] R.R. Netz, H. Orland, *Beyond Poisson-Boltzmann: Fluctuation effects and correlation functions*, The European Physical Journal E, 1 (2000) 203-214.

# CHAPTER 5

# Quantum Mechanics – II

### ❖ Schrodinger Wave Equation for a Particle in a Three Dimensional Box

In the first chapter of this book, we derived and discussed the Schrodinger wave equation for a particle in the one-dimensional box. In this chapter, we will extend that procedure to the particle in a three-dimensional box. In order to do so, consider a particle trapped in a 3-dimensional box of length, breadth, and height as $a$, $b$ and $c$, respectively. This means that this particle can travel in any direction i.e. along $x$-, $y$- *and* $z$-axis. The potential inside the box is 0, while outside to the box it is infinite.

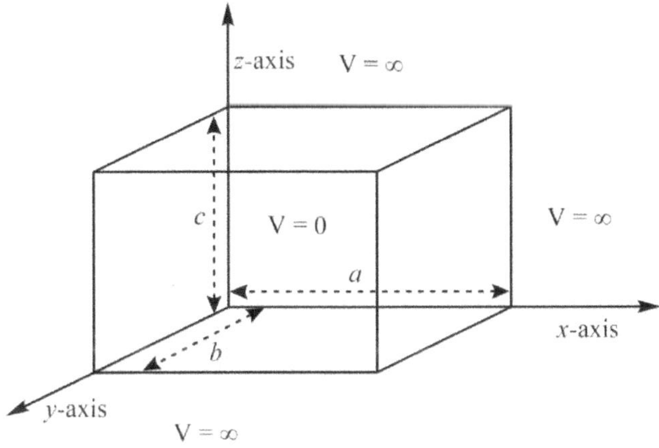


Figure 1. The particle in a three-dimensional box.

So far we have considered a quantum mechanical system of a particle trapped in a three-dimensional box. Now suppose that we need to find various physical properties associated with different states of this system. Had it been a classical system, we would use simple formulas from classical mechanics to determine the value of different physical properties. However, being a quantum mechanical system, we cannot use those expressions because they would give irrational results. Therefore, we need to use the postulates of quantum mechanics to evaluate various physical properties.

Let $\psi$ be the function that describes all the states of the particle in a three-dimensional box. At this point we have no information about the exact mathematical expression of $\psi$; nevertheless, we know that there is one operator that does not need the absolute expression of wave function but uses the symbolic form only, the Hamiltonian operator. The operation of Hamiltonian operator over this symbolic form can be rearranged to give to construct the Schrodinger wave equation; and we all know that the wave function as well the energy, both are obtained as this second-order differential equation is solved. Mathematically, we can say that

$$\hat{H}\psi = E\psi \tag{1}$$

After putting the value of three-dimensional Hamiltonian in equation (1), we get

$$\left[\frac{-h^2}{8\pi^2 m}\left(\frac{\partial^2}{\partial x^2} + \frac{\partial^2}{\partial y^2} + \frac{\partial^2}{\partial z^2}\right) + V\right]\psi = E\psi \tag{2}$$

or

$$\frac{-h^2}{8\pi^2 m}\left(\frac{\partial^2\psi}{\partial x^2} + \frac{\partial^2\psi}{\partial y^2} + \frac{\partial^2\psi}{\partial z^2}\right) + V\psi = E\psi \tag{3}$$

$$\frac{-h^2}{8\pi^2 m}\left(\frac{\partial^2\psi}{\partial x^2} + \frac{\partial^2\psi}{\partial y^2} + \frac{\partial^2\psi}{\partial z^2}\right) + V\psi - E\psi = 0 \tag{4}$$

or

$$\frac{\partial^2\psi}{\partial x^2} + \frac{\partial^2\psi}{\partial y^2} + \frac{\partial^2\psi}{\partial z^2} + \frac{8\pi^2 m}{h^2}(E - V)\psi = 0 \tag{5}$$

The above-mentioned second order differential equation is the Schrodinger wave equation for a particle moving along three dimensions. Since the conditions outside and inside the box are different, the equation (5) must be solved separately for both cases.

**1. The solution of Schrodinger wave equation for outside the box:** After putting the value of potential outside the box in equation (5) i.e. V = ∞, we get

$$\frac{\partial^2\psi}{\partial x^2} + \frac{\partial^2\psi}{\partial y^2} + \frac{\partial^2\psi}{\partial z^2} + \frac{8\pi^2 m}{h^2}(E - \infty)\psi = 0 \tag{6}$$

Since E is negligible in comparison to the ∞, the above equation becomes

$$\frac{\partial^2\psi}{\partial x^2} + \frac{\partial^2\psi}{\partial y^2} + \frac{\partial^2\psi}{\partial z^2} - \infty\psi = 0 \tag{7}$$

$$\infty\psi = \frac{\partial^2\psi}{\partial x^2} + \frac{\partial^2\psi}{\partial y^2} + \frac{\partial^2\psi}{\partial z^2} \tag{8}$$

$$\psi = \frac{1}{\infty}\left(\frac{\partial^2\psi}{\partial x^2} + \frac{\partial^2\psi}{\partial y^2} + \frac{\partial^2\psi}{\partial z^2}\right) = 0 \tag{9}$$

The physical significance of the equation (9) is that the particle cannot go outside the box, and is always reflected back when it strikes the boundaries. In other words, as the function describing the existence of particles is zero outside the box, the particle cannot exist outside the box.

**2. The solution of Schrodinger wave equation for inside the box:** After putting the value of potential inside the box in equation (5) i.e. V = 0, we get

$$\frac{\partial^2 \psi}{\partial x^2} + \frac{\partial^2 \psi}{\partial y^2} + \frac{\partial^2 \psi}{\partial z^2} + \frac{8\pi^2 m}{h^2}(E - 0)\psi = 0 \tag{10}$$

$$\frac{\partial^2 \psi}{\partial x^2} + \frac{\partial^2 \psi}{\partial y^2} + \frac{\partial^2 \psi}{\partial z^2} + \frac{8\pi^2 mE}{h^2}\psi = 0 \tag{11}$$

The above equation has three variables and is difficult to solve directly. Therefore, it is better to separate variable, we already know the steps to solve a one-variable equation. To do so, consider that the wave function $\psi$ is the multiplication of three individual functions as

$$\psi(x, y, z) = \psi(x) \times \psi(y) \times \psi(z) = XYZ \tag{12}$$

Using the above expression in equation (11), we get

$$\frac{\partial^2 XYZ}{\partial x^2} + \frac{\partial^2 XYZ}{\partial y^2} + \frac{\partial^2 XYZ}{\partial z^2} + \frac{8\pi^2 mE}{h^2}XYZ = 0 \tag{13}$$

From the rules of partial derivative, the equation (13) takes the form

$$YZ\frac{\partial^2 X}{\partial x^2} + XZ\frac{\partial^2 Y}{\partial y^2} + XY\frac{\partial^2 Z}{\partial z^2} + \frac{8\pi^2 mE}{h^2}XYZ = 0 \tag{14}$$

Now divide the above equation by XYZ on both side i.e.

$$\frac{1}{X}\frac{\partial^2 X}{\partial x^2} + \frac{1}{Y}\frac{\partial^2 Y}{\partial y^2} + \frac{1}{Z}\frac{\partial^2 Z}{\partial z^2} + \frac{8\pi^2 mE}{h^2} = 0 \tag{15}$$

Assuming

$$k^2 = \frac{8\pi^2 mE}{h^2} \tag{16}$$

The equation (15) becomes

$$\frac{1}{X}\frac{\partial^2 X}{\partial x^2} + \frac{1}{Y}\frac{\partial^2 Y}{\partial y^2} + \frac{1}{Z}\frac{\partial^2 Z}{\partial z^2} + k^2 = 0 \tag{17}$$

Also fragmenting the constant $k^2$ along three x-, y- and z-axis i.e. $k^2 = k_x^2 + k_y^2 + k_z^2$, the equation (17) can

$$\frac{1}{X}\frac{d^2 X}{\partial x^2} + \frac{1}{Y}\frac{d^2 Y}{\partial y^2} + \frac{1}{Z}\frac{\partial^2 Z}{\partial z^2} + k_x^2 + k_y^2 + k_z^2 = 0 \tag{18}$$

The above equation can be written as the sum of three equations with only one variable in each i.e.

$$\frac{\partial^2 X}{\partial x^2} + k_x^2 X = 0 \tag{19}$$

$$\frac{\partial^2 Y}{\partial y^2} + k_y^2 Y = 0 \tag{20}$$

$$\frac{\partial^2 Z}{\partial z^2} + k_z^2 Z = 0 \tag{21}$$

The equations (19-21) are simple one-dimensional differential equations whose solutions can be obtained just like in the one-dimensional box. The solution of equation (19) will give the $x$-dependent wave function as well the energy distribution along $x$-axis i.e.

$$\psi_{n_x}(x) = X = \sqrt{\frac{2}{a}}\, Sin\frac{n_x \pi x}{a} \quad and \quad E_{n_x} = \frac{n_x^2 h^2}{8ma^2} \tag{22}$$

Similarly, the solution of equation (20) will be

$$\psi_{n_y}(y) = Y = \sqrt{\frac{2}{b}}\, Sin\frac{n_y \pi y}{b} \quad and \quad E_{n_y} = \frac{n_y^2 h^2}{8mb^2} \tag{23}$$

Just like the above two, the solution of equation (21) will be

$$\psi_{n_z}(z) = Z = \sqrt{\frac{2}{c}}\, Sin\frac{n_z \pi z}{c} \quad and \quad E_{n_z} = \frac{n_z^2 h^2}{8mc^2} \tag{24}$$

After putting the expressions of individual wave functions from equation (22-24) in equation (12), the total wave function can be obtained i.e.

$$\psi_{n_x n_y n_z}(x, y, z) = \sqrt{\frac{8}{abc}}\, Sin\frac{n_x \pi x}{a}\, Sin\frac{n_y \pi y}{b}\, Sin\frac{n_z \pi z}{c} \tag{25}$$

Since $k^2 = k_x^2 + k_y^2 + k_z^2$, the total energy must be the sum of individual energies i.e.

$$E_{n_x n_y n_z} = \left(\frac{n_x^2}{a^2} + \frac{n_y^2}{b^2} + \frac{n_z^2}{c^2}\right)\frac{h^2}{8m} \tag{26}$$

Where $n_x$, $n_y$, $n_y$ are the discrete variable whose permitted values from boundary conditions can be 0, 1, 2, 3, 4….∞. Nevertheless, it is worthy to note that even though the $n = 0$ is permitted by the boundary conditions, we still don't use it in equation (25); which is obviously because it makes the whole function zero.

## ❖ The Concept of Degeneracy Among Energy Levels for a Particle in Three Dimensional Box

The solution of Schrodinger wave equation for a particle of mass '$m$' trapped in three dimensional of sides $a$, $b$ and $c$ with zero potential inside and infinite potential outside provide the total wave function $\psi$ as

$$\psi_{n_x n_y n_z}(x, y, z) = \sqrt{\frac{8}{abc}}\, Sin\frac{n_x \pi x}{a}\, Sin\frac{n_y \pi y}{b}\, Sin\frac{n_z \pi z}{c} \tag{27}$$

Where $n_x$, $n_y$, $n_y$ are the discrete variable whose permitted values from boundary conditions can be 1, 2, 3, 4….∞. The variable $x$, $y$ and $z$ represent the position of the particle along the corresponding axis. Besides, the expression for total energy is

$$E_{n_x n_y n_z} = \left(\frac{n_x^2}{a^2} + \frac{n_y^2}{b^2} + \frac{n_z^2}{c^2}\right)\frac{h^2}{8m} \tag{28}$$

For a cubical box, all the sides become equal ($a = b = c$). Using this condition in equation (27), the total wave function representing different quantum mechanical states take the following form.

$$\psi_{n_x n_y n_z}(x, y, z) = \sqrt{\frac{8}{a^3}}\, Sin\frac{n_x \pi x}{a}\, Sin\frac{n_y \pi y}{a}\, Sin\frac{n_z \pi z}{a} \tag{29}$$

Similarly, the energy expression also changes to

$$E_{n_x n_y n_z} = \left(n_x^2 + n_y^2 + n_z^2\right)\frac{h^2}{8ma^2} \tag{30}$$

Now, in order to define various quantum mechanical states, we need to put valid set quantum numbers. The expression for first quantum mechanical and corresponding energy can be obtained by putting $n_x = n_y = n_z = 1$ in equations (29–30) i.e.

$$\psi_{111} = \sqrt{\frac{8}{a^3}}\, Sin\frac{\pi x}{a}\, Sin\frac{\pi y}{a}\, Sin\frac{\pi z}{a} \quad \text{and} \quad E_{111} = \frac{3h^2}{8ma^2} \tag{31}$$

Similarly, the next state with energy can be obtained by putting $n_x = n_y = 1$ and $n_z = 2$ in equations (29–30) i.e.

$$\psi_{112} = \sqrt{\frac{8}{a^3}}\, Sin\frac{\pi x}{a}\, Sin\frac{\pi y}{a}\, Sin\frac{2\pi z}{a} \quad \text{and} \quad E_{112} = \frac{6h^2}{8ma^2} \tag{32}$$

If $n_x = n_z = 1$ and $n_y = 2$; the wavefunction and energy become

$$\psi_{121} = \sqrt{\frac{8}{a^3}}\; Sin\frac{\pi x}{a}\; Sin\frac{2\pi y}{a}\; Sin\frac{\pi z}{a} \quad \text{and} \quad E_{121} = \frac{6h^2}{8ma^2} \tag{33}$$

If $n_y = n_z = 1$ and $n_x = 2$, the state with energy becomes

$$\psi_{211} = \sqrt{\frac{8}{a^3}}\; Sin\frac{2\pi x}{a}\; Sin\frac{\pi y}{a}\; Sin\frac{\pi z}{a} \quad \text{and} \quad E_{211} = \frac{6h^2}{8ma^2} \tag{34}$$

It can be clearly seen that three quantum mechanical states $\psi_{112}$, $\psi_{121}$ and $\psi_{211}$ possess the same amount of energy (i.e. $6h^2/8ma^2$); and therefore, are said to be degenerate. In other words, there are three different ways of existence of the particle inside the box so that the particle possesses $6h^2/8ma^2$ energy as total.

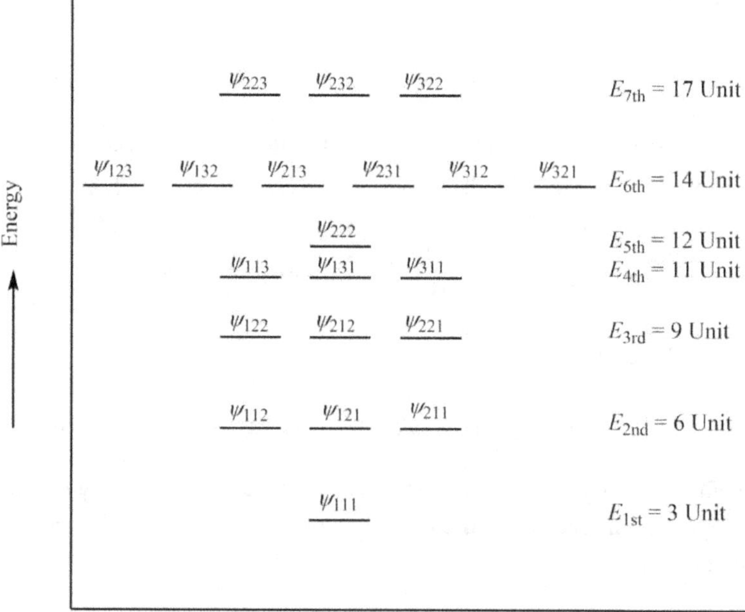

Figure 2. The energy level diagram representing different quantum mechanical states (in the units of $h^2/8ma^2$) for a particle trapped in a cubical box.

Hence, the degeneracy of the ground state is one i.e. there is only one way for the particle to exist in the box to create zero-point energy ($3h^2/8ma^2$). On the other hand, the degeneracy of first excited stated is 3 as $\psi_{112}$, $\psi_{121}$ and $\psi_{211}$, all have 6 units of energy. Moreover, after careful examination of energy diagram, it can be concluded that degeneracy is 1 if $n_x = n_y = n_z$, 3 if $n_x = n_y$ or $n_y = n_z$ or $n_x = n_z$, and 6 if $n_x \neq n_y \neq n_z$.

## ❖ Schrodinger Wave Equation for a Linear Harmonic Oscillator & Its Solution by Polynomial Method

A diatomic molecule is the quantum-mechanical analog of the classical version of the harmonic oscillator. It represents the vibrational motion and is one of the few quantum-mechanical systems for which an exact solution is available. In this section, we will discuss the classical and quantum mechanical oscillator and their comparative study.

### ➢ *The Classical Treatment of Simple Harmonic Oscillator*

In order to understand the vibrational states of a simple diatomic molecule, we must understand the classical oscillator first. In order to do so, consider a spring of length $r$ in which a displacement '$x$' is incorporated by expending or compressing it.

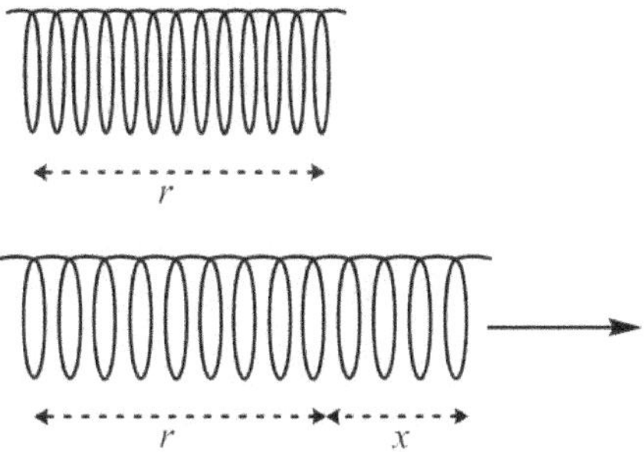

Figure 2. The pictorial representation of the displacement-inducing in a typical spiral.

For a moment, imagine that the spiral is extended by a displacement of '$x$'; then the restoring force ($F$) developed in the spiral can be obtained using Hook's law as

$$F = -kx \qquad (35)$$

Where $k$ is the constant of proportionality. The minus sign is because the restoring force and the displacement both are vector quantity but in the opposite direction. In other words, if we expend the spiral, the spiral will try to compress itself and vice-versa. From equation (35), it seems that the restoring force depends only upon displacement induced only, however, it is found that stronger spirals have larger restoring force than the weaker ones for the same magnitude of displacement, indicating a lager force constant. Therefore, the physical significance of the force constant lies in the fact that it can be used to comment on the strength of oscillator.

Since the potential energy in this expended state is simply the amount of work done in the process of incorporating the displacement '$x$', we need calculate the same for further analysis. The restoring force is proportional to the displacement, and therefore, is a variable quantity; suggesting that we need to carry out the integration force curve vs displacement. Suppose that the total displacement "$x$" is fragmented in very small

"$dx$" segments. The amount of work done in inducing '$dx$' displacement will be '$dw$' and can be given by the following relation

$$dw = F.dx \tag{36}$$

The total work from zero displacement to '$x$' displacement will be

$$W = \int_0^x F.dx \tag{37}$$

$$= \int_0^x -kx.dx \tag{38}$$

$$= -k\left[\frac{x^2}{2}\right]_0^x \tag{39}$$

$$W = -\frac{1}{2}kx^2 \tag{40}$$

Since there is no electrostatic attraction, the potential energy ($V$) of the system at displacement will simply be

$$V = \frac{1}{2}kx^2 \tag{41}$$

The above equation represents a parabolic behavior and shows that the potential energy varies continuously with the displacement. Larger the displacement, higher will be the potential energy.

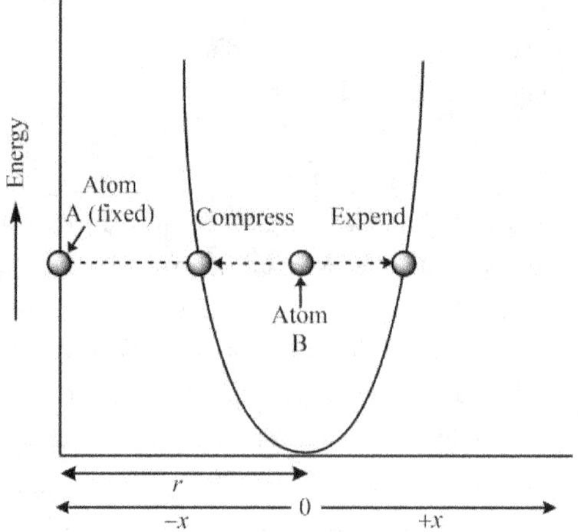


Figure 3. The variation of potential energy as a function of displacement in a classical oscillator.

If '$m$' is the reduced mass of the diatomic molecule, the equilibrium vibrational frequency '$v$' of the oscillator can be given as

$$v = \frac{1}{2\pi}\sqrt{\frac{k}{m}} \tag{42}$$

Where $m$ is the reduced mass defined by the ratio of the product to the sum of individual masses i.e. $m = m_1 m_2/(m_1 + m_2)$. It is obvious that the energy levels of a simple harmonic oscillator in classical mechanics are continuous (including zero), and have a limit over the expansion and compression for each value. Furthermore, the classical oscillator is bound to spend most of its time in the extreme state (fully compressed and fully expended) and the least time in the equilibrium position.

> ➤ *The Quantum Mechanical Treatment of Simple Harmonic Oscillator*

In order to find out the quantum mechanical behavior of a simple harmonic oscillator, assume that all the vibrational states can be described by a mathematical expression $\psi$. Since we don't know the exact nature of $\psi$, we need to follow the postulates of quantum mechanics. Therefore, after applying the Hamiltonian operator over this symbolic wave function, we have

$$H\psi = E\psi \tag{43}$$

$$\left(\frac{-h^2}{8\pi^2 m}\frac{\partial^2}{\partial x^2} + V\right)\psi = E\psi \tag{44}$$

or

$$\frac{-h^2}{8\pi^2 m}\frac{\partial^2\psi}{\partial x^2} + V\psi - E\psi = 0 \tag{45}$$

Rearranging, we have

$$\frac{\partial^2\psi}{\partial x^2} + \frac{8\pi^2 m}{h^2}(E - V)\psi = 0 \tag{46}$$

After putting the value of potential energy form equation (41) in equation (46), we get

$$\frac{\partial^2\psi}{\partial x^2} + \frac{8\pi^2 m}{h^2}\left(E - \frac{1}{2}kx^2\right)\psi = 0 \tag{47}$$

Now put the value of $k$ form equation (42) in equation (47) i.e.

$$\frac{\partial^2\psi}{\partial x^2} + \frac{8\pi^2 m}{h^2}(E - 2\pi^2 v^2 m x^2)\psi = 0 \tag{48}$$

or

$$\frac{\partial^2 \psi}{\partial x^2} + \left(\frac{8\pi^2 mE}{h^2} - \frac{16\pi^4 m^2 v^2 x^2}{h^2}\right)\psi = 0 \tag{49}$$

After defining constants

$$\alpha = \frac{8\pi^2 mE}{h^2} \quad and \quad \beta = \frac{4\pi^2 mv}{h} \tag{50}$$

The equation (49) takes the form

$$\frac{\partial^2 \psi}{\partial x^2} + (\alpha - \beta^2 x^2)\psi = 0 \tag{51}$$

or

$$\frac{\partial^2 \psi}{\partial x^2} + (\alpha - \beta\beta x^2)\psi = 0 \tag{52}$$

Now define a new variable $y = \sqrt{\beta}x$, then we have the derivative as

$$\frac{dy}{dx} = \sqrt{\beta} \tag{53}$$

Squaring both side of the equation (53), and then rearranging

$$\frac{d^2 y}{dx^2} = \beta \quad or \quad dx^2 = d^2 y/\beta \tag{54}$$

Now put the value of $dx^2$ and $\beta x^2$ in equation (52), we get

$$\beta \frac{\partial^2 \psi}{\partial y^2} + (\alpha - \beta y^2)\psi = 0 \tag{55}$$

Dividing the above equation by $\beta$, we get

$$\frac{\partial^2 \psi}{\partial y^2} + \left(\frac{\alpha}{\beta} - y^2\right)\psi = 0 \tag{56}$$

The equation (56) can be solved asymptotically i.e. at very large values of $y$. Thus, when $y \gg \alpha/\beta$, the equation (56) becomes

$$\frac{\partial^2 \psi}{\partial y^2} - y^2 \psi = 0 \tag{57}$$

The two possible solutions of the above equation are

$$\psi = e^{\pm y^2/2} \tag{58}$$

Nevertheless, only one of them is acceptable because for $\psi = e^{+y^2/2}$, the wavefunction becomes infinite as y tends to approach $\infty$. Therefore, the only single-valued, continuous and finite solution we left with is

$$\psi = e^{-y^2/2} \tag{59}$$

Since the acceptable solution given above is valid only at very large values of $y$, it is quite reasonable to think that the exact solution may also contain some pre-exponential part to attain validity at all values of $y$. Therefore, after incorporating some $y$-dependent unknown function '$F(y)$' in equation (59), we get

$$\psi = F(y)\, e^{-y^2/2} \tag{60}$$

In order to find the value of $F(y)$, differentiate the equation (60) first i.e.

$$\frac{d\psi}{dy} = -ye^{-y^2/2}.F + \frac{dF}{dy}e^{-y^2/2} \tag{61}$$

Differentiating again, we get

$$\frac{d^2\psi}{dy^2} = \left[ -y.(-y)e^{-\frac{y^2}{2}}.F(y) + \left(-1.e^{-y^2/2}.F(y)\right) + \left(-y.e^{-y^2/2}\frac{dF}{dy}\right) \right] \tag{62}$$
$$+ \left[ -y.e^{-y^2/2}\frac{dF}{dy} + e^{-y^2/2}\frac{d^2F}{dy^2} \right]$$

or

$$\frac{d^2\psi}{dy^2} = y^2 e^{-y^2/2}.F(y) - e^{-y^2/2}.F(y) - 2y.e^{-y^2/2}\frac{dF}{dy} + e^{-y^2/2}\frac{d^2F}{dy^2} \tag{63}$$

or

$$\frac{d^2\psi}{dy^2} = \left[ \frac{d^2F}{dy^2} - 2y.\frac{dF}{dy} + (y^2 - 1)\,F \right] e^{-y^2/2} \tag{64}$$

Now, after using equation (60) and equation (64) in equation (56), we get

$$\left[ \frac{d^2F}{dy^2} - 2y.\frac{dF}{dy} + (y^2 - 1)\,F \right] e^{-y^2/2} + \left( \frac{\alpha}{\beta} - y^2 \right) F(y)\, e^{-y^2/2} = 0 \tag{65}$$

or

$$\left[ \frac{d^2F}{dy^2} - 2y.\frac{dF}{dy} + \left( \frac{\alpha}{\beta} - 1 \right) F(y) \right] e^{-y^2/2} = 0 \tag{66}$$

Now because of the quantity $e^{-y^2/2}$ will be zero only at $y = \pm\infty$, the sum of the terms present in the bracket must be zero at normal $y$-values i.e.

$$\frac{d^2F}{dy^2} - 2y.\frac{dF}{dy} + \left(\frac{\alpha}{\beta} - 1\right) F(y) = 0 \tag{67}$$

The above differential equation is a "Hermit differential equation" and can be solved to find the expression for the "unknown" function $F(y)$. The solution of equation (67) can be obtained by the polynomial method by expressing the function $F(y)$ as a power series in terms of variable '$y$'.

$$F = a_0 + a_1 y + a_2 y^2 + a_3 y^3 + a_4 y^4 \dots \dots \dots \dots \tag{68}$$

Differentiating the above equation, we get

$$\frac{dF}{dy} = a_1 + 2a_2 y + 3a_3 y^2 + 4a_4 y^3 \dots \dots \dots \dots \tag{69}$$

Differentiating again, we get

$$\frac{d^2F}{dy^2} = 2a_2 + 6a_3 y + 12a_4 y^2 \dots \dots \dots \dots \tag{70}$$

Using equation (68-70) in equation (67), we get

$$[2a_2 + 6a_3 y + 12a_4 y^2 \dots] - 2y[a_1 + 2a_2 y + 3a_3 y^2 + 4a_4 y^3 \dots] \tag{71}$$
$$+ \left(\frac{\alpha}{\beta} - 1\right) [a_0 + a_1 y + a_2 y^2 + a_3 y^3 + a_4 y^4 \dots] = 0$$

or

$$[2a_2 + 6a_3 y + 12a_4 y^2 \dots] - [2a_1 y + 4a_2 y^2 + 6a_3 y^3 + 8a_4 y^4 \dots] \tag{72}$$
$$+ \left[\left(\frac{\alpha}{\beta} - 1\right) a_0 + \left(\frac{\alpha}{\beta} - 1\right) a_1 y + \left(\frac{\alpha}{\beta} - 1\right) a_2 y^2 + \cdots\right] = 0$$

After further rearranging

$$\left[2a_2 + \left(\frac{\alpha}{\beta} - 1\right) a_0\right] + \left[6a_3 y - 2a_1 y + \left(\frac{\alpha}{\beta} - 1\right) a_1 y\right] \tag{73}$$
$$+ \left[12a_4 y^2 - 4a_2 y^2 + \left(\frac{\alpha}{\beta} - 1\right) a_2 y^2\right] + \cdots = 0$$

or

$$\left[2a_2 + \left(\frac{\alpha}{\beta} - 1\right) a_0\right] + \left[6a_3 - 2a_1 + \left(\frac{\alpha}{\beta} - 1\right) a_1\right] y \tag{74}$$
$$+ \left[12a_4 - 4a_2 + \left(\frac{\alpha}{\beta} - 1\right) a_2\right] y^2 + \cdots = 0$$

The above equation is valid only when coefficients of the individual power of $y$ are zero i.e.

For $y^0$

---

$$2a_2 + \left(\frac{\alpha}{\beta} - 1\right)a_0 = 0 \tag{75}$$

For $y^1$

$$6a_3 + \left(\frac{\alpha}{\beta} - 1 - 2\right)a_1 = 0 \tag{76}$$

For $y^2$

$$12a_4 + \left(\frac{\alpha}{\beta} - 1 - 4\right)a_2 = 0 \tag{77}$$

Similarly, for $y^k$

$$(k+1)(k+2)a_{k+2} + \left(\frac{\alpha}{\beta} - 1 - 2k\right)a_k = 0 \tag{78}$$

The above equation can be rearranged for the coefficient $a_{k+2}$ i.e.

$$a_{k+2} = -\frac{\left(\frac{\alpha}{\beta} - 1 - 2k\right)a_k}{(k+1)(k+2)} \tag{79}$$

Where $k$ is an integer. The expression given above is popularly known as the recursion formula, and allows one to determine the coefficient $a_{k+2}$ of the term $y^{k+2}$ in terms of $a_k$-coefficient of the $y^k$ term. In simple words, we can calculate $a_2$, $a_4$, $a_6$ etc. in terms of $a_0$ if we set $a_1 = 0$; likewise, the coefficients $a_3$, $a_5$, $a_7$ etc. can be obtained in terms of $a_1$ if we set $a_0 = 0$.

However, the power series will still be made up of the infinite number of terms, making function $F(y)$ infinite at $y = \infty$. Therefore, we must restrict the number of terms so that the function remains acceptable. This can be made possible if, at a certain value of $k = n$, the numerator in equation (79) becomes zero i.e.

$$\frac{\alpha}{\beta} - 1 - 2k = \frac{\alpha}{\beta} - 1 - 2n = 0 \tag{80}$$

or

$$\frac{\alpha}{\beta} = 2n + 1 \tag{81}$$

Where $n = 0, 1, 2, 3$ .... etc. The series that is obtained so contains a finite number of terms and is called as "Hermit polynomial" i.e. $H_n(y)$. All these Hermit polynomials are generating-function defined and are given below.

$$H_n(y) = (-1)^n . e^{y^2} . \frac{d^n}{dy^n} . e^{-y^2} \tag{82}$$

For instance, some of the Hermit polynomials for $n = 0$, 1 and 2 are calculated as given below.

For $n = 0$, the equation (82) becomes

$$H_0(y) = (-1)^0 . e^{y^2} . \frac{d^0}{dy^0} . e^{-y^2} \tag{83}$$

$$H_0(y) = 1 . e^{y^2} . 1 . e^{-y^2} = e^{y^2 - y^2} = e^0 \tag{84}$$

$$H_0(y) = 1 \tag{85}$$

For $n = 1$, the equation (82) becomes

$$H_1(y) = (-1)^1 . e^{y^2} . \frac{d}{dy} . e^{-y^2} \tag{86}$$

$$H_1(y) = (-1) . e^{y^2} . (-2y) . e^{-y^2} \tag{87}$$

$$H_1(y) = (-1) . (-2y) . e^{y^2 - y^2} = 2y . e^0 \tag{88}$$

$$H_1(y) = 2y \tag{89}$$

For $n = 2$, the equation (82) becomes

$$H_2(y) = (-1)^2 . e^{y^2} . \frac{d^2}{dy^2} . e^{-y^2} \tag{90}$$

$$H_2(y) = (+1) . e^{y^2} . \frac{d}{dy} . -2y . e^{-y^2} \tag{91}$$

$$H_2(y) = e^{y^2} \left[ (-2y)e^{-y^2}(-2y) + (-2)e^{-y^2} \right] \tag{92}$$

$$H_2(y) = e^{y^2} \left[ (4y^2 - 2)e^{-y^2} \right] \tag{93}$$

$$H_2(y) = (4y^2 - 2)e^{y^2 - y^2} = (4y^2 - 2)e^0 \tag{94}$$

$$H_2(y) = 4y^2 - 2 \tag{95}$$

**The total wavefunction:** After knowing the unknown part $F(y)$, the complete eigenfunction for a simple harmonic oscillator in the quantum mechanical world can be written as

$$\psi_n(y) = N_n H_n(y)\, e^{-y^2/2} \tag{96}$$

Where $N_n$ is the normalization constant for $n$th state while the symbol $H_n(y)$ represents the Hermit polynomial of $n$th order in terms of $y$-variable. Once the wavefunctions are obtained in terms of $y$, they can easily be converted into $x$-dependent function by simply putting $y = \sqrt{\beta}x$.

Table 1. Eigenfunctions representing various quantum mechanical states of a simple harmonic oscillator.

| $\alpha/\beta$ | $n$ | $H_n(y)$ | $\psi_n(y)$ | $\psi_n(x)$ |
|---|---|---|---|---|
| 1 | 0 | 1 | $N_0 . e^{-y^2/2}$ | $N_0 . e^{-\beta x^2/2}$ |
| 3 | 1 | $2y$ | $N_1 . 2y . e^{-y^2/2}$ | $N_1(2\sqrt{\beta}x)e^{-\beta x^2/2}$ |
| 5 | 2 | $4y^2 - 2$ | $N_2(4y^2 - 2)e^{-y^2/2}$ | $N_2(4\beta x^2 - 2)e^{-\beta x^2/2}$ |
| 7 | 3 | $8y^3 - 12y$ | $N_3(8y^3 - 12y)e^{-y^2/2}$ | $N_3(8\beta^{3/2}x^3 - 12\sqrt{\beta}x)e^{-\beta x^2/2}$ |

The normalization constant can be obtained by recalling the fact that every wave function must describe the corresponding state completely. This means that square of wave function under consideration over the whole configurational space must be equal to unity i.e.

$$\int_{-\infty}^{+\infty} \psi_n^2 \, dx = \int_{-\infty}^{+\infty} \left[ N_n H_n(x) \, e^{-\frac{\beta x^2}{2}} \right]^2 = 1 \tag{97}$$

$$\frac{N_n^2}{\sqrt{\beta}} 2^n \, n! \, \sqrt{\pi} = 1 \tag{98}$$

$$N_n^2 = \frac{\sqrt{\beta}}{2^n \, n! \, \sqrt{\pi}} \tag{99}$$

$$N_n = \left( \frac{\sqrt{\beta}}{2^n \, n! \, \sqrt{\pi}} \right)^{1/2} \tag{100}$$

It can be clearly seen from the above equation that the normalization consents are different for different states. For instance, some of the normalization constants are given below.

$$N_0 = \left( \frac{\sqrt{\beta}}{2^0 \, 0! \, \sqrt{\pi}} \right)^{1/2} = \left( \frac{\beta}{\pi} \right)^{1/4} \tag{101}$$

$$N_1 = \left( \frac{\sqrt{\beta}}{2^1 \, 1! \, \sqrt{\pi}} \right)^{1/2} = \frac{1}{\sqrt{2}} \left( \frac{\beta}{\pi} \right)^{1/4} \tag{102}$$

$$N_2 = \left( \frac{\sqrt{\beta}}{2^2 \, 2! \, \sqrt{\pi}} \right)^{1/2} = \frac{1}{2\sqrt{2}} \left( \frac{\beta}{\pi} \right)^{1/4} \tag{103}$$

**The eigenvalues of energy:** Since we have already proved that the total wavefunction for a simple harmonic oscillator is acceptable only when the following condition of equation (81) is satisfied i.e.

$$\frac{\alpha}{\beta} = 2n + 1 \tag{104}$$

Furthermore, we also know that

$$\alpha = \frac{8\pi^2 mE}{h^2} \quad and \quad \beta = \frac{4\pi^2 mv}{h} \tag{105}$$

After using the value of $\alpha$ and $\beta$, the equation (104) take the form

$$\frac{8\pi^2 mE}{h^2} \times \frac{h}{4\pi^2 mv} = 2n + 1 \tag{106}$$

or

$$E = (2n + 1)\frac{4\pi^2 mv}{h} \times \frac{h^2}{8\pi^2 m} \tag{107}$$

or

$$E_n = (2n + 1)\frac{hv}{2} \tag{108}$$

or

$$E_n = \left(n + \frac{1}{2}\right) hv \tag{109}$$

Where $n$ is a discrete variable (vibrational quantum number) with values 0, 1, 2, 3…∞. The symbol $E_n$ represents the vibrational energies of different vibrational states.

> ### *The Classical and Quantum-Mechanical Interpretation of Vibrational States*

In order to have a relative interpretation of vibrational states in the classical and quantum-mechanical framework, recall the classical expression for potential energy curve of simple harmonic oscillator i.e.

$$V = \frac{1}{2} kx^2 \tag{110}$$

Where $k$ is the force constant and $x$ is the displacement induced. Also, the general expressions for all the vibrational states and corresponding energies are

$$\psi_n(x) = N_n H_n(x)\, e^{-\beta x^2} \quad and \quad E_n = \left(n + \frac{1}{2}\right) hv \tag{111}$$

Where $N_n$ and $H_n(x)$ are the normalization constant and Hermit polynomial for $n$th state.

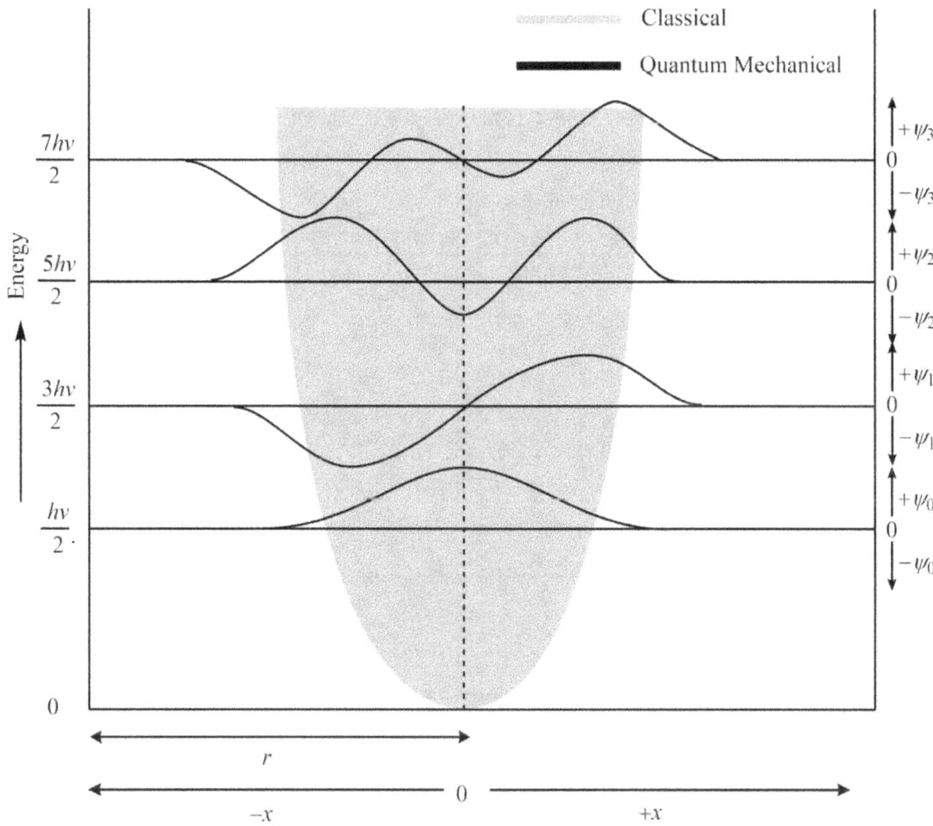


Figure 4. The depiction of various vibrational states of a simple harmonic oscillator in the classical and quantum mechanical framework.

After looking at the figure given above, the following points can be made about the differences and similarities in the classical and quantum mechanical oscillators.

*i*) It can be clearly seen that the energy levels of a classical oscillator are continuous including zero while the energy levels the quantum-mechanical analog is discontinuous with zero-point energy of $hv/2$. In other words, the classical oscillator can have zero vibrational energy but the vibrational motion cannot be ceased completely in case of the quantum mechanical version of the simple harmonic oscillator.

*ii*) There is always a limit over the compression as well as over the expansion in the classical oscillator to have a certain amount of energy. On the other hand, since the function becomes zero only at infinite displacement, there is no limit over the compression and expansion in the quantum oscillator theoretically.

*iii*) The classical oscillator spends more time in the extreme states i.e. fully compressed and fully expended, and spends the least time with equilibrium bond length. As far as the ground vibrational state of the quantum mechanical oscillator is concerned, it spends most time with equilibrium (because the function is maximum for $r_{equ}$ or $x = 0$), and probability to spend time in compressed and expended mode decreases as the magnitude of compression and expansion increases.

*iv*) If we plot the square of wavefunction vs displacement incorporated, it can be clearly seen that the most probable bond lengths shift towards the compressed and expanded states as the quantum number increases which is in accordance with the Bohr's correspondence principle.

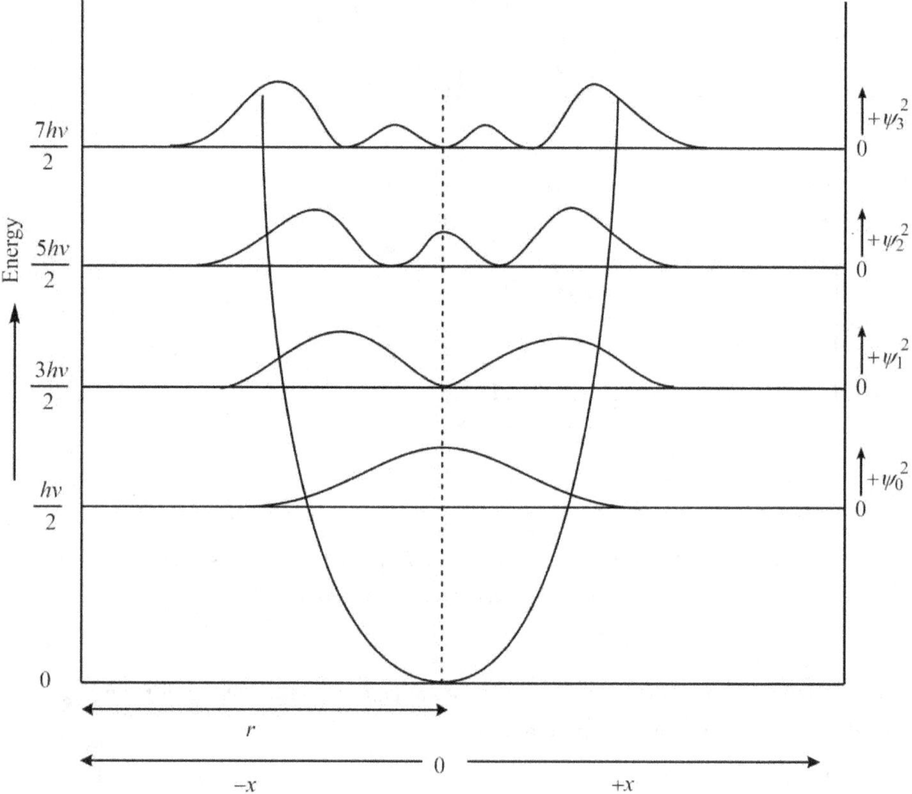


Figure 5. The variation of probability as a function of bond length or displacement in different vibrational states of a simple harmonic oscillator.

Furthermore, it also worthy to note that all $\psi_{even}$ states are symmetric while all $\psi_{odd}$ wavefunctions are asymmetric in nature with $n$ nodes.

## ❖ Zero Point Energy of a Particle Possessing Harmonic Motion and Its Consequence

In order to understand the minimum or the zero-point energy of a simple harmonic oscillator, recall the general wavefunction representing all the vibrational states of a simple harmonic oscillator i.e.

$$\psi_n(y) = N_n H_n(y)\, e^{-y^2/2} \tag{112}$$

Where $y$ is a displacement-based variable with a value equal to $\sqrt{\beta}x$. The constant $\beta$ depends upon the reduced mass of the oscillator $(m)$ and equilibrium vibrational frequency $(v)$ as

$$\beta = \frac{4\pi^2 m v}{h} \tag{113}$$

The symbol $N_n$ and $H_n(y)$ are the normalization constant and Hermit polynomial for $n$th state i.e.

$$N_n = \left(\frac{\sqrt{\beta}}{2^n\, n!\, \sqrt{\pi}}\right)^{1/2} \quad and \quad H_n(y) = (-1)^n. e^{y^2}. \frac{d^n}{dy^n}. e^{-y^2} \tag{114}$$

Also, the general expression for the energies is given below.

$$E_n = \left(n + \frac{1}{2}\right) hv \tag{115}$$

Now, for the ground vibrational state $(n = 0)$, $N_0$ and $H_0(y)$ can be obtained from equation (114) i.e.

$$N_0 = \left(\frac{\beta}{\pi}\right)^{1/4} \quad and \quad H_0(y) = 1 \tag{116}$$

After using the values of $y$, $N_0$ and $H_0(y)$ in equation (112), the ground state function becomes

$$\psi_0(x) = \left(\frac{\beta}{\pi}\right)^{1/4} 1\, e^{-\beta x^2/2} \tag{117}$$

Hence, the ground state wave function does not collapse at $n = 0$, which means that corresponding energy can also be obtained by putting $n = 0$ in equation (115) i.e.

$$E_0 = \left(0 + \frac{1}{2}\right) hv \tag{118}$$

or

$$E_0 = \frac{1}{2} hv \tag{119}$$

The above equation gives the minimum energy which is always possessed by a simple harmonic oscillator.

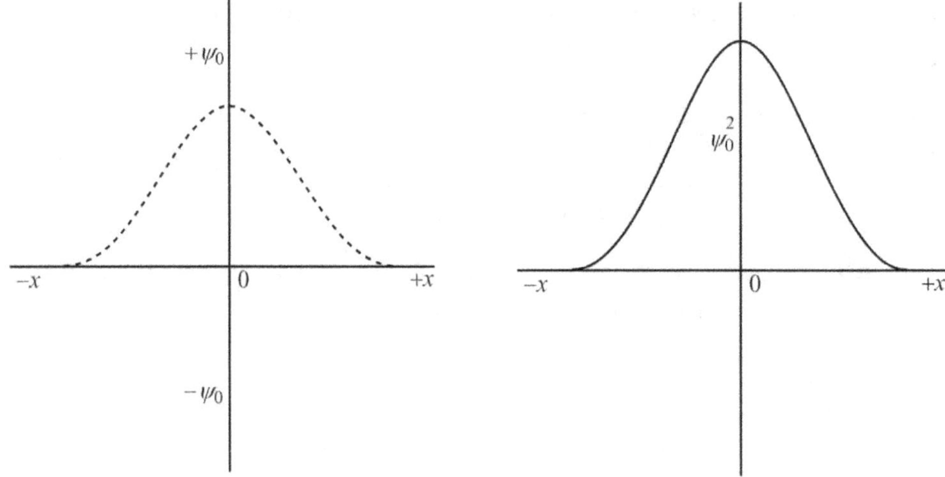

Figure 6. The variation of ground vibrational wavefunction and probability as a function of bond length or displacement in a simple harmonic oscillator.

It is well-known that the classical oscillator spends more time in the extreme states i.e. fully compressed and fully expended, and spends the least time with equilibrium bond length. However, as far as the ground vibrational state of the quantum mechanical oscillator is concerned, it spends the most time with equilibrium (because the function is maximum for $r_{equ}$ or $x = 0$), and probability to spend time in compressed and expended mode decreases as the magnitude of compression and expansion increases. Moreover, there is always a limit over the compression as well as over the expansion in the classical oscillator to have a certain amount of energy; however, since the function becomes zero only at infinite displacement, there is no limit over the compression and expansion in the quantum oscillator theoretically.

It is also worthy to note that the energy given by the equation (119) is in joules. However, in many textbooks or papers, it is also reported in terms of wavenumbers. To do so, we need first put the value of frequency as $v = c/\lambda$ and then $1/\lambda = \bar{v}$ in the equation (119) i.e.

$$E_0 = \frac{1}{2}h\frac{c}{\lambda} = \frac{1}{2}hc\bar{v} \tag{119}$$

Where $c$ is the velocity of light. Now, to convert the zero-point energy in wavenumbers, divide equation (119) by $hc$ i.e.

$$\bar{v}_0 = \frac{1}{2}\frac{hc\bar{v}}{hc} = \frac{\bar{v}}{2} \tag{120}$$

## ❖ Schrodinger Wave Equation for Three Dimensional Rigid Rotator

In order to study the rotational behavior of a diatomic molecule, consider a system two masses $m_1$ and $m_2$ joined by a rigid rod of length "$r$". Now assume that this dumbbell type geometry rotates about an axis that is perpendicular to $r$ and passes through the center of mass.

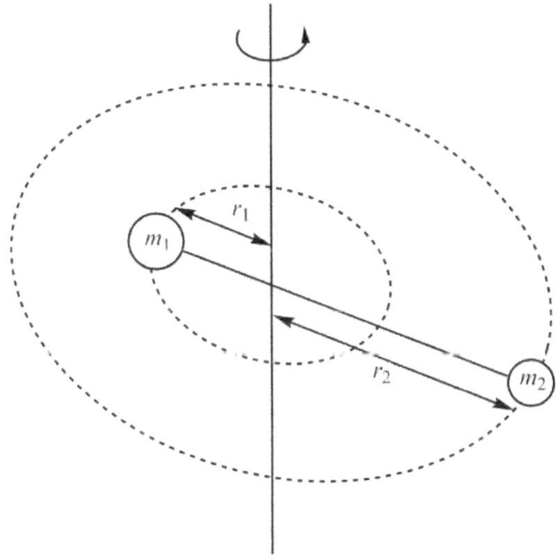


Figure 7. The pictorial representation of the diatomic rigid rotator in classical mechanics.

If $v_1$ and $v_2$ are the velocities of the mass $m_1$ and $m_2$ revolving about the axis of rotation, the total kinetic energy ($T$) of the rotator can be given by the following relation.

$$T = \frac{1}{2}m_1 v_1^2 + \frac{1}{2}m_2 v_2^2 \tag{121}$$

Since we know that linear velocity $v$ is simply equal to the angular velocity $\omega$ multiplied by the radius of rotation $r$ i.e. $v = \omega r$, the equation (121) takes the form

$$T = \frac{1}{2}m_1(r_1\omega)^2 + \frac{1}{2}m_2(r_2\omega)^2 \tag{122}$$

$$T = \frac{1}{2}(m_1 r_1^2 + m_2 r_2^2)\omega^2 \tag{123}$$

$$T = \frac{1}{2}I\omega^2 \tag{123}$$

Where $I$ is the moment of inertia with definition $I = \sum m_i r_i^2$. Furthermore, we know from mass-center that

 **DALAL INSTITUTE**

$$m_1 r_1 = m_2 r_2 \tag{124}$$

Now since $r = r_1 + r_2$, we rearrange equation (124) to give

$$r_1 = \frac{m_2}{m_1 + m_2} r \quad and \quad r_2 = \frac{m_1}{m_1 + m_2} r \tag{125}$$

In the two-mass system $I = m_1 r_1^2 + m_2 r_2^2$, so have

$$I = m_1 \left(\frac{m_2}{m_1 + m_2} r\right)^2 + m_2 \left(\frac{m_1}{m_1 + m_2} r\right)^2 \tag{126}$$

$$I = \left(\frac{m_1 m_2}{m_1 + m_2}\right) r^2 \tag{127}$$

$$I = \mu r^2 \tag{128}$$

Where $\mu = m_1 m_2 / m_1 + m_2$ is the reduced mass of the rigid diatomic system. Since we that the kinetic energy and linear moment of a particle of mass $m$ moving with velocity $v$ are

$$T = \frac{1}{2} m v^2 \quad and \quad p = mv \tag{129}$$

The counterparts in the angular motion can be written as

$$T = \frac{1}{2} I \omega^2 \quad and \quad L = I \omega \tag{130}$$

Multiplying and dividing the rotational kinetic energy by $I$, we have

$$T = \frac{I^2 \omega^2}{2I} = \frac{(I\omega)^2}{2I} = \frac{L^2}{2I} \tag{131}$$

It is clear from the above equation that the kinetic energy of a classical rotator can have any value because the value-domain of angular velocity is continuous. Moreover, as no external force is working on the rotator, the potential can be set to zero. In other words, the Hamiltonian for diatomic rigid rotator can be given as

$$\hat{H} = \hat{T} + \hat{V} \tag{132}$$

$$\hat{H} = \frac{\hat{L}^2}{2I} + 0 \tag{133}$$

The expression for the operator $\hat{L}^2$ in polar coordinates is

$$L^2 = -\frac{h^2}{4\pi^2} \left[\frac{1}{Sin\,\theta} \frac{\partial}{\partial \theta} \left(Sin\,\theta \frac{\partial}{\partial \theta}\right) + \frac{1}{Sin^2\,\theta} \frac{\partial}{\partial \phi}\right] \tag{134}$$

Using equation (134) in equation (133), the Hamiltonian operator takes the form

$$\widehat{H} = -\frac{h^2}{8\pi^2 I}\left[\frac{1}{Sin\,\theta}\frac{\partial}{\partial\theta}\left(Sin\,\theta\frac{\partial}{\partial\theta}\right) + \frac{1}{Sin^2\,\theta}\frac{\partial}{\partial\phi}\right] + 0 \tag{135}$$

Now, let $\psi$ be the function that describes all the rotational states of the diatomic rigid rotator. The operation of Hamiltonian operator over $\psi$ can be rearranged to give to construct the Schrodinger wave equation; and we all know that the wave function as well the energy, both are the obtained as this second-order differential equation is solved. Mathematically, we can say that

$$\widehat{H}\psi = E\psi \tag{136}$$

After putting the expression of the Hamiltonian operator from equation (135) in equation (136), we get

$$-\frac{h^2}{8\pi^2 I}\left[\frac{1}{Sin\,\theta}\frac{\partial}{\partial\theta}\left(Sin\,\theta\frac{\partial}{\partial\theta}\right) + \frac{1}{Sin^2\,\theta}\frac{\partial}{\partial\phi}\right]\psi = E\psi \tag{137}$$

or

$$-\frac{h^2}{8\pi^2 I}\left[\frac{1}{Sin\,\theta}\frac{\partial}{\partial\theta}\left(Sin\,\theta\frac{\partial\psi}{\partial\theta}\right) + \frac{1}{Sin^2\,\theta}\frac{\partial\psi}{\partial\phi}\right] = E\psi \tag{138}$$

or

$$\frac{1}{Sin\,\theta}\frac{\partial}{\partial\theta}\left(Sin\,\theta\frac{\partial\psi}{\partial\theta}\right) + \frac{1}{Sin^2\,\theta}\frac{\partial\psi}{\partial\phi} = -\frac{8\pi^2 IE\psi}{h^2} \tag{139}$$

or

$$\frac{1}{Sin\,\theta}\frac{\partial}{\partial\theta}\left(Sin\,\theta\frac{\partial\psi}{\partial\theta}\right) + \frac{1}{Sin^2\,\theta}\frac{\partial\psi}{\partial\phi} + \frac{8\pi^2 IE\psi}{h^2} = 0 \tag{140}$$

The above differential equation contains two variable $\phi$ and $\theta$, and therefore, is difficult to solve. Thus, we need to use the same mathematical technique we used to study particle in a 3-dimensional box i.e. the separation of variables. To do so, consider the total wavefunction as the product of two independent, one $\theta$ dependent and other as a $\phi$ dependent function only i.e.

$$\psi(\theta,\phi) = \psi(\theta) \times \psi(\phi) = \Theta \times \Phi \tag{141}$$

After putting the value of equation (141) in equation (140), we get

$$\frac{1}{Sin\,\theta}\frac{\partial}{\partial\theta}\left(Sin\,\theta\frac{\partial\Theta\Phi}{\partial\theta}\right) + \frac{1}{Sin^2\,\theta}\frac{\partial\Theta\Phi}{\partial\phi} + \frac{8\pi^2 IE\Theta\Phi}{h^2} = 0 \tag{142}$$

Since the first and second terms contain the partial derivatives w.r.t. $\theta$ and $\phi$, respectively; function $\Phi$ and $\Theta$ must be kept constant correspondingly, i.e.,

$$\Phi \frac{1}{Sin\,\theta} \frac{\partial}{\partial\theta}\left(Sin\,\theta\,\frac{\partial\Theta}{\partial\theta}\right) + \Theta\,\frac{1}{Sin^2\,\theta}\frac{\partial\Phi}{\partial\phi} + \frac{8\pi^2 I E \Theta \Phi}{h^2} = 0 \qquad (143)$$

Dividing the above equation by $\Theta\Phi$, the equation (143) takes the form

$$\frac{1}{\Theta}\frac{1}{Sin\,\theta}\frac{\partial}{\partial\theta}\left(Sin\,\theta\,\frac{\partial\Theta}{\partial\theta}\right) + \frac{1}{\Phi}\frac{1}{Sin^2\,\theta}\frac{\partial\Phi}{\partial\phi} + \frac{8\pi^2 I E}{h^2} = 0 \qquad (144)$$

After multiplying equation (144) by $Sin^2\theta$, we get

$$\frac{Sin\,\theta}{\Theta}\frac{\partial}{\partial\theta}\left(Sin\,\theta\,\frac{\partial\Theta}{\partial\theta}\right) + \frac{1}{\Phi}\frac{\partial\Phi}{\partial\phi} + \frac{8\pi^2 I E}{h^2}Sin^2\theta = 0 \qquad (145)$$

Rearranging

$$\frac{Sin\,\theta}{\Theta}\frac{\partial}{\partial\theta}\left(Sin\,\theta\,\frac{\partial\Theta}{\partial\theta}\right) + \frac{8\pi^2 I E}{h^2}Sin^2\theta = -\frac{1}{\Phi}\frac{\partial\Phi}{\partial\phi} \qquad (146)$$

At this point, we can set both sides equal to constant $m^2$ i.e.

$$\frac{Sin\,\theta}{\Theta}\frac{\partial}{\partial\theta}\left(Sin\,\theta\,\frac{\partial\Theta}{\partial\theta}\right) + \frac{8\pi^2 I E}{h^2}Sin^2\theta = m^2 = -\frac{1}{\Phi}\frac{\partial\Phi}{\partial\phi} \qquad (147)$$

The equation (147) can be fragmented into two equations, each containing a single variable i.e.

$$\frac{\partial\Phi}{\partial\phi} + m^2\Phi = 0 \qquad (148)$$

And

$$Sin\,\theta\,\frac{\partial}{\partial\theta}\left(Sin\,\theta\,\frac{\partial\Theta}{\partial\theta}\right) + \Theta\,\frac{8\pi^2 I E}{h^2}Sin^2\theta - m^2\Theta = 0 \qquad (149)$$

Now dividing the above equation by $Sin^2\theta$, we get

$$\frac{1}{Sin\,\theta}\frac{\partial}{\partial\theta}\left(Sin\,\theta\,\frac{\partial\Theta}{\partial\theta}\right) + \Theta\,\frac{8\pi^2 I E}{h^2} - \frac{m^2\Theta}{Sin^2\theta} = 0 \qquad (150)$$

or

$$\frac{1}{Sin\,\theta}\frac{\partial}{\partial\theta}\left(Sin\,\theta\,\frac{\partial\Theta}{\partial\theta}\right) + \left(\frac{8\pi^2 I E}{h^2} - \frac{m^2}{Sin^2\theta}\right)\Theta = 0 \qquad (151)$$

or

$$\frac{1}{Sin\,\theta}\frac{\partial}{\partial\theta}\left(Sin\,\theta\,\frac{\partial\Theta}{\partial\theta}\right) + \left(\beta - \frac{m^2}{Sin^2\theta}\right)\Theta = 0 \qquad (152)$$

Where the constant $\beta$ is defined as

$$\beta = \frac{8\pi^2 IE}{h^2} \tag{153}$$

**The solution of Φ equation:** Recall the differential equation obtained after separation of variables having $\phi$ dependence i.e.

$$\frac{\partial \Phi}{\partial \phi} + m^2 \Phi = 0 \tag{154}$$

The general solution of such an equation is

$$\Phi(\phi) = N e^{im\phi} \tag{155}$$

Where $N$ represents the normalization constant. The wavefunction given above will be acceptable only if $m$ has integer value i.e. $0, \pm 1, \pm 2$, etc. This can be understood in terms of single-valued, continuous and finite nature of quantum states.

*i) The boundary condition for function Φ:* If we replace the angle "$\phi$" with "$\phi + 2\pi$", the position of the point under consideration should remain the same i.e.

$$\Phi(\phi + 2\pi) = \Phi(\phi) \tag{156}$$

Therefore

$$N e^{im(\phi + 2\pi)} = N e^{im\phi} \tag{157}$$

or

$$e^{im(\phi + 2\pi)} = e^{im\phi} \tag{158}$$

$$e^{im\phi} . e^{im2\pi} = e^{im\phi} \tag{159}$$

$$e^{im2\pi} = e^{im\phi} e^{-im\phi} \tag{160}$$

$$e^{im2\pi} = e^{im\phi - im\phi} = e^0 \tag{161}$$

$$e^{im2\pi} = 1 \tag{162}$$

Since we know from the Euler's expansion $e^{ix} = Cos\, x + i\, Sin\, x$, the equation (162) takes the form

$$e^{im2\pi} = Cos\, 2\pi m + i\, Sin\, 2\pi m \tag{163}$$

After putting the value of equation (163) in equation (162), we get

$$Cos\, 2\pi m + i\, Sin\, 2\pi m = 1 \tag{164}$$

The relation holds true only when we use $m = 0, \pm 1, \pm 2, \pm 3, \pm 4$, etc.

Copyright © *Mandeep Dalal*

*ii) The normalization constant for function $\Phi$:* In order to determine the normalization constant for the $\Phi$ function, we must put the squared-integral over whole configuration space as unity i.e.

$$\int_{0}^{2\pi} \Phi^*\Phi \, d\phi = 1 \tag{165}$$

or

$$N^2 \int_{0}^{2\pi} e^{im\phi} . e^{-im\phi} \, d\phi = 1 \tag{166}$$

$$N^2 \int_{0}^{2\pi} e^{im\phi - im\phi} \, d\phi = N^2 \int_{0}^{2\pi} e^{0} \, d\phi = 1 \tag{167}$$

$$N^2 [\phi]_{0}^{2\pi} = N^2 [2\pi] = 1 \tag{168}$$

$$N = \sqrt{\frac{1}{2\pi}} \tag{169}$$

After using the value of normalization constant in equation (155), we get

$$\Phi(\phi) = \sqrt{\frac{1}{2\pi}} e^{\pm im\phi} \tag{170}$$

**Solution of $\Theta$ equation:** Recall the differential equation obtained after separation of variables having $\theta$ dependence i.e.

$$\frac{1}{Sin\,\theta} \frac{\partial}{\partial\theta} \left( Sin\,\theta \, \frac{\partial\Theta}{\partial\theta} \right) + \left( \beta - \frac{m^2}{Sin^2\theta} \right) \Theta = 0 \tag{171}$$

After defining a new variable $x = Cos\,\theta$, we have

$$Sin^2\theta + Cos^2\theta = 1 \tag{172}$$

$$Sin^2\theta = 1 - Cos^2\theta \tag{173}$$

$$Sin\,\theta = \sqrt{1 - Cos^2\theta} \tag{174}$$

$$Sin\,\theta = \sqrt{1 - x^2} \tag{175}$$

Also, since we assumed $x = Cos\,\theta$, the first derivative w.r.t. $\theta$ will be

$$\frac{\partial x}{\partial \theta} = -Sin\ \theta \tag{176}$$

The derivative of $\Theta$ function w.r.t. $\theta$ can be rewritten as

$$\frac{\partial \Theta}{\partial \theta} = \frac{\partial \Theta}{\partial x}\cdot\frac{\partial x}{\partial \theta} \tag{177}$$

After putting the values of $\partial x/\partial \theta$ from equation (176) in equation (177), we get

$$\frac{\partial \Theta}{\partial \theta} = -Sin\ \theta\,\frac{\partial \Theta}{\partial x} \tag{178}$$

After removing $\Theta$ from both sides

$$\frac{\partial}{\partial \theta} = -Sin\ \theta\,\frac{\partial}{\partial x} \tag{179}$$

Multiplying both sides of equation (178) by $Sin\ \theta$, we have

$$Sin\ \theta\,\frac{\partial \Theta}{\partial \theta} = -Sin^2\ \theta\,\frac{\partial \Theta}{\partial x} \tag{180}$$

$$Sin\ \theta\,\frac{\partial \Theta}{\partial \theta} = -(1 - x^2)\frac{\partial \Theta}{\partial x} \tag{181}$$

Now, after putting the values of equation (179) and (181) in equation (171), we get

$$\frac{1}{Sin\ \theta}\left(-Sin\ \theta\,\frac{\partial}{\partial x}\right)\left[-(1 - x^2)\frac{\partial \Theta}{\partial x}\right] + \left(\beta - \frac{m^2}{1 - x^2}\right)\Theta = 0 \tag{182}$$

or

$$\frac{\partial}{\partial x}\left[(1 - x^2)\frac{\partial \Theta}{\partial x}\right] + \left(\beta - \frac{m^2}{1 - x^2}\right)\Theta = 0 \tag{183}$$

The equation given above is a Legendre's polynomial and has physical significance only in the range of $x$ = $-1$ to $+1$. Therefore, consider that one more form of $\Theta$ function so that this condition is satisfied i.e.

$$\Theta(\theta) = (1 - x^2)^{\frac{m}{2}}.X(x) \tag{184}$$

Where $X$ is a function depending upon variable $x$. The differentiation of the above equation w.r.t. $x$ yields

$$\frac{\partial \Theta}{\partial x} = -mx(1 - x^2)^{\frac{m}{2}-1}.X + (1 - x^2)^{\frac{m}{2}}.\frac{dX}{dx} \tag{185}$$

After multiplying the above equation by $1 - x^2$ and $\partial/\partial x$, we get

$$\frac{\partial}{\partial x}\left[(1-x^2)\frac{\partial\Theta}{\partial x}\right] = \frac{\partial}{\partial x}\left[-mx(1-x^2)^{\frac{m}{2}}.X + (1-x^2)^{\frac{m}{2}+1}.\frac{dX}{dx}\right] \tag{186}$$

$$= \left[-m(1-x^2)^{m/2} + m^2x^2(1-x^2)^{\frac{m}{2}-1}\right]X - \left[2x(m+1)(1-x^2)^{\frac{m}{2}}\right]X' \tag{187}$$

$$+ \left[(1-x^2)^{\frac{m}{2}+1}\right]X''$$

Where $\partial/\partial x$ and $\partial^2/\partial x^2$ are represented by the symbol $X'$ and $X''$, respectively. Now, after using the value of equation (184) and equation (187) in equation (183), we get

$$\left[-m(1-x^2)^{m/2} + m^2x^2(1-x^2)^{\frac{m}{2}-1}\right]X - \left[2x(m+1)(1-x^2)^{\frac{m}{2}}\right]X' \tag{188}$$

$$+ \left[(1-x^2)^{\frac{m}{2}+1}\right]X'' + \left(\beta - \frac{m^2}{1-x^2}\right)(1-x^2)^{\frac{m}{2}}.X = 0$$

Dividing above expression by $(1-x^2)^{m/2}$, we have

$$(1-x^2)X'' - 2(m+1)xX' + [\beta - m(m+1)]X = 0 \tag{189}$$

or

$$(1-x^2)X'' - 2\alpha xX' + \lambda X = 0 \tag{190}$$

Where $\alpha = m + 1$ and $\lambda = \beta - m(m+1)$. Now assume that the function $X$ can be expressed as a power series expansion as given below.

$$X = a_0 + a_1x + a_2x^2 + a_3x^3 \dots\dots\dots\dots \tag{191}$$

$$X' = a_1 + 2a_2x + 3a_3x^2 \dots\dots\dots\dots \tag{192}$$

$$X'' = 2a_2 + 6a_3x + 12a_4x^2 \dots\dots\dots\dots \tag{193}$$

Putting values of equation (191-193) in equation (190), we get

$$(1-x^2)(2a_2 + 6a_3x + 12a_4x^2 + 20a_5x^3) - 2\alpha x(a_1 + 2a_2x + 3a_3x^2 + 4a_4x^3) \tag{194}$$

$$+ \lambda(a_0 + a_1x + a_2x^2 + a_3x^3) = 0$$

or

$$(2a_2 + \lambda a_0) + [6a_3 + (\lambda - 2\alpha)a_1]x + [12a_4 + (\lambda - 2\alpha - 2)a_2]x^2 \dots\dots\dots = 0 \tag{195}$$

The above equation is satisfied only if each term on the left-hand side is individually equal to zero i.e. coefficients of each power of $x$ are vanish. The general expression for the coefficients must follow the condition given below.

$$(n+1)(n+2)a_{n+2} + [\lambda - 2n\alpha - n(n-1)]a_n = 0 \tag{196}$$

Where $n = 0, 1, 2, 3$ etc. Summarizing the result, we can write

$$a_{n+2} = \frac{2n\alpha + n(n-1) - \lambda}{(n+1)(n+2)} a_n \tag{197}$$

After putting values of $\alpha$ and $\lambda$ in equation (197), we get

$$\frac{a_{n+2}}{a_n} = \frac{(n+m)(n+m+1) - \beta}{(n+1)(n+2)} \tag{198}$$

Which is the Recursion formula for the coefficients of the power of $x$. Now, in order to obtain a valid wavefunction, the power series must contain a finite number of terms which is possible only if numerator becomes zero i.e.

$$(n+m)(n+m+1) - \beta = 0 \tag{199}$$

$$\beta = (n+m)(n+m+1) \tag{200}$$

Since we know that $m$ as well $n$ both are the whole numbers, their sum must also be a whole number. Therefore, the sum of $n$ and $m$ can be replaced by another whole number symbolized by $l$ i.e.

$$\beta = l(l+1) \tag{201}$$

Where $l = 0, 1, 2, 3$ etc. After putting the value of $\beta$ from equation (201) in equation (183), we get

$$\frac{\partial}{\partial x}\left[(1 - x^2)\frac{\partial \Theta}{\partial x}\right] + \left[l(l+1) - \frac{m^2}{1 - x^2}\right]\Theta = 0 \tag{202}$$

The general solution of equation (202) is

$$\Theta = NP_l^m(x) = NP_l^m(Cos\,\theta) \tag{203}$$

Where N is the normalization constant and $P_l^m(x)$ is the associated "Legendre function" which is defined as given below.

$$P_l^m(x) = (1 - x^2)^{m/2}\frac{d^m P_l(x)}{dx^m} \tag{204}$$

Where $P_l(x)$ is the Legendre polynomial given by

$$P_l(x) = \frac{1}{2^l\, l!}\frac{d^l(x^2 - 1)^l}{dx^l} \tag{205}$$

In order to proceed further, we must discuss the concept of orthogonality and the normalization of the "Legendre's function".

*i) Orthogonality of associated Legendre's function:* The orthogonality of the associated Legendre's polynomial follows the conditions given below.

$$\int_{-1}^{+1} P_k^m(x)\, P_l^m(x) = 0 \qquad if\ k \neq l \tag{206}$$

$$\int_{-1}^{+1} P_k^m(x)\, P_l^m(x) = \frac{2}{(2l+1)}\frac{(l+m)!}{(l-m)!} \qquad if\ k = l \tag{207}$$

*ii) Normalization of associated Legendre's function:* The normalization of the associated Legendre's polynomial follows the conditions given below.

$$\int_{-1}^{+1} \Theta_{m,l}\, \Theta_{m,l}^{*}(d\theta) = 1 \tag{208}$$

$$N^2 \int_{-1}^{+1} P_k^m(x)\, P_l^m(x)\, dx = 1 \tag{209}$$

After solving the integral, we get

$$N^2 \cdot \frac{2}{(2l+1)}\frac{(l+m)!}{(l-m)!} = 1 \tag{210}$$

$$N = \sqrt{\frac{(2l+1)(l-m)!}{2(l+m)!}} \tag{211}$$

Using the value of normalization constant in equation (203), we get

$$\Theta(\theta) = \sqrt{\frac{(2l+1)(l-m)!}{2(l+m)!}} \cdot P_l^m(Cos\ \theta) \tag{212}$$

**The complete eigenfunction of rigid rotator:** The total eigenfunction for the rigid rotator now can be obtained by simply multiplying the solution of $\phi$-dependent and $\theta$-dependent differential equations i.e. equation (170) and equation (203).

$$\psi_{l,m}(\theta,\phi) = \Theta_{l,m}(\theta)\Phi_m(\phi) = \sqrt{\frac{(2l+1)(l-m)!}{2(l+m)!}} \cdot P_l^m(Cos\ \theta) \cdot \sqrt{\frac{1}{2\pi}}\, e^{\pm im\phi} \tag{213}$$

$$\psi_{l,m}(\theta,\phi) = \sqrt{\frac{1}{2\pi}}\sqrt{\frac{(2l+1)(l-m)!}{2(l+m)!}} \cdot P_l^m(Cos\ \theta) \cdot e^{\pm im\phi} \tag{214}$$

## ❖ Energy of Rigid Rotator

The energy of a rigid rotator can be understood only after considering its classical and quantum mechanical aspects. In the previous section of this chapter, we discussed the classical and quantum mechanical nature of the rigid rotator. consider a system two masses $m_1$ and $m_2$ joined by a rigid rod of length "$r$". Now assume that this dumbbell type geometry rotates about an axis that is perpendicular to $r$ and passes through the center of mass.

> ### The energy of Classical Rigid Rotator

If $v_1$ and $v_2$ are the velocities of the mass $m_1$ and $m_2$ revolving about the axis of rotation, the total kinetic energy ($T$) of the rotator can be given by the following relation.

$$T = \frac{1}{2}m_1 v_1^2 + \frac{1}{2}m_2 v_2^2 \tag{215}$$

Since we know that linear velocity $v$ is simply equal to the angular velocity $\omega$ multiplied by the radius of rotation $r$ i.e. $v = \omega r$, the equation (215) takes the form

$$T = \frac{1}{2}m_1(r_1\omega)^2 + \frac{1}{2}m_2(r_2\omega)^2 \tag{216}$$

$$T = \frac{1}{2}(m_1 r_1^2 + m_2 r_2^2)\omega^2 \tag{217}$$

$$T = \frac{1}{2}I\omega^2 \tag{218}$$

Where $I$ is the moment of inertia equal with definition $I = \sum m_i r_i^2$. Furthermore, the value of I can also be written as

$$I = \left(\frac{m_1 m_2}{m_1 + m_2}\right)r^2 \tag{219}$$

$$I = \mu r^2 \tag{220}$$

Where $\mu - m_1 m_2/m_1 + m_2$ is the reduced mass of the rigid diatomic system  After multiplying and dividing the rotational kinetic energy by $I$ i.e. equation (218), we have

$$T = \frac{I^2\omega^2}{2I} = \frac{(I\omega)^2}{2I} = \frac{L^2}{2I} \tag{221}$$

Where $L$ is the angular momentum of the rotator. It is clear from the above equation that the kinetic energy of a classical rotator can have any value because the value-domain of angular velocity is continuous. Moreover, as now the external force is working on the rotator, the potential can be set to zero. Therefore, we can conclude that the total energy of a classical diatomic rigid rotator is given by equation (221).

> ### *The energy of Quantum Mechanical Rigid Rotator*

In order to understand the energy of a quantum mechanical rigid rotator, recall the Schrodinger wave equation for the same first i.e.

$$\frac{1}{Sin\,\theta}\frac{\partial}{\partial\theta}\left(Sin\,\theta\,\frac{\partial\psi}{\partial\theta}\right) + \frac{1}{Sin^2\,\theta}\frac{\partial\psi}{\partial\phi} + \frac{8\pi^2 IE\psi}{h^2} = 0 \tag{222}$$

Where $\psi$ is the mathematical expression defining various quantum mechanical states depending upon two variables $\theta$ and $\phi$. During the course of the solution of the above equation, a constant $\beta$ is defined for simplicity as given below.

$$\beta = \frac{8\pi^2 IE}{h^2} \tag{223}$$

However, the boundary conditions that keep the function single-valued, continuous and finite; also proved that the constant $\beta$ must satisfy the following condition also.

$$\beta = l(l+1) \tag{224}$$

Where $l = 0, 1, 2, 3, 4$ etc. After equating the value of $\beta$ from equation (223) and equation (224), we get

$$\frac{8\pi^2 IE}{h^2} = l(l+1) \tag{225}$$

$$E_l = \frac{h^2}{8\pi^2 I}\,l(l+1) \tag{226}$$

Hence, unlike the classical counterpart, the energy levels of quantum mechanical rigid rotators are discontinuous.

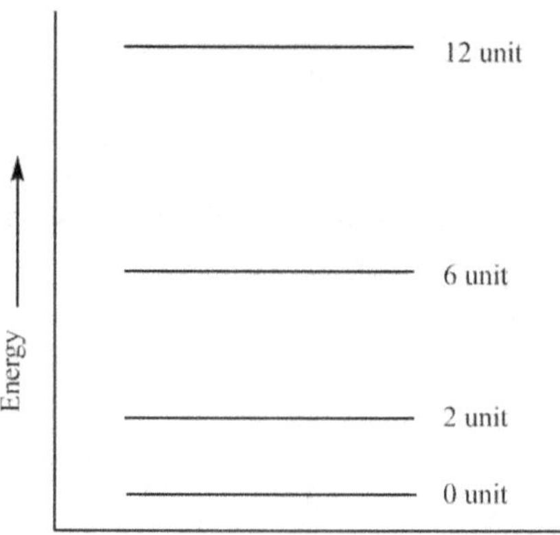


Figure 8. The energy level diagram of the diatomic rigid rotator in units of $h^2/8\pi^2 I$.

## ❖ Space Quantization

The solution of the Schrodinger wave equation for the diatomic rigid rotator provided the mathematical descriptions of all the rotational states along with their corresponding energies. The general form of total eigenfunction for the rigid rotator is given below.

$$\psi_{l,m}(\theta, \phi) = \sqrt{\frac{1}{2\pi}} \sqrt{\frac{(2l+1)(l-m)!}{2(l+m)!}} \cdot P_l^m(Cos\ \theta) \cdot e^{\pm im\phi} \tag{227}$$

Where $\psi$ is the mathematical expression defining various quantum mechanical states depending upon two variables $\theta$ and $\phi$. Furthermore, the most important property of a rigid rotator after energy is the angular momentum which can be obtained using the last postulate of quantum mechanics i.e.

$$<L> = \oint \psi_{l,m}(\theta, \phi)\ \hat{L}\ \psi_{l,m}(\theta, \phi) \tag{228}$$

$$L_l = \sqrt{l(l+1)}\ \frac{h}{2\pi} \tag{229}$$

Alternatively, we know that the energies of various rotational states of rigid rotators are given by the following relation.

$$E = \frac{h^2}{8\pi^2 I}\ l(l+1) \tag{230}$$

Where $l = 0, 1, 2, 3, 4$ etc. Also, we know that the angular momentum and energy are related classically as

$$E = \frac{1}{2}\ I\omega^2 = \frac{(I\omega)^2}{2I} = \frac{L^2}{2I} \tag{231}$$

or

$$L - \sqrt{2EI} \tag{232}$$

After using the value of energy from equation (230) into equation (232), we get

$$L_l = \sqrt{2I \cdot \frac{h^2}{8\pi^2 I}\ l(l+1)} \tag{233}$$

or

$$L_l = \sqrt{\frac{h^2}{4\pi^2}\ l(l+1)} \tag{234}$$

or

$$L_l = \sqrt{l(l+1)}\,\frac{h}{2\pi} \tag{235}$$

Which is exactly the same as given by equation (229). Since $l = 0, 1, 2, 3, 4$ etc., the quantum mechanically allowed values of angular momentum (in the units of $h/2\pi$) are given below.

$$L_0 = \sqrt{0(0+1)}\ \text{unit} = 0\ \text{unit} \tag{236}$$

$$L_1 = \sqrt{1(1+1)}\ \text{unit} = \sqrt{2}\ \text{unit} \tag{237}$$

$$L_2 = \sqrt{2(2+1)}\ \text{unit} = \sqrt{6}\ \text{unit} \tag{238}$$

$$L_3 = \sqrt{3(3+1)}\ \text{unit} = \sqrt{12}\ \text{unit} \tag{239}$$

However, there is boundary condition in quantum mechanics that says that only integral effects are allowed reference direction if the angular momentum is generated by integral quantum number and half-integral effects are allowed in reference direction if the momentum is generated by half-integral quantum number.

This can be understood by taking the example of a diatomic molecule rotating in the first excited rotational state i.e. $l = 1$. The angular momentum of such a molecule will be $\sqrt{2}$ or 1.414 units. However, since this angular momentum is obtained using an integral quantum number ($l = 1$), only integral effects (i.e. $+1, 0, -1$) are allowed in reference direction. Now if $z$-axis is the reference direction, the effect of any vector $\vec{A}$ in the reference direction is calculated by multiplying its magnitude with the cosine of the angle it makes with reference direction.

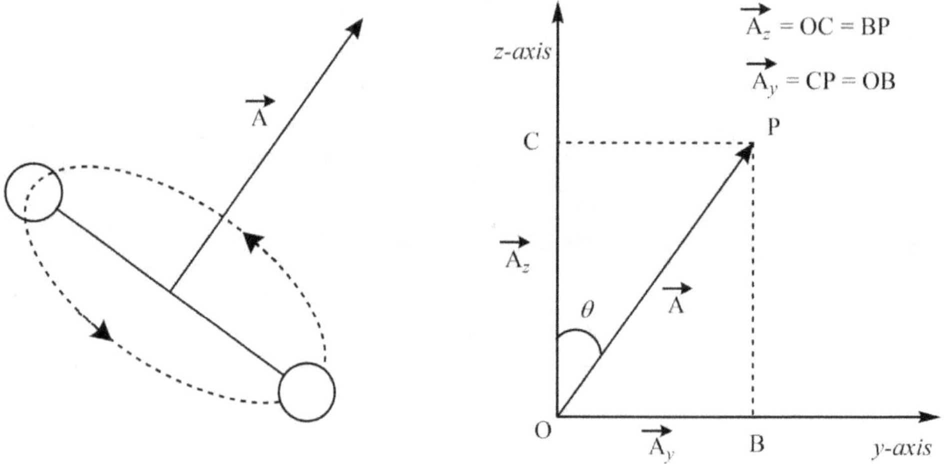


Figure 9. The angular momentum of the diatomic rigid rotator (left) and its rectangular resolution.

In tringle OPC, the side OC represents the effect of the angular momentum vector $\vec{A}$ along $z$-axis, can be calculated as given below.

$$\frac{OC}{OP} = Cos\,\theta \tag{240}$$

$$OC = OP.\,Cos\,\theta \tag{241}$$

$$\vec{A}_z = A\,Cos\,\theta \tag{242}$$

Hence, a diatomic molecule in its first rotational state cannot rotate in $xy$-plane since it will generate $\sqrt{2}$ or 1.414 units of angular momentum along the $z$-axis (from right-hand thumb rule). In other words, the $\sqrt{2}$ units of angular momentum cannot orient itself along $z$-axis because this makes $\theta = 0°$ and since $Cos\,0 = 1$, $\vec{A}_z = A$ i.e. angular momentum effect along the z-axis is also 1.414 unit which is not allowed quantum mechanically. The effects of angular momentum allowed in the $z$-direction are $+1, 0, -1$; for which angles required are determined as follows.

$$+1 = \sqrt{2}\,Cos\,\theta \quad \Rightarrow \quad \theta = Cos^{-1}\frac{1}{\sqrt{2}} = 45° \tag{243}$$

$$0 = \sqrt{2}\,Cos\,\theta \quad \Rightarrow \quad \theta = Cos^{-1}\frac{0}{\sqrt{2}} = 90° \tag{244}$$

$$-1 = \sqrt{2}\,Cos\,\theta \quad \Rightarrow \quad \theta = Cos^{-1}\frac{-1}{\sqrt{2}} = 135° \tag{245}$$

Hence, we can say that in order to be allowed, the 1.414 units of angular momentum must orient itself only at $45°$, $90°$ and $135°$ in space from reference direction ($z$-axis in this case).

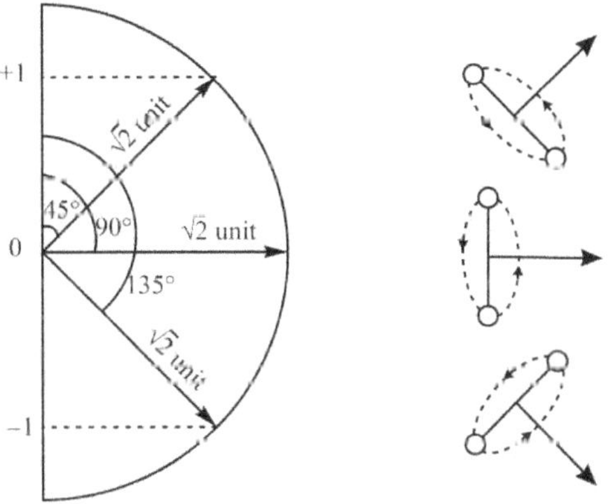


Figure 10. The space quantization of angular momentum of rigid rotator in $l = 1$ rotational state.

Since the orientation of angular momentum can orient itself in any direction from the $z$-axis as far as the effective angular momentum +1 unit along $z$-direction; therefore, we should use a cone around the same at 45°. The same is true for 0 and −1 effects with 90° and 135°, respectively.

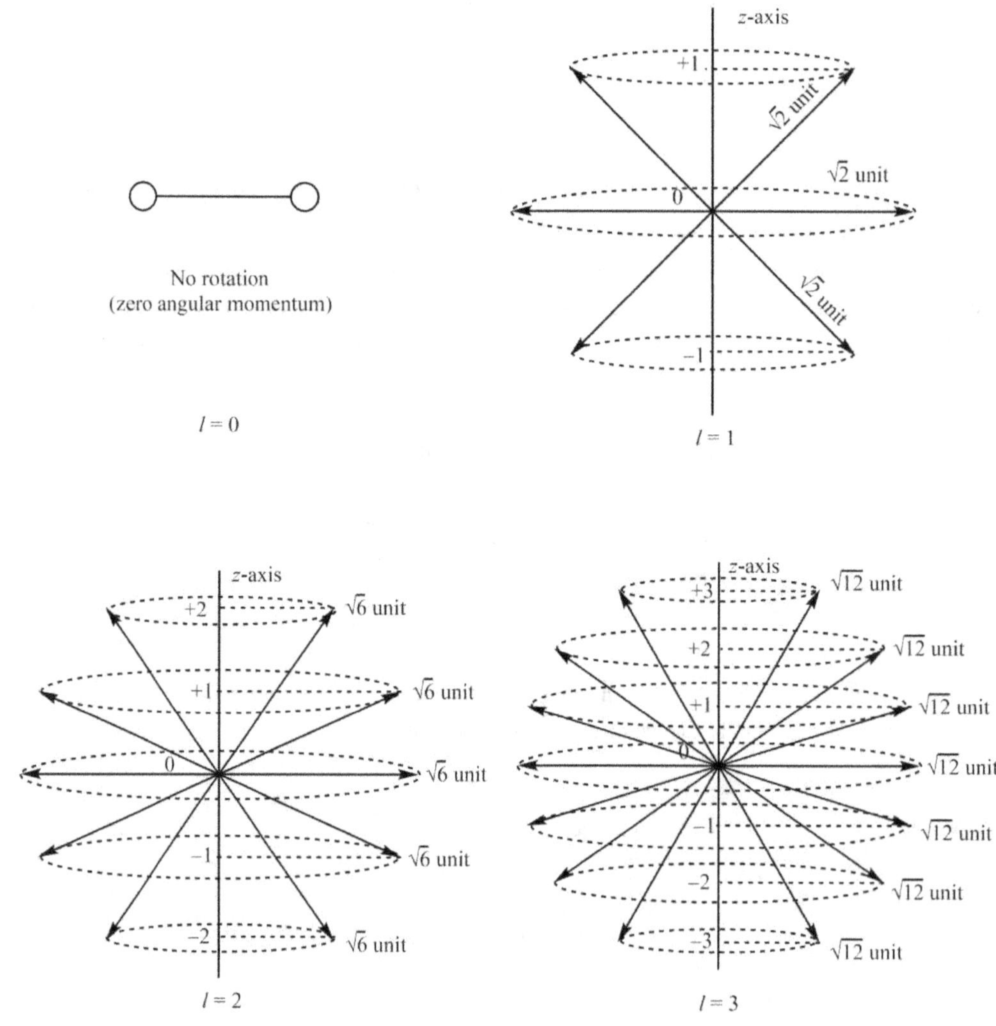

Figure 11. The space quantization of angular momentum of the rigid rotator in $l = 0$, 1, 2 and 3 states.

It is also worthy to mention that the concept of space quantization is equally applicable to the angular momentums of all other systems also like orbital or spin angular momentum of electrons or nuclei.

### ❖ Schrodinger Wave Equation for Hydrogen Atom: Separation of Variable in Polar Spherical Coordinates and Its Solution

In the first section of this chapter, we derived and discussed the Schrodinger wave equation for a particle in a three-dimensional box. In this section, we will apply the procedure to an electron that exits around the nucleus. In order to do so, consider an electron at a distance $r$ from the center of the nucleus, and this electron can travel in any direction i.e. along $x$-, $y$- and $z$-axis. The potential energy of such an electron-nucleus system will be $-Ze^2/r$; where $Ze$ and $e$ are charges on nucleus and electron respectively.

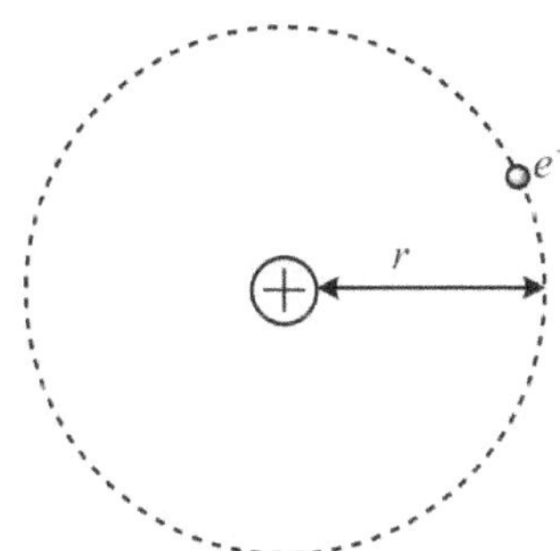


Figure 12. An electron around nucleus at $r$ distance.

So far we have considered a quantum mechanical system of an electron around the nucleus. Now suppose that we need to find various physical properties associated with different states of this system. Had it been a classical system, we would use simple formulas from classical mechanics to determine the value of different physical properties. However, being a quantum mechanical system, we cannot use those expressions because they would give irrational results. Therefore, we need to use the postulates of quantum mechanics to evaluate various physical properties.

Let $\psi$ be the function that describes all the states of the electron around the nucleus. At this point we have no information about the exact mathematical expression of $\psi$; nevertheless, we know that there is one operator that does not need the absolute expression of wave function but uses the symbolic form only, the Hamiltonian operator. The operation of Hamiltonian operator over this symbolic form can be rearranged to give to construct the Schrodinger wave equation; and we all know that the wave function as well the energy, both are the obtained as this second-order differential equation is solved. Mathematically, we can say that

$$\hat{H}\psi = E\psi \tag{246}$$

After putting the value of three-dimensional Hamiltonian in equation (1), we get

$$\left[\frac{-h^2}{8\pi^2 m}\left(\frac{\partial^2}{\partial x^2} + \frac{\partial^2}{\partial y^2} + \frac{\partial^2}{\partial z^2}\right) + V\right]\psi = E\psi \tag{247}$$

or

$$\frac{-h^2}{8\pi^2 m}\left(\frac{\partial^2 \psi}{\partial x^2}+\frac{\partial^2 \psi}{\partial y^2}+\frac{\partial^2 \psi}{\partial z^2}\right)+V\psi=E\psi \tag{248}$$

$$\frac{-h^2}{8\pi^2 m}\left(\frac{\partial^2 \psi}{\partial x^2}+\frac{\partial^2 \psi}{\partial y^2}+\frac{\partial^2 \psi}{\partial z^2}\right)+V\psi-E\psi=0 \tag{249}$$

$$\frac{\partial^2 \psi}{\partial x^2}+\frac{\partial^2 \psi}{\partial y^2}+\frac{\partial^2 \psi}{\partial z^2}+\frac{8\pi^2 m}{h^2}(E-V)\psi=0 \tag{250}$$

After putting the value of potential energy of the electron-nucleus system in equation (250), we get

$$\frac{\partial^2 \psi}{\partial x^2}+\frac{\partial^2 \psi}{\partial y^2}+\frac{\partial^2 \psi}{\partial z^2}+\frac{8\pi^2 m}{h^2}\left(E+\frac{Ze^2}{r}\right)\psi=0 \tag{251}$$

The above-mentioned second order differential equation is the Schrodinger wave equation for an electron around the nucleus. However, since it is neither completely in cartesian nor completely in polar coordinates (contains $x$, $y$, $z$ as well as $r$ variable), the solution is very much difficult. Therefore, recall the transformation of cartesian coordinates to polar coordinates in three dimensions as given below.

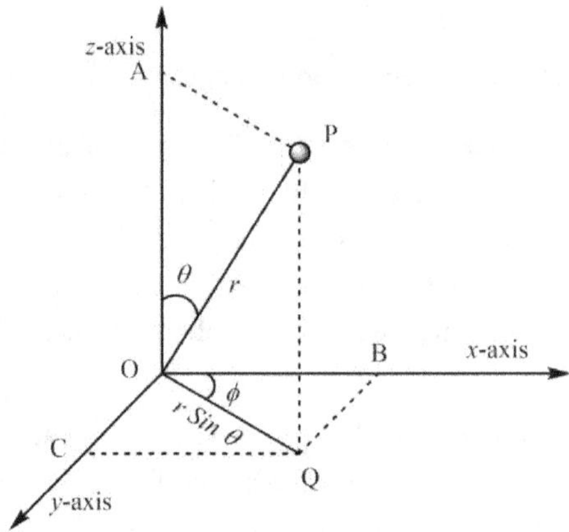


Figure 13. Correlation between cartesian and polar coordinates in three dimensions.

In tringle AOP, the side OA is simply the z-coordinate and can be obtained as

$$\frac{OA}{OP}=Cos\,\theta \quad \Rightarrow \quad OA=OP\,Cos\,\theta \quad \Rightarrow \quad z=r\,Cos\,\theta \tag{252}$$

Similarly, in AOP

$$\frac{AP}{OP}=Sin\,\theta \quad \Rightarrow \quad AP=OP\,Sin\,\theta \quad \Rightarrow \quad OQ=r\,Sin\,\theta \tag{253}$$

In tringle BOQ, the side OB is simply the $x$-coordinate and can be obtained as

$$\frac{OB}{OQ} = Cos\,\phi \quad \Rightarrow \quad OB = OQ\,Cos\,\phi \quad \Rightarrow \quad x = r\,Sin\,\theta\,Cos\,\phi \tag{254}$$

Since the side BQ equal to OC, BQ also represents the $y$-coordinate and can be obtained as

$$\frac{BQ}{OQ} = Sin\,\phi \quad \Rightarrow \quad BQ = OQ\,Sin\,\phi \quad \Rightarrow \quad y = r\,Sin\,\theta\,Sin\,\phi \tag{255}$$

Now using equation (252-254), the equation (251) can be transformed to polar coordinates as given below.

$$\frac{1}{r^2}\frac{\partial}{\partial r}\left(r^2\frac{\partial\psi}{\partial r}\right) + \frac{1}{r^2 Sin\theta}\frac{\partial}{\partial\theta}\left(Sin\theta\frac{\partial\psi}{\partial\theta}\right) + \frac{1}{r^2\,Sin^2\theta}\frac{\partial^2\psi}{\partial\phi^2} + \frac{8\pi^2\mu}{h^2}\left(E + \frac{Ze^2}{r}\right)\psi = 0 \tag{256}$$

or

$$\frac{1}{r^2}\left[\frac{\partial}{\partial r}\left(r^2\frac{\partial\psi}{\partial r}\right) + \frac{1}{Sin\theta}\frac{\partial}{\partial\theta}\left(Sin\theta\frac{\partial\psi}{\partial\theta}\right) + \frac{1}{Sin^2\theta}\frac{\partial^2\psi}{\partial\phi^2}\right] + \frac{8\pi^2\mu}{h^2}\left(E + \frac{Ze^2}{r}\right)\psi = 0 \tag{257}$$

Which is the Schrodinger wave equation for hydrogen and hydrogen-like species in polar coordinates.

> ### ➤ *Separation of Variables*

The wave function representing quantum mechanical states, in this case, is actually a function of three variable $r$, $\theta$ and $\phi$. Now, we know that it is easier to solve three differential equations with one variable in each rather a single differential equation with three variables. Therefore, in order to separate variables, consider that the wave function $\psi$ is the multiplication of three individual functions as

$$\psi(r,\theta,\phi) = \psi(r) \times \psi(\theta) \times \psi(\phi) = R.\,\Theta.\,\Phi \tag{258}$$

After putting the value of equation (258) in equation (257) and then multiplying throughout by $r^2$, we get

$$\Theta\Phi\frac{\partial}{\partial r}\left(r^2\frac{\partial R}{\partial r}\right) + \frac{\Phi R}{Sin\theta}\frac{\partial}{\partial\theta}\left(Sin\theta\frac{\partial\Theta}{\partial\theta}\right) + \frac{R\Theta}{Sin^2\theta}\frac{\partial^2\Phi}{\partial\psi^2} + \frac{8\pi^2\mu r^2}{h^2}\left(E + \frac{Ze^2}{r}\right)\Theta\Phi R = 0 \tag{259}$$

Furthermore, divide equation (259) throughout $\Theta\Phi R$ i.e.

$$\frac{1}{R}\frac{\partial}{\partial r}\left(r^2\frac{\partial R}{\partial r}\right) + \frac{1}{\Theta}\frac{1}{Sin\theta}\frac{\partial}{\partial\theta}\left(Sin\theta\frac{\partial\Theta}{\partial\theta}\right) + \frac{1}{\Phi}\frac{1}{Sin^2\theta}\frac{\partial^2\Phi}{\partial\phi^2} + \frac{8\pi^2\mu r^2}{h^2}\left(E + \frac{Ze^2}{r}\right) = 0 \tag{260}$$

or

$$\frac{1}{R}\frac{\partial}{\partial r}\left(r^2\frac{\partial R}{\partial r}\right) + \frac{8\pi^2\mu r^2}{h^2}\left(E + \frac{Ze^2}{r}\right) = -\frac{1}{\Theta}\frac{1}{Sin\theta}\frac{\partial}{\partial\theta}\left(Sin\theta\frac{\partial\Theta}{\partial\theta}\right) - \frac{1}{\Phi}\frac{1}{Sin^2\theta}\frac{\partial^2\Phi}{\partial\phi^2} \tag{261}$$

The above equation holds true if we put both sides equal to a constant $\beta$ i.e.

$$\frac{1}{R}\frac{\partial}{\partial r}\left(r^2\frac{\partial R}{\partial r}\right) + \frac{8\pi^2\mu r^2}{h^2}\left(E + \frac{Ze^2}{r}\right) = \beta \tag{262}$$

and

$$\frac{1}{\Theta}\frac{1}{Sin\theta}\frac{\partial}{\partial\theta}\left(Sin\theta\frac{\partial\Theta}{\partial\theta}\right) + \frac{1}{\Phi}\frac{1}{Sin^2\theta}\frac{\partial^2\Phi}{\partial\phi^2} = -\beta \tag{263}$$

The equation (262) contains only $r$ variable, and therefore, is called as the "radial equation". However, the equation (263) still contains two variable, and thus, needs further separation. To do so, first multiply equation (263) throughout by $Sin^2\theta$ i.e.

$$\frac{Sin\theta}{\Theta}\frac{\partial}{\partial\theta}\left(Sin\theta\frac{\partial\Theta}{\partial\theta}\right) + \frac{1}{\Phi}\frac{\partial^2\Phi}{\partial\phi^2} = -\beta\,Sin^2\theta \tag{264}$$

or

$$\frac{Sin\theta}{\Theta}\frac{\partial}{\partial\theta}\left(Sin\theta\frac{\partial\Theta}{\partial\theta}\right) + \beta\,Sin^2\theta = -\frac{1}{\Phi}\frac{\partial^2\Phi}{\partial\phi^2} \tag{265}$$

The above equation also holds true if we put both sides equal to a constant $m^2$ i.e.

$$\frac{Sin\theta}{\Theta}\frac{\partial}{\partial\theta}\left(Sin\theta\frac{\partial\Theta}{\partial\theta}\right) + \beta\,Sin^2\theta = m^2 \tag{266}$$

and

$$\frac{1}{\Phi}\frac{\partial^2\Phi}{\partial\phi^2} = -m^2 \tag{267}$$

The equation (266) contains only $\theta$ variable, and therefore, is called as "theta equation". Likewise, the equation (267) contains only $\phi$ variable, and therefore, is called as "phi equation".

> ### Solutions of R(r), Θ(θ) and Φ(φ) Equations

The single variable equations obtained after separation of variables can be solved separately to yield r, θ and φ-dependent functions which then are multiplied give total wave function.

**1. The solution of Φ(φ) equation:** Recall and rearrange the differential equation obtained after separation of variables having φ dependence i.e.

$$\frac{1}{\Phi}\frac{\partial^2\Phi}{\partial\phi^2} = -m^2 \qquad \Rightarrow \qquad \frac{\partial\Phi}{\partial\phi} + m^2\Phi = 0 \tag{268}$$

The general solution of such an equation is

$$\Phi(\phi) = Ne^{im\phi} \tag{269}$$

Where $N$ represents the normalization constant. The wavefunction given above will be acceptable only if $m$ has integer value i.e. $0, \pm1, \pm2$, etc. This can be understood in terms of single-valued, continuous and finite nature of quantum states.

*i) The boundary condition for function $\Phi$:* If we replace the angle "$\phi$" with "$\phi + 2\pi$", the position of point under consideration should remain the same i.e.

$$\Phi(\phi + 2\pi) = \Phi(\phi) \tag{270}$$

Therefore

$$N e^{im(\phi+2\pi)} = N e^{im\phi} \tag{271}$$

$$e^{im(\phi+2\pi)} = e^{im\phi} \tag{272}$$

$$e^{im\phi}.e^{im2\pi} = e^{im\phi} \tag{273}$$

$$e^{im2\pi} = e^{im\phi} e^{-im\phi} \tag{274}$$

$$e^{im2\pi} = e^{im\phi - im\phi} = e^{0} \tag{275}$$

$$e^{im2\pi} = 1 \tag{276}$$

Since we know from the Euler's expansion $e^{ix} = Cos\ x + i\ Sin\ x$, the equation (276) takes the form

$$e^{im2\pi} = Cos\ 2\pi m + i\ Sin\ 2\pi m \tag{277}$$

After putting the value of equation (277) in equation (276), we get

$$Cos\ 2\pi m + i\ Sin\ 2\pi m = 1 \tag{278}$$

The relation holds true only when we use $m = 0, \pm1, \pm2, \pm3, \pm4$, etc.

*ii) The normalization constant for function $\Phi$:* In order to determine the normalization constant for the $\Phi$ function, we must put the squared-integral over whole configuration space as unity i.e.

$$\int_{0}^{2\pi} \Phi^* \Phi\ d\phi = 1 \tag{279}$$

$$N^2 \int_{0}^{2\pi} e^{im\phi}.e^{-im\phi}\ d\phi = 1 \tag{280}$$

$$N^2 \int_{0}^{2\pi} e^{im\phi - im\phi}\ d\phi = N^2 \int_{0}^{2\pi} e^{0}\ d\phi = 1 \tag{281}$$



**D** **DALAL INSTITUTE**

$$N^2[\phi]_0^{2\pi} = N^2[2\pi] = 1 \qquad (282)$$

or

$$N = \sqrt{\frac{1}{2\pi}} \qquad (283)$$

After using the value of normalization constant in equation (269), we get

$$\Phi_m(\phi) = \sqrt{\frac{1}{2\pi}}\, e^{im\phi} \qquad (284)$$

Which is the complete solution of $\phi$-equation.

Table 1. Complex and real forms of some normalized $\Phi$-functions.

| $|m|$ | Complex form | Real form |
|---|---|---|
| 0 | $\Phi_0(\phi) = \sqrt{\dfrac{1}{2\pi}}$ | $\Phi_0(\phi) = \sqrt{\dfrac{1}{2\pi}}$ |
| 1 | $\Phi_{+1}(\phi) = \sqrt{\dfrac{1}{2\pi}}\, e^{i\phi}$ | $\Phi_{+1}(\phi) = \sqrt{\dfrac{1}{\pi}}\, Cos\,\phi$ |
|   | $\Phi_{-1}(\phi) = \sqrt{\dfrac{1}{2\pi}}\, e^{-i\phi}$ | $\Phi_{-1}(\phi) = \sqrt{\dfrac{1}{\pi}}\, Sin\,\phi$ |
| 2 | $\Phi_{+2}(\phi) = \sqrt{\dfrac{1}{2\pi}}\, e^{i2\phi}$ | $\Phi_{+2}(\phi) = \sqrt{\dfrac{1}{\pi}}\, Cos\,(2\phi)$ |
|   | $\Phi_{-2}(\phi) = \sqrt{\dfrac{1}{2\pi}}\, e^{-i2\phi}$ | $\Phi_{-2}(\phi) = \sqrt{\dfrac{1}{\pi}}\, Sin\,(2\phi)$ |
| 3 | $\Phi_{+3}(\phi) = \sqrt{\dfrac{1}{2\pi}}\, e^{i3\phi}$ | $\Phi_{+3}(\phi) = \sqrt{\dfrac{1}{\pi}}\, Cos\,(3\phi)$ |
|   | $\Phi_{-3}(\phi) = \sqrt{\dfrac{1}{2\pi}}\, e^{-i3\phi}$ | $\Phi_{-3}(\phi) = \sqrt{\dfrac{1}{\pi}}\, Sin\,(3\phi)$ |

**2. The solution of Θ(θ) equation:** Recall and rearrange the differential equation obtained after separation of variables having θ-dependence i.e.

$$\frac{Sin\theta}{\Theta}\frac{\partial}{\partial\theta}\left(Sin\theta\frac{\partial\Theta}{\partial\theta}\right) + \beta\,Sin^2\theta = m^2 \tag{285}$$

or

$$Sin\,\theta\frac{\partial}{\partial\theta}\left(Sin\,\theta\frac{\partial\Theta}{\partial\theta}\right) + \Theta\beta Sin^2\theta - m^2\Theta = 0 \tag{286}$$

Now dividing the above equation by $Sin^2\theta$, we get

$$\frac{1}{Sin\,\theta}\frac{\partial}{\partial\theta}\left(Sin\,\theta\frac{\partial\Theta}{\partial\theta}\right) + \Theta\beta - \frac{m^2\Theta}{Sin^2\theta} = 0 \tag{287}$$

or

$$\frac{1}{Sin\,\theta}\frac{\partial}{\partial\theta}\left(Sin\,\theta\frac{\partial\Theta}{\partial\theta}\right) + \left(\beta - \frac{m^2}{Sin^2\theta}\right)\Theta = 0 \tag{288}$$

After defining a new variable $x = Cos\,\theta$, we have

$$Sin^2\theta + Cos^2\theta = 1 \tag{289}$$

$$Sin^2\theta = 1 - Cos^2\theta \tag{290}$$

$$Sin\,\theta = \sqrt{1 - Cos^2\theta} \tag{291}$$

$$Sin\,\theta = \sqrt{1 - x^2} \tag{292}$$

Also, since we assumed $x = Cos\,\theta$, the first derivative w.r.t. $\theta$ will be

$$\frac{\partial x}{\partial\theta} = -\,Sin\,\theta \tag{293}$$

The derivative of Θ function w.r.t. $\theta$ can be rewritten as

$$\frac{\partial\Theta}{\partial\theta} = \frac{\partial\Theta}{\partial x}\cdot\frac{\partial x}{\partial\theta} \tag{294}$$

After putting the values of $\partial x/\partial\theta$ from equation (293) in equation (294), we get

$$\frac{\partial\Theta}{\partial\theta} = -Sin\,\theta\frac{\partial\Theta}{\partial x} \tag{295}$$

After removing Θ from both sides

$$\frac{\partial}{\partial\theta} = -Sin\,\theta\,\frac{\partial}{\partial x} \tag{296}$$

Multiplying both sides of equation (295) by $Sin\,\theta$, we have

$$Sin\,\theta\,\frac{\partial\Theta}{\partial\theta} = -Sin^2\,\theta\,\frac{\partial\Theta}{\partial x} \tag{297}$$

$$Sin\,\theta\,\frac{\partial\Theta}{\partial\theta} = -(1-x^2)\frac{\partial\Theta}{\partial x} \tag{298}$$

Now, after putting the values of equation (296) and (298) in equation (288), we get

$$\frac{1}{Sin\,\theta}\left(-Sin\,\theta\,\frac{\partial}{\partial x}\right)\left[-(1-x^2)\frac{\partial\Theta}{\partial x}\right] + \left(\beta - \frac{m^2}{1-x^2}\right)\Theta = 0 \tag{299}$$

$$\frac{\partial}{\partial x}\left[(1-x^2)\frac{\partial\Theta}{\partial x}\right] + \left(\beta - \frac{m^2}{1-x^2}\right)\Theta = 0 \tag{300}$$

The equation given above is a Legendre's polynomial and has physical significance only in the range of $x = +1$ to $-1$. Therefore, consider that one more form of $\Theta$ function so that this condition is satisfied i.e.

$$\Theta(\theta) = (1-x^2)^{\frac{m}{2}}.X(x) \tag{301}$$

Where $X$ is a function depending upon variable $x$. The differentiation of the above equation w.r.t. $x$ yields

$$\frac{\partial\Theta}{\partial x} = -mx(1-x^2)^{\frac{m}{2}-1}.X + (1-x^2)^{\frac{m}{2}}.\frac{dX}{dx} \tag{302}$$

After multiplying the above equation by $1-x^2$ and $\partial/\partial x$, we get

$$\frac{\partial}{\partial x}\left[(1-x^2)\frac{\partial\Theta}{\partial x}\right] = \frac{\partial}{\partial x}\left[-mx(1-x^2)^{\frac{m}{2}}.X + (1-x^2)^{\frac{m}{2}+1}.\frac{dX}{dx}\right] \tag{303}$$

$$= \left[-m(1-x^2)^{m/2} + m^2x^2(1-x^2)^{\frac{m}{2}-1}\right]X - \left[2x(m+1)(1-x^2)^{\frac{m}{2}}\right]X' \tag{304}$$

$$+ \left[(1-x^2)^{\frac{m}{2}+1}\right]X''$$

Where $\partial/\partial x$ and $\partial^2/\partial x^2$ are represented by the symbol $X'$ and $X''$, respectively. Now, after using the value of equation (301) and equation (304) in equation (300), we get

$$\left[-m(1-x^2)^{m/2} + m^2x^2(1-x^2)^{\frac{m}{2}-1}\right]X - \left[2x(m+1)(1-x^2)^{\frac{m}{2}}\right]X' \tag{305}$$

$$+ \left[(1-x^2)^{\frac{m}{2}+1}\right]X'' + \left(\beta - \frac{m^2}{1-x^2}\right)(1-x^2)^{\frac{m}{2}}.X = 0$$

Dividing the above expression by $(1-x^2)^{m/2}$, we have

$$(1 - x^2)X'' - 2(m + 1)xX' + [\beta - m(m + 1)]X = 0 \tag{306}$$

or

$$(1 - x^2)X'' - 2\alpha xX' + \lambda X = 0 \tag{307}$$

Where $\alpha = m + 1$ and $\lambda = \beta - m(m + 1)$. Now assume that the function $X$ can be expressed as a power series expansion as given below.

$$X = a_0 + a_1 x + a_2 x^2 + a_3 x^3 \ldots \ldots \ldots \ldots \tag{308}$$

$$X' = a_1 + 2a_2 x + 3a_3 x^2 \ldots \ldots \ldots \ldots \tag{309}$$

$$X'' = 2a_2 + 6a_3 x + 12a_4 x^2 \ldots \ldots \ldots \ldots \tag{310}$$

Putting values of equation (308-310) in equation (307), we get

$$(1 - x^2)(2a_2 + 6a_3 x + 12a_4 x^2 + 20a_5 x^3) - 2\alpha x(a_1 + 2a_2 x + 3a_3 x^2 + 4a_4 x^3) \tag{311}$$
$$+ \lambda(a_0 + a_1 x + a_2 x^2 + a_3 x^3) = 0$$

or

$$(2a_2 + \lambda a_0) + [6a_3 + (\lambda - 2\alpha)a_1]x + [12a_4 + (\lambda - 2\alpha - 2)a_2]x^2 \ldots \ldots \ldots = 0 \tag{312}$$

The above equation is satisfied only if each term on the left-hand side is individually equal to zero i.e. coefficients of each power of $x$ are vanish. The general expression for the coefficients must follow the condition given below.+

$$(n + 1)(n + 2)a_{n+2} + [\lambda - 2n\alpha - n(n - 1)]a_n = 0 \tag{313}$$

Where $n = 0, 1, 2, 3$ etc. Summarizing the result, we can write

$$a_{n+2} = \frac{2n\alpha + n(n - 1) - \lambda}{(n + 1)(n + 2)} a_n \tag{314}$$

After putting values of $\alpha$ and $\lambda$ in equation (314), we get

$$\frac{a_{n+2}}{a_n} = \frac{(n + m)(n + m + 1) - \beta}{(n + 1)(n + 2)} \tag{315}$$

Which is the Recursion formula for the coefficients of the power of $x$. Now, in order to obtain a valid wavefunction, the power series must contain a finite number of terms which is possible only if numerator becomes zero i.e.

$$(n + m)(n + m + 1) - \beta = 0 \tag{316}$$

$$\beta = (n + m)(n + m + 1) \tag{317}$$

Since we know that $m$ as well $n$ both are the whole numbers, their sum must also be a whole number. Therefore, the sum of $n$ and $m$ can be replaced by another whole number symbolized by $l$ i.e.

$$\beta = l(l+1) \tag{318}$$

Where $l = 0, 1, 2, 3$ etc. After putting the value of $\beta$ from equation (318) in equation (300), we get

$$\frac{\partial}{\partial x}\left[(1-x^2)\frac{\partial \Theta}{\partial x}\right] + \left[l(l+1) - \frac{m^2}{1-x^2}\right]\Theta = 0 \tag{319}$$

The general solution of equation (319) is

$$\Theta = NP_l^m(x) = NP_l^m(Cos\,\theta) \tag{320}$$

Where N is the normalization constant and $P_l^m(x)$ is the associated "Legendre function" which is defined as given below.

$$P_l^m(x) = (1-x^2)^{m/2}\frac{d^m P_l(x)}{dx^m} \tag{321}$$

Where $P_l(x)$ is the Legendre polynomial given by

$$P_l(x) = \frac{1}{2^l\,l!}\frac{d^l(x^2-1)^l}{dx^l} \tag{322}$$

In order to proceed further, we must discuss the concept of orthogonality and the normalization of the "Legendre's function".

*i) Orthogonality of associated Legendre's function:* The orthogonality of the associated Legendre's polynomial follows the conditions given below.

$$\int_{-1}^{+1} P_k^m(x)\,P_l^m(x) = 0 \qquad if\ k \neq l \tag{323}$$

$$\int_{-1}^{+1} P_k^m(x)\,P_l^m(x) = \frac{2}{(2l+1)}\frac{(l+m)!}{(l-m)!} \qquad if\ k = l \tag{324}$$

*ii) Normalization of associated Legendre's function:* The normalization of the associated Legendre's polynomial follows the conditions given below.

$$\int_{-1}^{+1} \Theta_{m,l}\,\Theta_{m,l}^*(d\theta) = 1 \tag{325}$$

$$N^2 \int_{-1}^{+1} P_k^m(x) \, P_l^m(x) \, dx = 1 \tag{326}$$

$$N^2 . \frac{2}{(2l+1)} \frac{(l+m)!}{(l-m)!} = 1 \tag{327}$$

$$N = \sqrt{\frac{(2l+1)(l-m)!}{2(l+m)!}} \tag{328}$$

Using the value of normalization constant in equation (320), we get

$$\Theta_{l,m}(\theta) = \sqrt{\frac{(2l+1)(l-m)!}{2(l+m)!}} . P_l^m(Cos\,\theta) \tag{329}$$

Which is the complete solution of $\Theta$-equation.

Table 2. Some normalized $\Theta$-functions and corresponding spherical harmonics.

| $\Theta$-functions | Spherical harmonics |
|---|---|
| $\Theta_{0,0} = \dfrac{1}{\sqrt{2}}$ | $Y_{0,0} = \dfrac{1}{\sqrt{2}} . \sqrt{\dfrac{1}{2\pi}}$ |
| $\Theta_{1,0} = \sqrt{\dfrac{3}{2}} \, Cos\,\theta$ | $Y_{1,0} = \sqrt{\dfrac{3}{2}} \, Cos\,\theta . \dfrac{1}{\sqrt{2\pi}}$ |
| $\Theta_{1,\pm 1} = \sqrt{\dfrac{3}{4}} \, Sin\,\theta$ | $Y_{1,\pm 1} = \sqrt{\dfrac{3}{4}} \, Sin\,\theta . \sqrt{\dfrac{1}{2\pi}} e^{\pm i\phi}$ |
| $\Theta_{2,0} = \sqrt{\dfrac{5}{8}} \, (3Cos^2\,\theta - 1)$ | $Y_{2,0} = \sqrt{\dfrac{5}{8}} \, (3Cos^2\,\theta - 1) \dfrac{1}{\sqrt{2\pi}}$ |
| $\Theta_{2,\pm 1} = \sqrt{\dfrac{15}{4}} \, Sin\,\theta \, Cos\,\theta$ | $Y_{2,\pm 1} = \sqrt{\dfrac{15}{4}} \, Sin\,\theta \, Cos\,\theta . \sqrt{\dfrac{1}{2\pi}} e^{\pm i\phi}$ |
| $\Theta_{2,\pm 2} = \sqrt{\dfrac{15}{16}} \, Sin^2\,\theta$ | $Y_{2,\pm 2} = \sqrt{\dfrac{15}{16}} \, Sin^2\,\theta . \sqrt{\dfrac{1}{2\pi}} e^{\pm i2\phi}$ |

**3. The solution of R(r) equation:** Recall and rearrange the differential equation obtained after separation of variables having $r$-dependence i.e.

$$\frac{1}{R}\frac{\partial}{\partial r}\left(r^2\frac{\partial R}{\partial r}\right) + \frac{8\pi^2\mu r^2}{h^2}\left(E + \frac{Ze^2}{r}\right) = \beta \tag{330}$$

After putting $\hbar = h/2\pi$ and rearranging, we get

$$\frac{1}{R}\frac{\partial}{\partial r}\left(r^2\frac{\partial R}{\partial r}\right) + \frac{2\mu r^2}{\hbar^2}(E - V) = \beta \tag{331}$$

After multiplying by R on both sides and then dividing by $r^2$ throughout, we get

$$\frac{1}{r^2}\frac{\partial}{\partial r}\left(r^2\frac{\partial R}{\partial r}\right) + \frac{2\mu}{\hbar^2}(E - V)R = \frac{\beta R}{r^2} \tag{332}$$

Now, as we know from the solution of $\Theta$-equation that $\beta = l(l+1)$, the above equation takes the form

$$\frac{1}{r^2}\frac{\partial}{\partial r}\left(r^2\frac{\partial R}{\partial r}\right) + \frac{2\mu}{\hbar^2}(E - V)R = \frac{l(l+1)R}{r^2} \tag{333}$$

$$\frac{1}{r^2}\frac{\partial}{\partial r}\left(r^2\frac{\partial R}{\partial r}\right) + \frac{2\mu}{\hbar^2}(E - V)R - \frac{l(l+1)R}{r^2} = 0 \tag{334}$$

or

$$\frac{1}{r^2}\left[r^2\frac{\partial^2 R}{\partial r^2} + 2r\frac{\partial R}{\partial r}\right] + \left[\frac{2\mu}{\hbar^2}(E - V) - \frac{l(l+1)}{r^2}\right]R = 0 \tag{335}$$

$$\left[\frac{\partial^2 R}{\partial r^2} + \frac{2}{r}\frac{\partial R}{\partial r}\right] + \left[\frac{2\mu}{\hbar^2}(E - V) - \frac{l(l+1)}{r^2}\right]R = 0 \tag{336}$$

Putting the value of potential energy for atomic hydrogen or hydrogen-like species again in the above equation, we get

$$\frac{\partial^2 R}{\partial r^2} + \frac{2}{r}\frac{\partial R}{\partial r} + \left[\frac{2\mu E}{\hbar^2} + \frac{2\mu Ze^2}{\hbar^2 r} - \frac{l(l+1)}{r^2}\right]R = 0 \tag{337}$$

As we know from the classical mechanics that elliptical orbits represent bound states have energies less than zero whereas hyperbolic orbits represent unbound states have energies greater than zero. Now assume that electron around the nucleus is bound somehow i.e.

$$-\frac{2\mu E}{\hbar^2} = \alpha^2 \quad and \quad \frac{\mu Ze^2}{\hbar^2\alpha} = \lambda \tag{338}$$

Using equation (338) in equation (337), we get

$$\frac{\partial^2 R}{\partial r^2} + \frac{2}{r}\frac{\partial R}{\partial r} + \left[-\alpha^2 + \frac{2\alpha\lambda}{r} - \frac{l(l+1)}{r^2}\right]R = 0 \tag{339}$$

At this stage, we need to define a new variable $\rho = 2\alpha r$, so that

$$\frac{\partial \rho}{\partial r} = 2\alpha \tag{340}$$

Which follows

$$\frac{\partial R}{\partial r} = \frac{\partial R}{\partial \rho}\cdot\frac{\partial \rho}{\partial r} = 2\alpha\frac{\partial R}{\partial \rho} \tag{341}$$

Also

$$\frac{\partial^2 R}{\partial r^2} = \frac{\partial}{\partial r}\left[\frac{\partial R}{\partial r}\right] = \frac{\partial}{\partial r}\left[2\alpha\frac{\partial R}{\partial \rho}\right] = \frac{\partial}{\partial r}\frac{\partial \rho}{\partial \rho}\left[2\alpha\frac{\partial R}{\partial \rho}\right] = \frac{\partial \rho}{\partial r}\frac{\partial}{\partial \rho}\left[2\alpha\frac{\partial R}{\partial \rho}\right] \tag{341}$$

$$\frac{\partial^2 R}{\partial r^2} = 2\alpha\frac{\partial}{\partial \rho}\left[2\alpha\frac{\partial R}{\partial \rho}\right] = 4\alpha^2\frac{\partial^2 R}{\partial \rho^2} \tag{342}$$

After using the values of $\partial R/\partial r$ and $\partial^2 R/\partial r^2$ from equation (341) and equation (342) in equation (339), we get the following.

$$4\alpha^2\frac{\partial^2 R}{\partial \rho^2} + \frac{2}{r}2\alpha\frac{\partial R}{\partial \rho} + \left[-\alpha^2 + \frac{2\alpha\lambda}{r} - \frac{l(l+1)}{r^2}\right]R = 0 \tag{343}$$

Now divide the above equation by $4\alpha^2$ i.e.

$$\frac{\partial^2 R}{\partial \rho^2} + \frac{1}{\alpha r}\frac{\partial R}{\partial \rho} + \left[-\frac{1}{4} + \frac{\lambda}{2\alpha r} - \frac{l(l+1)}{4\alpha^2 r^2}\right]R = 0 \tag{344}$$

Using $\rho = 2\alpha r$, we get

$$\frac{\partial^2 R}{\partial \rho^2} + \frac{2}{\rho}\frac{\partial R}{\partial \rho} + \left[-\frac{1}{4} + \frac{\lambda}{\rho} - \frac{l(l+1)}{\rho^2}\right]R = 0 \tag{345}$$

When $\rho \to \infty$, the above equation takes the form

$$\frac{\partial^2 R}{\partial \rho^2} - \frac{1}{4}R = 0 \tag{346}$$

The general solutions of the differential equation given above are

$$R(\rho) = e^{+\rho/2} \qquad and \qquad R(\rho) = e^{-\rho/2} \tag{347}$$

The function $R(\rho) = e^{+\rho/2}$ becomes $\infty$ when $\rho = \infty$, and hence, is not acceptable. Therefore, we are left with

$$R(\rho) = e^{-\rho/2} \tag{348}$$

Since the acceptable solution given above is valid only at very large values of $\rho$, it is quite reasonable to think that the exact solution may also contain some pre-exponential part to attain validity at all values of $\rho$. Therefore, after incorporating some $\rho$-dependent unknown function '$F(\rho)$' in equation (348), we get

$$R(\rho) = F(\rho)\, e^{-\rho/2} \tag{349}$$

Differentiating above equation with w.r.t $\rho$ at first and second order and then putting the values of $R(\rho)$, $\partial R/\partial \rho$ and $\partial^2 R/\partial \rho^2$ in equation (345), we get

$$\frac{\partial^2 F}{\partial \rho^2} + \left(\frac{2}{\rho} - 1\right)\frac{\partial F}{\partial \rho} + \left[-\frac{1}{\rho} + \frac{\lambda}{\rho} - \frac{l(l+1)}{\rho^2}\right] F = 0 \tag{350}$$

For simplification, put $\partial^2 R/\partial \rho^2 = F''$ and $\partial R/\partial \rho = F'$ i.e.

$$F'' + \left(\frac{2}{\rho} - 1\right) F' + \left[-\frac{1}{\rho} + \frac{\lambda}{\rho} - \frac{l(l+1)}{\rho^2}\right] F = 0 \tag{351}$$

Hence, the problem has been reduced to the determination of the solution of $F$ which can be assumed as

$$F(\rho) = \rho^s G(\rho) \tag{352}$$

Where $G(\rho)$ represents a power series expansion of $\rho$ i.e.

$$G(\rho) = a_0 + a_1\rho + a_2\rho^2 + a_3\rho^3 \ldots \tag{353}$$

Or we can say that

$$G(\rho) = \sum_{k=0}^{k=\infty} a_k\rho^k \tag{354}$$

It is also worthy to mention that $a_0 \neq 0$. Now differentiating equation (352) w.r.t. $\rho$, we get

$$F'(\rho) = s\rho^{s-1}G + \rho^s G' \tag{355}$$

The double derivative of the same will be

$$F''(\rho) = s(s-1)\rho^{s-2}G + 2s\rho^{s-1}G' + \rho^s G'' \tag{356}$$

After putting the values of $F(\rho)$, $F'(\rho)$ and $F''(\rho)$ from equation (352, 355, 356) into equation (351), we get

$$s(s-1)\rho^{s-2}G + 2s\rho^{s-1}G' + \rho^s G'' + \left(\frac{2}{\rho} - 1\right)[s\rho^{s-1}G + \rho^s G'] \tag{357}$$

$$+ \left[-\frac{1}{\rho} + \frac{\lambda}{\rho} - \frac{l(l+1)}{\rho^2}\right]\rho^s G = 0$$

Multiplying throughout by $4\rho^2$, we get

$$4\rho^2 s(s-1)\rho^{s-2}G + 4\rho^2.2s\rho^{s-1}G' + 4\rho^2.\rho^s G'' + (8\rho - 4\rho^2)[s\rho^{s-1}G + \rho^s G'] \tag{358}$$
$$+ [-4\rho + 4\rho\lambda - 4l(l+1)]\rho^s G = 0$$

or

$$4s(s-1)\rho^s G + 8s\rho^{s+1}G' + 4\rho^{s+2}G'' + 8s\rho^s G - 4s\rho^{s+1}G + 8\rho^{s+1}G' \tag{359}$$
$$- 4\rho^{s+2}G' - 4\rho^{s+1}G + 4\lambda\rho^{s+1}G - 4l(l+1)\rho^s G = 0$$

or

$$4s(s-1)\rho^s G + 8s\rho^s G - 4s\rho^{s+1}G - 4\rho^{s+1}G + 4\lambda\rho^{s+1}G - 4l(l+1)\rho^s G \tag{360}$$
$$+ 8s\rho^{s+1}G' + 8\rho^{s+1}G' - 4\rho^{s+2}G' + 4\rho^{s+2}G'' = 0$$

or

$$[4s(s-1)\rho^s + 8s\rho^s - 4s\rho^{s+1} - 4\rho^{s+1} + 4\lambda\rho^{s+1} - 4l(l+1)\rho^s]G \tag{361}$$
$$+ [8s\rho^{s+1} + 8\rho^{s+1} - 4\rho^{s+2}]G' + 4\rho^{s+2}G'' = 0$$

Dividing throughout by $\rho^s$, we get

$$[4s(s-1) + 8s - 4s\rho - 4\rho + 4\lambda\rho - 4l(l+1)]G + [8s\rho + 8\rho - 4\rho^2]G' \tag{362}$$
$$+ 4\rho^2 G'' = 0$$

If $\rho = 0$, the function $G(\rho) = a_0$ and the above equation takes the form

$$[4s(s-1) + 8s - 4l(l+1)]a_0 = 0 \tag{363}$$

Since $a_0 \neq 0$, the quantity that must be equal to zero to satisfy the above result is

$$4s(s-1) + 8s - 4l(l+1) = 0 \tag{364}$$

$$s(s-1) + 2s - l(l+1) = 0 \tag{365}$$

$$s(s+1) - l(l+1) = 0 \tag{366}$$

$$s(s+1) = l(l+1)$$

Which implies that

$$s = l \qquad or \qquad s = -(l+1) \tag{367}$$

Now, if we put $s = -(l+1)$ the first term in the function $F(\rho)$ becomes $a_0/0^{l+1}$ at $\rho = 0$ which infinite, and hence is not an acceptable solution. Thus, the only we are left with is $s = l$; after using the same in equation (362), we get

$$[4l(l-1) + 8l - 4l\rho - 4\rho + 4\lambda\rho - 4l(l+1)]G + [8l\rho + 8\rho - 4\rho^2]G' + 4\rho^2 G'' \tag{368}$$
$$= 0$$

$$[-4l\rho - 4\rho + 4\lambda\rho]G + [8l\rho + 8\rho - 4\rho^2]G' + 4\rho^2 G'' = 0 \tag{369}$$

Dividing the above equation by $4\rho$, we get

$$[-l - 1 + \lambda]G + [2l + 2 - \rho]G' + \rho G'' = 0 \tag{370}$$

Now differentiating equation (353) at first and second order, we get

$$G'(\rho) = a_1 1\rho^{1-1} + a_2 2\rho^{2-1} + a_3 3\rho^{3-1} \dots = \sum_{k=0}^{k=\infty} a_k . k . \rho^{k-1} \tag{371}$$

Similarly

$$G''(\rho) = a_2 . 2 . (2-1)\rho^{2-2} + a_3 . 3 . (3-1)\rho^{3-2} \dots = \sum_{k=0}^{k=\infty} a_k . k . (k-1) . \rho^{k-2} \tag{372}$$

After using the values equation (354, 371, 372) into equation (370), we get

$$[-l - 1 + \lambda] \sum_{k=0}^{k=\infty} a_k \rho^k + [2l + 2 - \rho] \sum_{k=0}^{k=\infty} a_k . k . \rho^{k-1} + \rho \sum_{k=0}^{k=\infty} a_k . k . (k-1) . \rho^{k-2} \tag{373}$$
$$= 0$$

The above equation holds true only if the coefficients of individual powers of $\rho$ become zero. So, simplifying equation (373) for two summation terms ($a_k$ and $a_{k+1}$), we have

$$[-l - 1 + \lambda][a_k\rho^k + a_{k+1}\rho^{k+1}] + [2l + 2 - \rho][a_k . k . \rho^{k-1} + a_{k+1} . (k+1) . \rho^k] \tag{374}$$
$$+ \rho[a_k . k . (k-1) . \rho^{k-2} + a_{k+1} . (k+1) . k . \rho^{k-1}] = 0$$

$$-la_k\rho^k - a_k\rho^k + \lambda a_k\rho^k - la_{k+1}\rho^{k+1} - a_{k+1}\rho^{k+1} + \lambda a_{k+1}\rho^{k+1} + 2la_k . k . \rho^{k-1} \tag{375}$$
$$+ 2a_k . k . \rho^{k-1} - \rho a_k . k . \rho^{k-1} + 2la_{k+1} . (k+1) . \rho^k + 2a_{k+1} . (k$$
$$+ 1) . \rho^k - \rho a_{k+1} . (k+1) . \rho^k + \rho . a_k . k . (k-1) . \rho^{k-2}$$
$$+ \rho . a_{k+1} . (k+1) . k . \rho^{k-1} = 0$$

Now putting a coefficient of $\rho^k$ equal to zero, we get

$$-la_k\rho^k - a_k\rho^k + \lambda a_k\rho^k - a_k . k . \rho^k + 2la_{k+1} . (k+1) . \rho^k + 2a_{k+1} . (k+1) . \rho^k \tag{376}$$
$$+ a_{k+1} . (k+1) . k . \rho^k = 0$$

$$-la_k - a_k + \lambda a_k - a_k k + 2la_{k+1}(k+1) + 2a_{k+1}(k+1) + a_{k+1}(k+1)k = 0 \tag{377}$$

or

$$[-l - 1 + \lambda - k]a_k + [2l(k + 1) + 2(k + 1) + (k + 1)k]a_{k+1} = 0 \tag{378}$$

$$[2l(k + 1) + 2(k + 1) + (k + 1)k]a_{k+1} = -[-l - 1 + \lambda - k]a_k \tag{379}$$

or

$$a_{k+1} = \frac{l + 1 - \lambda + k}{2l(k + 1) + 2(k + 1) + (k + 1)k}\, a_k \tag{380}$$

or

$$a_{k+1} = \frac{l + 1 - \lambda + k}{(k + 1)(2l + k + 2)}\, a_k \tag{381}$$

The equation (384) is the recursion formula where $k$ is an integer. This expression allows one to determine the coefficient $a_{k+1}$ in terms of $a_k$ which is arbitrary.

Now, since the series $G(\rho)$ consists of the infinite number of terms, the function $F(\rho)$ becomes infinite at a very large value of $k$ i.e. infinite. Consequently, the function $R(\rho)$ will also become infinite if the number of terms is not limited to a finite value. Therefore, we must break off the series to a finite number of terms which is possible only if the numerator becomes zero i.e.

$$l + 1 - \lambda + k = 0 \tag{382}$$

Define a new quantum number "$n$" at this stage as

$$\lambda = l + 1 + k = n \tag{383}$$

Since $l$ and $k$ are integers, $n$ can be 1, 2, 3. 4 …. and so on. Moreover, as $n \geq l + 1$, the largest value that l can have is $n - 1$. Hence, the value of $l$ has a domain ranging from 0 to $n - 1$. Putting $\lambda = n$ in equation (370)

$$[-l - 1 + n]G + [2l + 2 - \rho]G' + \rho G'' = 0 \tag{384}$$

Defining $2l + 1 = p$ and $n + l = q$, we get

$$[q - p]G + [p + 1 - \rho]G' + \rho G'' = 0 \tag{385}$$

The solution of the equation given above is the "associated Laguerre polynomial" multiplied by a constant factor i.e.

$$G(p) = CL_q^p(\rho) = CL_{n+l}^{2l+1}(\rho) \tag{385}$$

The constant C can be set as normalization constant and "associated Laguerre polynomial" is

$$L_{n+l}^{2l+1}(\rho) = \sum_{k=0}^{k=n-l-1} \frac{(-1)^{k+1}[(n + l)!]^2 \rho^k}{(n - l - 1 - k)!\,(2l + 1 + k)!\,k!} \tag{385}$$

After using the value of $F(\rho)$ from equation (352) in equation (349), we get radial wavefunction as

$$R(\rho) = \rho^s G(\rho)\, e^{-\rho/2} \tag{386}$$

Since $s = l$ and also using $G(p)$ from equation (385), the above equation takes the form

$$R_{n,l}(\rho) = C\, e^{-\rho/2} \rho^l L_{n+l}^{2l+1}(\rho) \tag{387}$$

Now, after using the value of $L_{n+l}^{2l+1}(\rho)$ from equation (385) in equation (387), we get

$$R_{n,l}(\rho) = C\, e^{-\rho/2} \rho^l \sum_{k=0}^{k=n-l-1} \frac{(-1)^{k+1}[(n+l)!]^2 \rho^k}{(n-l-1-k)!\,(2l+1+k)!\,k!} \tag{388}$$

*i) The normalization constant for function R(r):* In order to determine the normalization constant for the $R$ function, we must put the squared-integral over whole configuration space as unity i.e.

$$\int_0^\infty R_{n,l}^2(r).\,r^2.\,dr = 1 \tag{389}$$

The factor $r^2$ is introduced to convert the length $dr$ into a volume around the center of the nucleus. At this point, recall the value of $\rho$ again but in terms of equation (338, 383) i.e.

$$\rho = 2ar = \frac{2\mu Z e^2 r}{\hbar^2 \lambda} = \frac{2\mu Z e^2 r}{\hbar^2 n} = \frac{2Zr}{n} \cdot \frac{\mu e^2}{\hbar^2} \tag{390}$$

Since $a_0 = \hbar^2/\mu e^2$ i.e. the "Bohr radius", the equation (390) takes the form

$$\rho = \frac{2Zr}{n} \cdot \frac{1}{a_0} \tag{391}$$

So that

$$r = \frac{na_0}{2Z}\rho \tag{392}$$

Also

$$dr = \frac{na_0}{2Z}d\rho \tag{393}$$

After using the values of $R_{n,l}(\rho)$, $r$ and $dr$ from equation (388, 392, 393) in equation (389), we get

$$C^2 \int_0^\infty e^\rho.\,\rho^{2l}.\,\left[L_{n+l}^{2l+1}(\rho)\right]^2 .\,\left[\frac{na_0}{2Z}\rho\right]^2 .\,\left[\frac{na_0}{2Z}\right] d\rho = 1 \tag{394}$$

or

$$C^2 \left(\frac{na_0}{2Z}\right)^3 \left[\frac{2n\{(n+l)!\}^3}{(n-l-1)!}\right] = 1 \tag{395}$$

or

$$C = \sqrt{\left(\frac{2Z}{na_0}\right)^3 \left[\frac{(n-l-1)!}{2n\{(n+l)!\}^3}\right]} \tag{395}$$

After using the value of normalization constant from above equation into equation (388), we get

$$R_{n,l}(\rho) = \sqrt{\left(\frac{2Z}{na_0}\right)^3 \left[\frac{(n-l-1)!}{2n\{(n+l)!\}^3}\right]} \, e^{-\rho/2} \rho^l \sum_{k=0}^{k=n-l-1} \frac{(-1)^{k+1}[(n+l)!]^2 \rho^k}{(n-l-1-k)!\,(2l+1+k)!\,k!} \tag{396}$$

$$-\sqrt{\left(\frac{2Z}{na_0}\right)^3 \left[\frac{(n-l-1)!}{2n\{(n+l)!\}^3}\right]} \cdot \exp\left(-\frac{Zr}{na_0}\right) \cdot \left(\frac{2Zr}{na_0}\right)^l \cdot \sum_{k=0}^{k=n-l-1} \frac{(-1)^{k+1}[(n+l)!]^2 \left(\frac{2Zr}{na_0}\right)^k}{(n-l-1-k)!\,(2l+1+k)!\,k!} \tag{397}$$

Which is the complete solution of $R$-equation.

Table 3. Some of the initial radial wave functions in terms of distance from the center of the nucleus for the hydrogen atom and other hydrogen-like species.

| $n$ | $l$ | Radial wave function ($R_{n,l}$) |
|-----|-----|----------------------------------|
| 1 | 0 | $R_{1,0} = 2\left(\dfrac{Z}{a_0}\right)^{3/2} e^{-Zr/a_0}$ |
| 2 | 0 | $R_{2,0} = \dfrac{1}{2\sqrt{2}}\left(\dfrac{Z}{a_0}\right)^{3/2}\left(2 - \dfrac{Zr}{a_0}\right) e^{-Zr/2a_0}$ |
| 2 | 1 | $R_{2,1} = \dfrac{1}{2\sqrt{6}}\left(\dfrac{Z}{a_0}\right)^{3/2}\left(\dfrac{Zr}{a_0}\right) e^{-Zr/2a_0}$ |
| 3 | 0 | $R_{3,0} = \dfrac{2}{81\sqrt{3}}\left(\dfrac{Z}{a_0}\right)^{3/2}\left(27 - 18\dfrac{Zr}{a_0} - 2\left(\dfrac{Zr}{a_0}\right)^2\right) e^{-Zr/3a_0}$ |
| 3 | 1 | $R_{3,1} = \dfrac{4}{81\sqrt{6}}\left(\dfrac{Z}{a_0}\right)^{3/2}\left(6\left(\dfrac{Zr}{a_0}\right) - \left(\dfrac{Zr}{a_0}\right)^2\right) e^{-Zr/3a_0}$ |
| 3 | 2 | $R_{3,2} = \dfrac{1}{81\sqrt{30}}\left(\dfrac{Z}{a_0}\right)^{3/2}\left(\dfrac{Zr}{a_0}\right)^{3/2} - \left(\dfrac{Zr}{a_0}\right)^2 e^{-Zr/3a_0}$ |

**The total wavefunction:** After solving the $\phi$-, $\theta$- and $r$-dependent equations, we have $\Phi_m(\phi)$, $\Theta_{l,m}(\theta)$ and $R_{n,l}(r)$ functions. Now, recall the total wave function that depends upon all the three variable i.e.

$$\psi_{n,l,m}(r,\theta,\phi) = \psi_{n,l}(r) \times \psi_{l,m}(\theta) \times \psi_m(\phi) \tag{398}$$

After putting the values of $\Phi_m(\phi)$, $\Theta_{l,m}(\theta)$ and $R_{n,l}(r)$ from equation (397) in equation (398), we get

$$\psi_{n,l,m}(r,\theta,\phi) = R_{n,l}.\,\Theta_{l,m}.\,\Phi_m \tag{399}$$

$$
\begin{aligned}
= & \sqrt{\left(\frac{2Z}{na_0}\right)^3 \left[\frac{(n-l-1)!}{2n\{(n+l)!\}^3}\right]} \cdot \exp\left(-\frac{Zr}{na_0}\right).\left(\frac{2Zr}{na_0}\right)^l \cdot \sum_{k=0}^{k=n-l-1} \frac{(-1)^{k+1}[(n+l)!]^2 \left(\frac{2Zr}{na_0}\right)^k}{(n-l-1-k)!\,(2l+1+k)!\,k!} \\
& \times \sqrt{\frac{(2l+1)(l-m)!}{2(l+m)!}} \cdot P_l^m(Cos\,\theta) \times \sqrt{\frac{1}{2\pi}}\, e^{im\phi}
\end{aligned}
\tag{400}
$$

Which is the complete expression for all the quantum mechanical states of a single electron around the nucleus.

Table 4. Some of the initial total wave functions for the hydrogen atom and other hydrogen-like species.

| $n$ | $l$ | $m$ | Total wave function ($\psi_{n,l,m}$) |
|---|---|---|---|
| 1 | 0 | 0 | $\psi_{1,0,0} = \dfrac{1}{\sqrt{\pi}}\left(\dfrac{Z}{a_0}\right)^{3/2} e^{-Zr/a_0}$ |
| 2 | 0 | 0 | $\psi_{2,0,0} = \dfrac{1}{4\sqrt{\pi}}\left(\dfrac{Z}{a_0}\right)^{3/2}\left(2 - \dfrac{Zr}{a_0}\right) e^{-Zr/2a_0}$ |
| 2 | 1 | 0 | $\psi_{2,1,0} = \dfrac{1}{4\sqrt{\pi}}\left(\dfrac{Z}{a_0}\right)^{5/2} e^{-Zr/2a_0}\, r\, Cos\theta$ |
| 2 | 1 | $\pm 1$ | $\psi_{2,1,\pm 1} = \dfrac{1}{4\sqrt{\pi}}\left(\dfrac{Z}{a_0}\right)^{5/2} e^{-Zr/2a_0}\, Sin\theta\, e^{-i\phi}$ |
| 3 | 0 | 0 | $\psi_{3,0,0} = \dfrac{2}{81\sqrt{3}}\left(\dfrac{Z}{a_0}\right)^{3/2}\left(27 - 18\dfrac{Zr}{a_0} - 2\left(\dfrac{Zr}{a_0}\right)^2\right) e^{-Zr/3a_0}.\dfrac{1}{\sqrt{4\pi}}$ |
| 3 | 1 | 0 | $\psi_{3,1,0} = \dfrac{4}{81\sqrt{6}}\left(\dfrac{Z}{a_0}\right)^{3/2}\left(6\left(\dfrac{Zr}{a_0}\right) - \left(\dfrac{Zr}{a_0}\right)^2\right) e^{-Zr/3a_0}.\sqrt{\dfrac{3}{2}}\, Cos\,\theta.\dfrac{1}{\sqrt{2\pi}}$ |
| 3 | 1 | $\pm 1$ | $\psi_{3,1,\pm 1} = \dfrac{4}{81\sqrt{6}}\left(\dfrac{Z}{a_0}\right)^{3/2}\left(6\left(\dfrac{Zr}{a_0}\right) - \left(\dfrac{Zr}{a_0}\right)^2\right) e^{-Zr/3a_0}.\sqrt{\dfrac{3}{2}}\, Sin\,\theta.\sqrt{\dfrac{1}{2\pi}}\, e^{\pm i\phi}$ |

**The eigenvalues of energy:** Since the series $G(\rho)$ consists of infinite number of terms, the function $F(\rho)$ becomes infinite at a very large value of $k$ i.e. infinite. Consequently, the function $R(\rho)$ will also become infinite if the number of terms are not limited to a finite value. Therefore, we must break off the series to a finite number of terms which is possible only if the numerator in equation (381) becomes zero i.e.

$$l + 1 - \lambda + k = 0 \tag{401}$$

or

$$\lambda = l + 1 + k = n \tag{402}$$

Where n is the principal quantum number and can have values 1, 2, 3. 4 …. because $l$ and $k$ are integers always. Now recall the value of $\lambda$ from equation (338) and then squaring both sides, we get

$$\lambda^2 = \frac{\mu^2 Z^2 e^4}{\hbar^4 \alpha^2} \tag{403}$$

Also putting the value of $\alpha^2$ from equation (338) in equation (403), we get

$$\lambda^2 = \frac{\mu^2 Z^2 e^4}{\hbar^4 \alpha^2} = -\frac{\mu^2 Z^2 e^4}{\hbar^4} \cdot \frac{\hbar^2}{2\mu E} = -\frac{\mu Z^2 e^4}{2 E \hbar^2} \tag{404}$$

$$E_n = -\frac{\mu Z^2 e^4}{2 \lambda^2 \hbar^2} == -\frac{\mu Z^2 e^4}{2 n^2 \hbar^2} \tag{405}$$

Which is the same as given by the pre-wave-mechanical quantum theory.

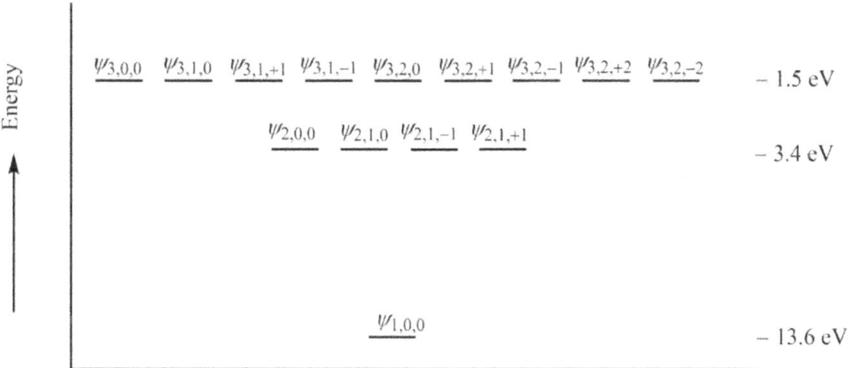


Figure 14. The energy level for various quantum mechanical states of the hydrogen atom.

It is also worthy to note that the total number of wave functions that can be written for a given value of $n$ are $n^2$, and therefore, we can say that the degeneracy of any energy level is also $n^2$.

## ❖ Principal, Azimuthal and Magnetic Quantum Numbers and the Magnitude of Their Values

The Schrodinger wave equation for hydrogen and hydrogen-like species in the polar coordinates can be written as:

$$\frac{1}{r^2}\left[\frac{\partial}{\partial r}\left(r^2\frac{\partial\psi}{\partial r}\right) + \frac{1}{Sin\theta}\frac{\partial}{\partial\theta}\left(Sin\theta\frac{\partial\psi}{\partial\theta}\right) + \frac{1}{Sin^2\theta}\frac{\partial^2\psi}{\partial\phi^2}\right] + \frac{8\pi^2\mu}{h^2}\left(E + \frac{Ze^2}{r}\right)\psi = 0 \tag{406}$$

After separating the variables present in the equation given above, the solution of the differential equation was found to be

$$\psi_{n,l,m}(r,\theta,\phi) = R_{n,l}.\Theta_{l,m}.\Phi_m \tag{407}$$

$$
= \sqrt{\left(\frac{2Z}{na_0}\right)^3\left[\frac{(n-l-1)!}{2n\{(n+l)!\}^3}\right]} . \exp\left(-\frac{Zr}{na_0}\right).\left(\frac{2Zr}{na_0}\right)^l . \sum_{k=0}^{k=n-l-1} \frac{(-1)^{k+1}[(n+l)!]^2\left(\frac{2Zr}{na_0}\right)^k}{(n-l-1-k)!\,(2l+1+k)!\,k!} \tag{408}
$$

$$
\times \sqrt{\frac{(2l+1)(l-m)!}{2(l+m)!}} . P_l^m(Cos\,\theta) \times \sqrt{\frac{1}{2\pi}}\,e^{im\phi}
$$

It is obvious that the solution of equation (406) contains three discrete ($n$, $l$, $m$) and three continuous ($r$, $\theta$, $\phi$) variables. In order to be a well-behaved function, there are some conditions over the values of discrete variables that must be followed i.e. boundary conditions. Therefore, we can conclude that principal ($n$), azimuthal ($l$) and magnetic ($m$) quantum numbers are obtained as a solution of the Schrodinger wave equation for hydrogen atom; and these quantum numbers are used to define various quantum mechanical states. In this section, we will discuss the properties and significance of all these three quantum numbers one by one.

### ➢ *Principal Quantum Number*

The principal quantum number is denoted by the symbol $n$; and can have value 1, 2, 3, 4, 5…..∞. The label "principal" is allotted because valid values of $l$ and $m$ can be defined only after defining an acceptable value of $n$. Some of the most important significances of the principal quantum number are given below.

**1. The energy of an electron in hydrogen-like systems:** The principal quantum number gives the energy of the electron in all hydrogen and hydrogen-like species by the following relation.

$$E_n = -\frac{\mu Z^2 e^4}{2\,n^2\,\hbar^2} \tag{409}$$

Where $\mu$ is the reduced mass of the system while $e$ represents the electronic charge. The symbol $Z$ represents the nuclear charge of the one-electron system. Now since main shells are nothing but the classification of different quantum mechanical states of electron on the basis of energy only, we can also say that n tells about the main shells in the modern wave mechanical model of the atom.

**2. Degeneracy in hydrogen-like systems:** Since the total number of wave functions that can be written for a given value of $n$ are $n^2$, we can say that the degeneracy of any energy level is also $n^2$. In other words, we can say that because the energy depends only upon the value of $n$, all wave functions with the same value of $n$ must possess the same energy. For instance, if we $n = 2$, a total of four wave-functions can be written i.e. $\psi_{2,0,0}, \psi_{2,1,0}, \psi_{2,1,+1}$ and $\psi_{2,1,-1}$. Owing to the same value of $n$, all of these states are bound to have the same energy, and thus, are degenerate.

**3. The maximum number of electrons per unit cell:** Since two electrons can have the same set of principal, orbital and magnetic quantum numbers via opposite spins, the maximum number of electrons per unit cell will be $2n^2$ i.e. the double of the degeneracy. For instance, if we $n = 2$, the maximum number of electrons that can be filled in the second main shell is $2 \times 2^2 = 8$. Similarly, if we $n = 3$, the maximum number of electrons that can be filled in the third main shell is $2 \times 3^2 = 18$.

**4. Spectra of elemental hydrogen:** In order to understand this concept, recall the energy expression for the hydrogen atom i.e.

$$E_n = -\frac{m\,e^4}{2\,n^2\,\hbar^2} = -\frac{4\pi^2 m\,e^4}{2\,n^2 h^2} \tag{410}$$

In the SI system, the above equation needs to be corrected for permittivity factor ($4\pi\varepsilon_0$) i.e.

$$E_n = -\frac{4\pi^2 m\,e^4}{2\,n^2 h^2 (4\pi\varepsilon_0)^2} Joules = -\frac{4\pi^2 m\,e^4}{32\,\pi^2 n^2 h^2 \varepsilon_0^2} Joules \tag{411}$$

or

$$E_n = -\frac{m\,e^4}{8\,n^2 h^2 \varepsilon_0^2} Joules \tag{412}$$

Converting Joules into cm$^{-1}$ (dividing by $hc$) the above equation takes the form

$$\bar{v}_n = -\frac{m\,e^4}{8\,n^2 h^3 c\,\varepsilon_0^2}\,cm^{-1} = -\frac{R}{n^2}\,cm^{-1} \tag{413}$$

Now the selection rules for electronic transitions are

$$\Delta n = anything \quad and \quad \Delta l = \pm 1 \tag{414}$$

This means that electron can move from $s$ to $p$-orbital only; and all $s$–$s$, $p$–$p$, $d$–$d$ and $f$–$f$ transitions are Laporte forbidden. Now assume that electron shows a transition from an initial quantum mechanical state ($n_1$) to the final quantum mechanical state ($n_2$). The energy of absorption can be formulated as:

$$\Delta\bar{v} = \bar{v}_{n_2} - \bar{v}_{n_1} = \left(-\frac{R}{n_2^2}\right) - \left(-\frac{R}{n_1^2}\right)cm^{-1} \tag{415}$$

or

$$\Delta\bar{v} = R\left(\frac{1}{n_1^2} - \frac{1}{n_2^2}\right) cm^{-1} \tag{416}$$

The above equation can also be used to determine the wavenumber of the emission spectral line of the hydrogen atom. All the possibilities and corresponding series are given below.

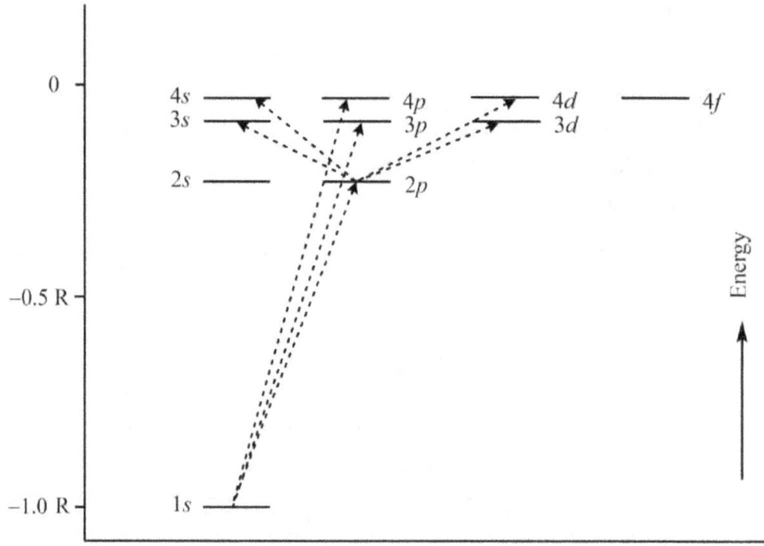


Figure 15. Energy level diagram of the hydrogen atom in the units of Rydberg constant.

At this stage, we must consider all the possibilities that may arise from the transitioning of the electron from different quantum mechanical states. These transitions are grouped in various series labeled as Lyman, Balmer, Paschen, Brackett and Pfund series.

Table 5. Different spectral series in the hydrogen atom.

| Series name | Lower state ($n_1$) | Higher state ($n_2$) | Region |
| --- | --- | --- | --- |
| Lyman | 1 | 2, 3, 4, 5, .... $\infty$ | UV |
| Balmer | 2 | 3, 4, 5, 6, .... $\infty$ | Visible |
| Paschen | 3 | 4, 5, 6, 7, .... $\infty$ | Near IR |
| Brackett | 4 | 5, 6, 7, 8, .... $\infty$ | Mid IR |
| Pfund | 5 | 6, 7, 8, 9, .... $\infty$ | Far IR |

> ### *Azimuthal Quantum Number*

The azimuthal quantum number is denoted by the symbol $l$; and can have value $n–1, n–2, n–3…..0$. The label azimuthal quantum number is also called as "angular momentum quantum number" because the values of $l$ also govern the orbital angular momentum of the electron in a particular quantum mechanical state. Some of the most important significances of the principal quantum number are given below.

**1. Orbital angular momentum of the electron:** The azimuthal quantum number gives the angular momentum of the electron in all hydrogen and hydrogen-like species by the following relation.

$$< L >= \oint \psi_{n,l,m}(r,\theta,\phi)\,\hat{L}\,\psi_{n,l,m}(r,\theta,\phi) \tag{417}$$

$$= \sqrt{l(l+1)}\,\frac{h}{2\pi} \tag{418}$$

After looking at the equation (418), it is obvious that it's only the '$l$' quantum number that controls the magnitude of the orbital angular momentum quantum number. Furthermore, owing to the quantized nature of '$l$' quantum number, the angular momentum of an electron in an atom is also quantized. For instance, if we use $l = 0, 1, 2, 3$ in equation (418), we will get $0, \sqrt{2}, \sqrt{6}$ and $\sqrt{12}$ units of angular momentum, respectively.

**2. Subshells in the main shell:** The azimuthal quantum number can also be used to classify different quantum mechanical states on the basis of orbital angular momentum. In other words, subshells are nothing but the classification degenerate quantum mechanical states on the basis of angular momentum.

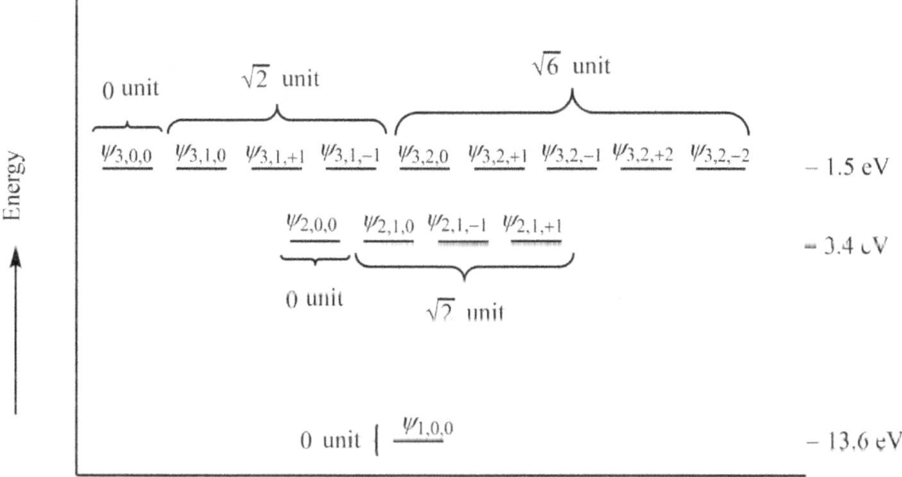


Figure 16. Energy level diagram of the hydrogen atom with further classification.

**3. Shape and number of angular nodes in atomic orbital:** The angular momentum quantum number, $l$, also controls the number of angular nodes that pass through the nucleus. An angular node (planar or conical) is observed when the angular part of the wave function passes through zero and changes sign.

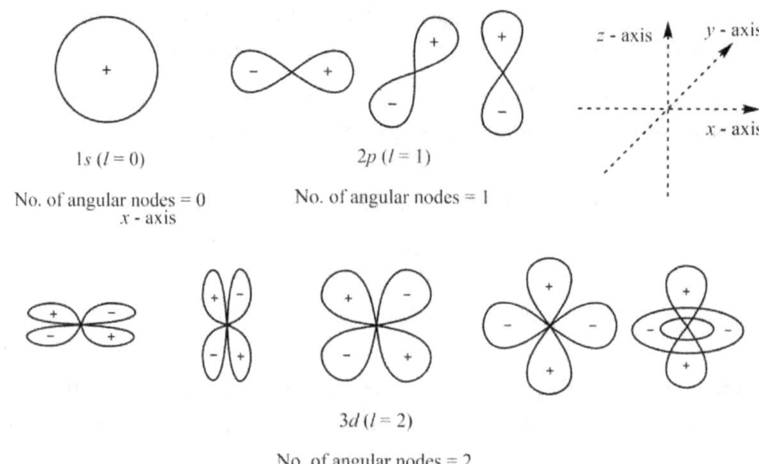

Figure 17. Different orbitals and the corresponding angular nodes.

**4. The energy of different subshells in multi-electron atoms:** In hydrogen and H-like atoms (i.e. one-electron systems), the energy levels depend only upon the principal quantum number. However, these energy levels also split according to the magnitude of $l$ as well. Quantum states of higher $l$ are placed above than the states with lower $l$. For instance, the energy of $2s$ orbital is lower than $2p$, $3d$ exists at higher position than $3p$.

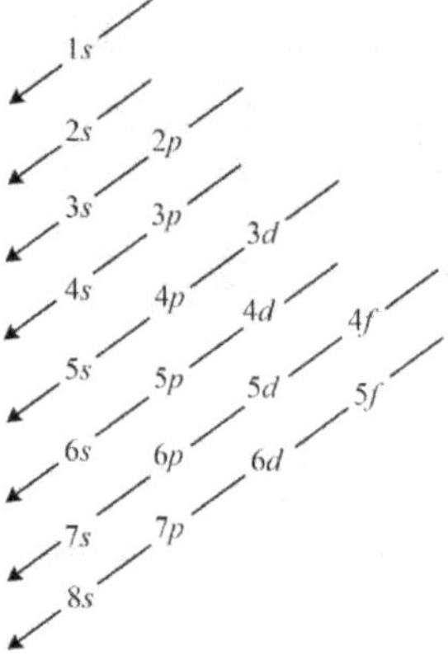

Figure 17. Different orbitals and the corresponding angular nodes.

> ### *Magnetic Quantum Number*

The magnetic quantum number is denoted by the symbol $m$, and can have values $+l$ to $–l$ in unit steps. In other words, the quantum number $m$ is nothing but the allowed effects of orbital angular momentum in the $z$-direction. The label "magnetic quantum number" arises because $m$ affects the energy of the electron in an externally applied magnetic field. In the absence of such a field, all spherical harmonics corresponding to the different arbitrary values of $m$ will be equivalent. Some of the most important significances of the principal quantum number are given below.

**1. The orientation of orbital angular momentum:** The azimuthal quantum number gives the angular momentum of the electron in all hydrogen and hydrogen-like species by the following relation:

$$L_l = \sqrt{l(l+1)}\,\frac{h}{2\pi} \tag{419}$$

Since $l = 0, 1, 2, 3, 4 \ldots \ldots (n-1)$ etc., the quantum mechanically allowed values of orbital angular momentum (in the units of $h/2\pi$) are given below.

$$L_0 = \sqrt{0(0+1)}\ \text{unit} = 0\ \text{unit} \tag{420}$$

$$L_1 = \sqrt{1(1+1)}\ \text{unit} = \sqrt{2}\ \text{unit} \tag{421}$$

$$L_2 = \sqrt{2(2+1)}\ \text{unit} = \sqrt{6}\ \text{unit} \tag{422}$$

$$L_3 = \sqrt{3(3+1)}\ \text{unit} = \sqrt{12}\ \text{unit} \tag{423}$$

However, there is boundary condition in quantum mechanics that says that only integral effects are allowed reference direction if the angular momentum is generated by integral quantum number and half-integral effects are allowed in reference direction if the momentum is generated by half-integral quantum number.

Since, $L_z = L\ Cos\ \theta$, $\sqrt{2}$ units of orbital angular momentum cannot orient itself along $z$-axis because this makes $\theta = 0°$, and since $Cos\ 0 = 1$, $\vec{L}_z = L$ i.e. orbital angular momentum effect along the $z$-axis is also 1.414 unit which is not allowed quantum mechanically. The effects of angular momentum allowed in the $z$-direction are $+1, 0, -1$; for which angles required are determined as follows.

$$+1 = \sqrt{2}\ Cos\ \theta \quad \Rightarrow \quad \theta = Cos^{-1}\frac{1}{\sqrt{2}} = 45° \tag{424}$$

$$0 = \sqrt{2}\ Cos\ \theta \quad \rightarrow \quad \theta = Cos^{-1}\frac{0}{\sqrt{2}} = 90° \tag{425}$$

$$-1 = \sqrt{2}\ Cos\ \theta \quad \Rightarrow \quad \theta = Cos^{-1}\frac{-1}{\sqrt{2}} = 135° \tag{426}$$

Hence, we can say that in order to be allowed, the 1.414 units of orbital angular momentum must orient itself only at 45°, 90° and 135° in space from reference direction ($z$-axis in this case). Since the orientation of angular momentum can orient itself in any direction from the $z$-axis as far as the effective orbital angular momentum +1 unit along $z$-direction; therefore, we should use a cone around the same at 45°. The same is true for 0 and −1 effects with 90° and 135°, respectively.

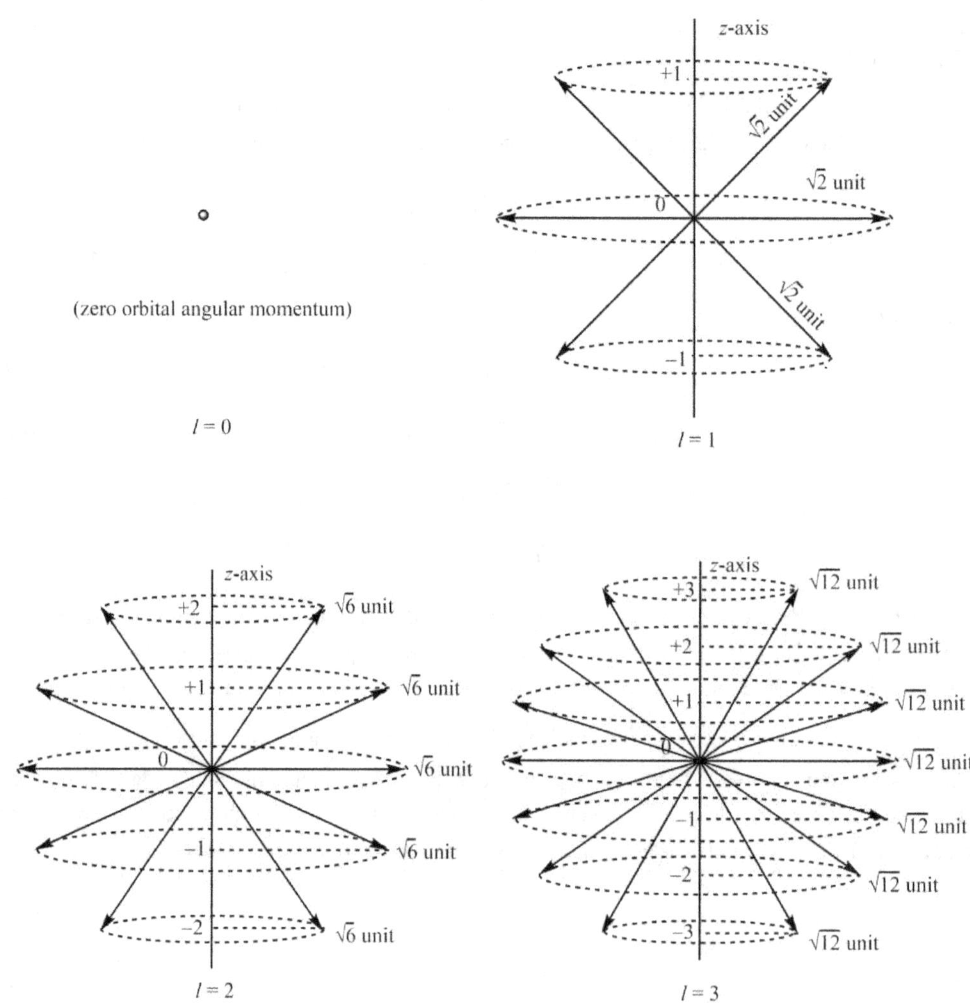


Figure 18. The space quantization of orbital angular momentum of an electron for $l$ = 0, 1, 2 and 3 states.

It is also worthy to mention that all the orientations of orbital angular momentum are degenerate in the absence of any externally applied magnetic field.

**2. The energy of different orientations:** The probability of different orientations of orbital angular momentum, and also the corresponding magnetic fields, are same. However, after applying the magnetic field along the $z$-direction, the situation will not be the same.

The orientation with a maximum negative value of orbital angular momentum along the z-axis will have magnetic dipole aligned along the applied magnetic field, and thus, will be most stable. Conversely, the orientations with the maximum positive value of orbital angular momentum along the z-axis will have magnetic dipole aligned opposite to the applied magnetic field, and thus, will be least stable. Likewise, the orientations with angular momentum component in between will also have intermediary energies in this case.

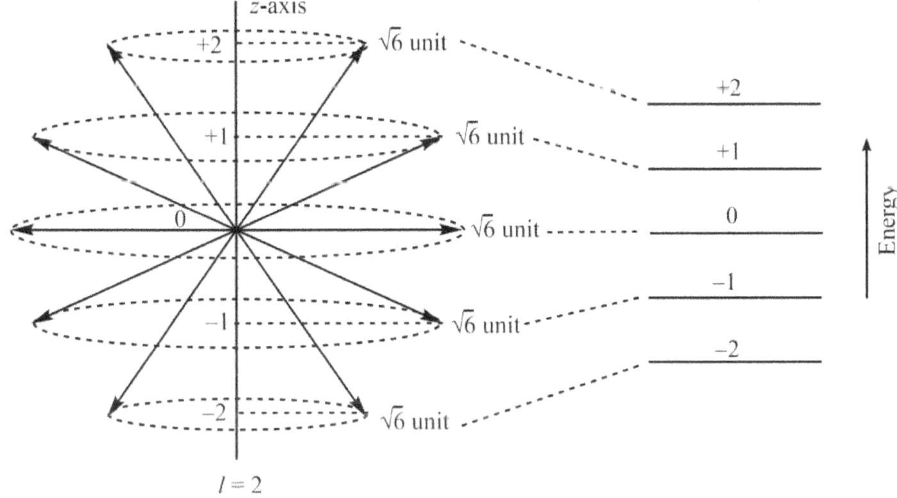


Figure 19. Different orientations of the orbital angular momentum of an electron in a $d$-subshell and corresponding energies in the applied magnetic field.

To understand this, consider the typical case of $d$-subshell. The orbital angular momentum for the corresponding electron can be obtained using equation (419) i.e.

$$L = \sqrt{2(2+1)}\,\frac{h}{2\pi} \tag{427}$$

$$L = \sqrt{6}\,\frac{h}{2\pi} \tag{428}$$

The allowed effects of orbital angular momentum in the $z$-direction are $+2, +1, 0, -1, -2$ in the units of $h/2\pi$. Therefore, each spectral line arising from the transition to the $d$-subshell will split in a quintet in the presence of the externally applied magnetic field.

## ❖ Probability Distribution Function

The probability distribution function is the behavior of $\psi^2$ at various points around the nucleus as a function of distance $r$ from the nucleus. The plots of such functions are also called as the probability distribution curves. Nevertheless, since it is only the radial part ($R_{n,l}$) that varies with the distance from the nucleus, the graphs of $\psi^2$ must behave in the same manner. To understand this more precisely, consider the plot of the first two quantum mechanical states of an electron in a hydrogen atom.

### ➢ Probability Distribution of $\psi_{1,0,0}$ State (1s Orbital)

In order to understand the probability distribution function of the electron in the ground state of the hydrogen atom, recall the mathematical expression for the same i.e.

$$\psi_{1,0,0} = \frac{1}{\sqrt{\pi}}\left(\frac{1}{a_0}\right)^{3/2} e^{-r/a_0} \tag{429}$$

Squaring both sides, we get

$$\psi_{1,0,0}^2 = \frac{1}{\pi a_0^3} e^{-2r/a_0} \tag{430}$$

It is obvious from the equation (429, 430) that the pre-exponential part is simply a constant and variation depends only upon the exponential part.

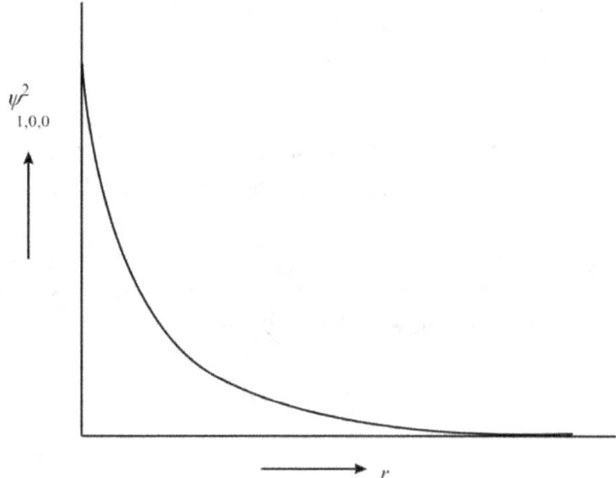


Figure 20. The variation of electron density vs distance from the center of the nucleus in 1$s$ orbital.

Hence, it can be concluded that the density of the electron wave is highest at the center of the nucleus and decreases as the distance from the center of the nucleus increases, and becomes zero only at infinite.

**DALAL INSTITUTE**

> ➤ *Probability Distribution of* $\psi_{2,0,0}$ *State (2s Orbital)*

In order to understand the probability distribution function of an electron in the $2s$ state of the hydrogen atom, recall the mathematical expression for the same i.e.

$$\psi_{2,0,0} = \frac{1}{4\sqrt{\pi}}\left(\frac{1}{a_0}\right)^{3/2}\left(2 - \frac{r}{a_0}\right)e^{-r/2a_0} \tag{431}$$

Squaring both sides, we get

$$\psi_{2,0,0}^2 = \frac{1}{16\pi a_0^3}\left(2 - \frac{r}{a_0}\right)^2 e^{-r/a_0} \tag{432}$$

It is obvious from the equation (431, 432) that the pre-exponential part is simply a constant and variation depends only upon the exponential part.

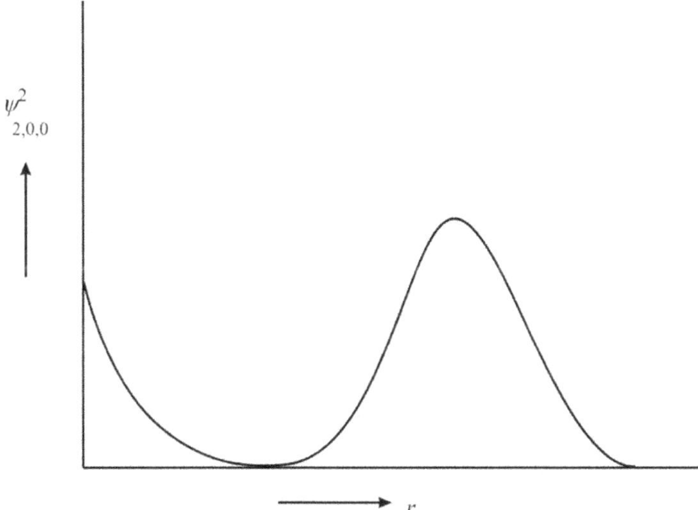

Figure 21. The variation of electron density vs distance from the center of the nucleus in 2s orbital.

Hence, it can be concluded that the density of electron wave is non zero at the center of the nucleus and decreases as the distance from the center of the nucleus increases, and becomes zero at $r = 2a_0$. Now since the wave function changes sign after $2a_0$, the density of electron wave after that increases first and then decreases exponentially and finally becomes zero at infinite distance. Now it's quite confusing because we have been told that the electron cannot reside within the nucleus and the probability of finding the electron inside the nucleus is almost zero, meaning that there must be something else that also governs the probability.

## ❖ Radial Distribution Function

The radial distribution function is the behavior of $R_{n,l}^2 \cdot 4\pi r^2 dr$ as a function of distance $r$ from the center of the nucleus. These plots solve the problem posed by the simple "probability distribution curves" which suggested that the probability of finding the electron must be highest at the center of the nucleus in the ground electronic state. In the radial distribution plots, we assume that the probability of finding the particle at a distance $r$ from the nucleus depends not only upon the density of electron wave but also varies with the magnitude of the volume of the spherical shell of $dr$ thickness at the same distance. This is quite rational because the $r$ can be in any direction around the nucleus.

Consider that the space around the nucleus is divided into an infinite number of concentric shells of thickness $dr$. Now though the electron density will show a decrease with increasing $r$, the volume of the concentric shells will increase. More volume at distance $r$ means more the chances of finding the electron at same. The two effects will try to counter each other, and therefore, the resultant probability at distance $r$ must be the multiplication of the two effects i.e.

$$Radial\ probability = \psi_{n,l,m}^2 \times dV_{shell} \tag{433}$$

Nevertheless, since it is only the radial part $(R_{n,l})$ that varies with the distance from the nucleus, the above expression for simplicity can be reduced to

$$Radial\ probability = R_{n,l}^2 \times dV_{shell} \tag{434}$$

Now as we have already derived the mathematical expression of radial wavefunction hydrogen atom already in this previously, the only thing we need is the mathematical expression of the volume element also.

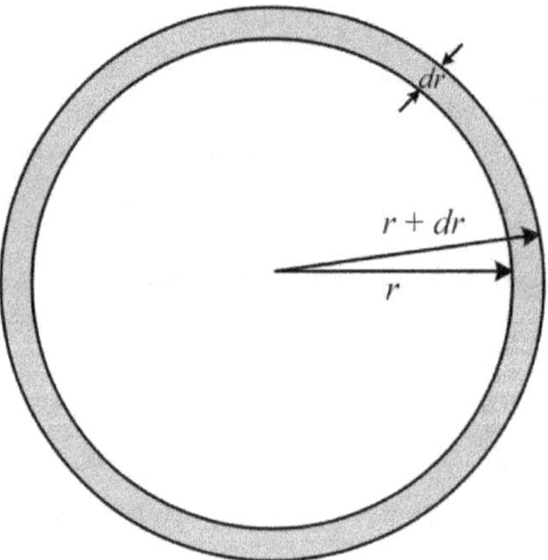


Figure 22. The depiction of a concentric shell of thickness $dr$ around the nucleus of a hydrogen atom at a distance $r$.

The volume of the shaded portion (spherical shell of thickness *dr*) can be obtained subtracting the volume of the inner sphere from the outer sphere i.e.

$$dV = \frac{4}{3}\pi(r+dr)^3 - \frac{4}{3}\pi r^3 \tag{435}$$

$$= \frac{4}{3}\pi(r^3 + dr^3 + 3r^2 dr + 3r\,dr^2) - \frac{4}{3}\pi r^3 \tag{436}$$

$$= \frac{4}{3}\pi r^3 + \frac{4}{3}\pi dr^3 + 4\pi r^2 dr + 4\pi r\,dr^2 - \frac{4}{3}\pi r^3 \tag{437}$$

$$dV = \frac{4}{3}\pi dr^3 + 4\pi r^2 dr + 4\pi r\,dr^2 \tag{438}$$

Since *dr* is very small, the terms involving square and cube of *dr* can be neglected for simplicity. All this leaves us with only one term i.e. $dV = 4\pi r^2 dr$. After using the value of *dV* in equation (434), we get

$$Radial\ probability = R_{n,l}^2 \times 4\pi r^2 dr \tag{439}$$

To understand this more precisely, consider the plot for the ground quantum mechanical state of an electron in a hydrogen atom i.e. 1*s* orbital.

> ### ➤ *Radial Probability Distribution Curve for Ground State of Hydrogen Atom*

The valid values of *n*, *l* and *m* that can be put in the general form of the hydrogenic wavefunction to obtain ground state are 1, 0 and 0, respectively. Therefore, we can start by writing the mathematical expression for the same i.e.

$$R_{1,0} = 2\left(\frac{1}{a_0}\right)^{3/2} e^{-r/a_0} \tag{440}$$

The probability distribution function can be obtained by squaring equation (440) i.e.

$$R_{1,0}^2 = \frac{1}{a_0^3} e^{-2r/a_0} \tag{441}$$

After multiplying the "probability distribution function" with "volume element", the expression for the "radial distribution function" can be formulated. Mathematically, we can say that

$$P(r) = \frac{4}{a_0^3} e^{-2r/a_0} \times 4\pi r^2 dr \tag{442}$$

It is obvious from the equation (442) that probability will become zero if we put $r = 0$ $(4\pi r^2 dr = 0)$. Now, if we increase the *r*, the radial probability will first increase due to increasing volume element, attaining maxima; and then it will start declining due to the dominance of $R_{1,0}^2$ part. In other words, the density of electron-wave decreases exponentially but the volume of the concentric shell increases continuously.

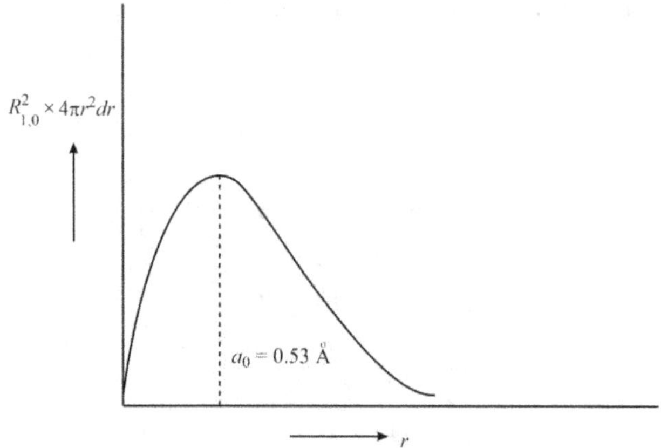

Figure 23. The variation of radial probability as a function of $r$ (1$s$ orbital).

In order to find the radius of maximum probability, we need to put $dP/dr$ equal to zero. It has been found that the radius of maximum probability will come out to be $0.53 \times 10^{-10}$ m, which is exactly equal to the radius of the first Bohr orbit ($a_0$).

> ### Radial Probability Distribution Curves for Other Hydrogenic Wavefunctions

The other valid sets of $n$, $l$ can be put in the general form of radial part of the wavefunction, to obtain $R_{n,l}^2$, and hence the corresponding "radial distribution functions".

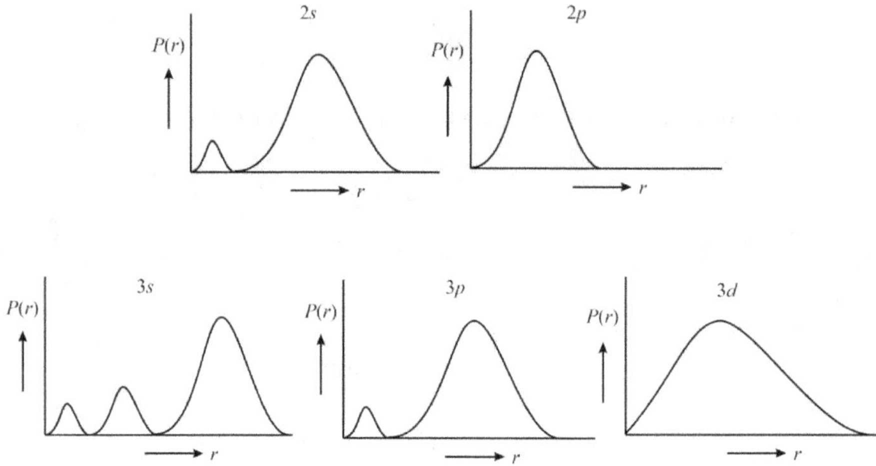

Figure 24. The variation of radial probability as a function of distance from the center of the nucleus.

### ❖ Shape of Atomic Orbitals (*s*, *p* & *d*)

The wave mechanical model of atom says that there is a non-zero probability of finding the electron almost everywhere in space excepting the angular and radial nodes. This means that primitive diagrams that depict the orbital shapes are intended to describe the region encompassing 90–95% probability density. In a typical drawing of orbital, we first plot the radial wave function and the angular part is superimposed. The shapes of some typical orbitals are discussed below.

#### ➤ *Shape of s-Orbitals*

In order to draw the shape of s-orbital, we first need to recall the radial part of the same and then we will have to superimpose the angular part. For instance, the radial part of $1s$ orbital is

$$R_{1,0} = 2\left(\frac{1}{a_0}\right)^{3/2} e^{-r/a_0} \tag{443}$$

It is obvious from the equation (443) that the radial part of the wave function has the largest magnitude when $r = 0$, and it decreases as we move away from the nucleus. The function will become zero only at infinite distance and will never change its sign. All this leads to a spherical-shaped cloud without any radial node. The angular part of every $s$-orbital is

$$Y_{0,0} = \frac{1}{\sqrt{4\pi}} = 0.28 \tag{444}$$

Hence, after multiplying radial wave function by a constant value of the angular part, the magnitude of function at all the points in space will reduce to 28% of the initial value.

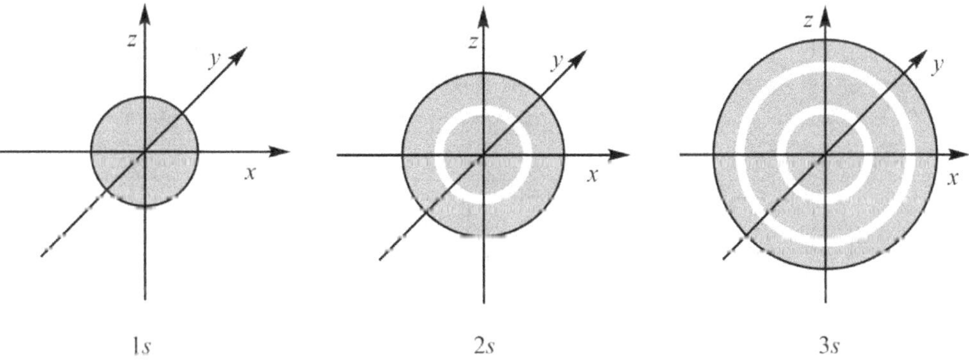


Figure 25. The shape of some lower energy $s$-orbitals.

Similarly, the shapes of some other s-orbitals are also given below to explain the concept more precisely. It is worthy to mention that the plots are easy to draw if we treat radial and angular parts consequently.

> ### *Shape of p-Orbitals*

In order to draw the shape of *p*-orbital, we first need to recall the radial part of the same and then we will have to superimpose the angular part. For instance, the radial part of $2p$ orbital is

$$R_{2,1} = \frac{1}{2\sqrt{6}}\left(\frac{1}{a_0}\right)^{3/2}\left(\frac{r}{a_0}\right)e^{-r/2a_0} \tag{445}$$

It is obvious from the equation (445) that the radial part of the wave function has the zero magnitudes when $r = 0$; and it increases as we move away from the nucleus, reaches a maximum, and decreases afterward. The function becomes zero only at infinite distance and will never change its sign. All this leads to a spherical-shaped cloud without any radial node. The angular part of *p*-orbitals are

$$Y_{1,0} = \sqrt{\frac{3}{2}}\; Cos\,\theta.\frac{1}{\sqrt{2\pi}} \tag{446}$$

$$Y_{1,\pm1} = \sqrt{\frac{3}{2}}\; Sin\,\theta.\sqrt{\frac{1}{2\pi}}\,e^{\pm i\phi} \tag{447}$$

Hence, the full plot for $p_z$-orbital is obtained after multiplying radial wave function by angular part given by equation (446); and the sign of function above the *xy*-plane will remain positive whereas a negative sign will be obtained below *xy*-plane. Similarly, we can obtain the three-dimensional plots for $p_x$ and $p_y$ by multiplying equation (445) by equation (447).

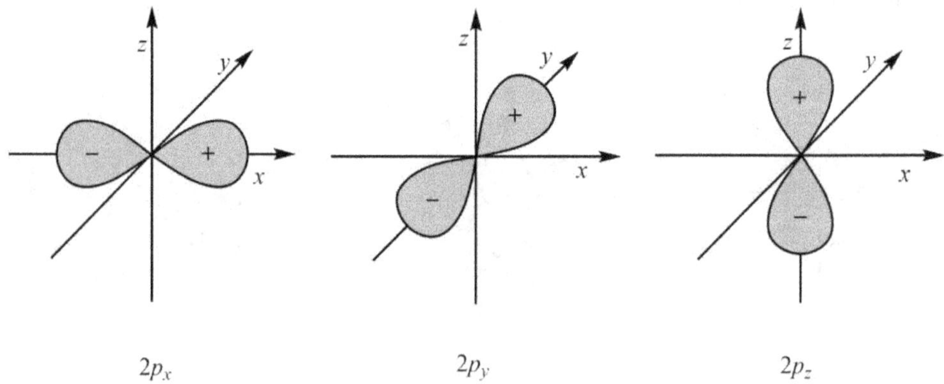


Figure 26. The shape of 2*p*-orbitals.

Similarly, the shapes of some other *p*-orbitals are also given below to explain the concept more precisely. It is worthy to mention that the plots are easy to draw if we treat radial and angular parts consequently.

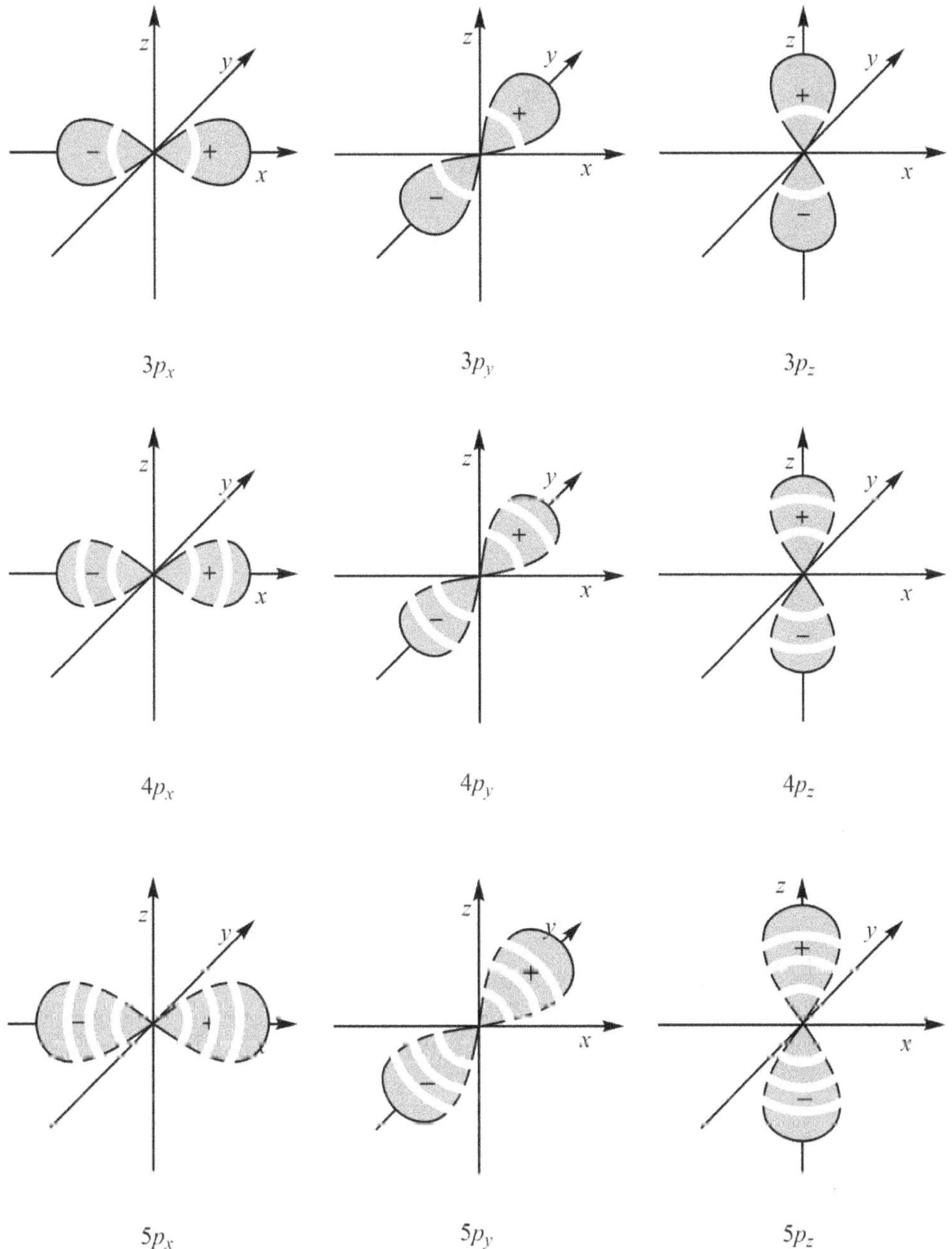


Figure 27. The shape of some $3p$, $4p$ and $5p$ orbitals.

> ### *Shape of d-Orbitals*

In order to draw the shape of *d*-orbital, we first need to recall the radial part of the same and then we will have to superimpose the angular part. For instance, the radial part of $3d$ orbital is

$$R_{3,2} = \frac{1}{81\sqrt{30}}\left(\frac{Z}{a_0}\right)^{3/2}\left(\frac{Zr}{a_0}\right)^{3/2} - \left(\frac{Zr}{a_0}\right)^{2} e^{-Zr/3a_0} \tag{448}$$

It is obvious from the equation (448) that the radial part of the wave function has the zero magnitudes when $r = 0$; and it increases as we move away from the nucleus, reaches a maximum, and decreases afterward. The function becomes zero only at infinite distance and will never change its sign. All this leads to a spherical-shaped cloud without any radial node. The angular part of *d*-orbitals are

$$Y_{2,0} = \sqrt{\frac{5}{8}}\,(3Cos^2\,\theta - 1).\frac{1}{\sqrt{2\pi}} \tag{449}$$

$$Y_{2,\pm 1} = \sqrt{\frac{15}{4}}\,Sin\,\theta\,Cos\,\theta.\sqrt{\frac{1}{2\pi}}\,e^{\pm i\phi} \tag{450}$$

$$Y_{2,\pm 2} = \sqrt{\frac{15}{16}}\,Sin^2\,\theta.\sqrt{\frac{1}{2\pi}}\,e^{\pm i2\phi} \tag{451}$$

Hence, the full plot for $d_{z^2}$-orbital is obtained after multiplying radial wave function by angular part given by equation (449); and the sign of function in the $xy$-plane will become negative whereas a positive sign will be obtained in two opposite lobes along $z$-axis. It is also worthy to note that the function becomes completely zero in two conical surfaces (above and below) before changing its sign. Similarly, we can obtain the three-dimensional plots for $p_x$ and $p_y$ by multiplying equation (448) by equation (450, 451).

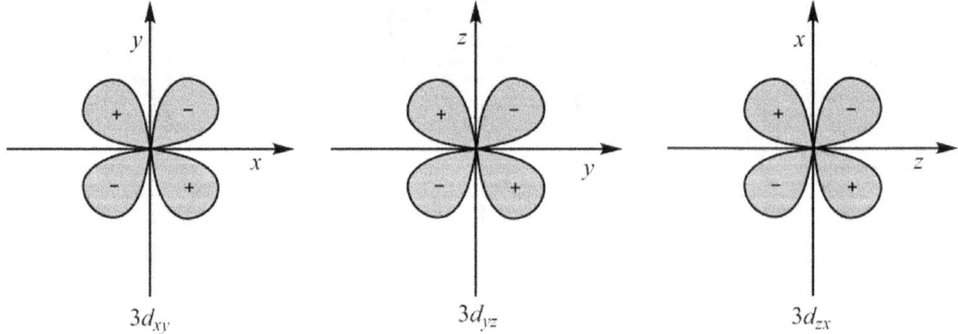


*Figure 28. Continued on the next page…*

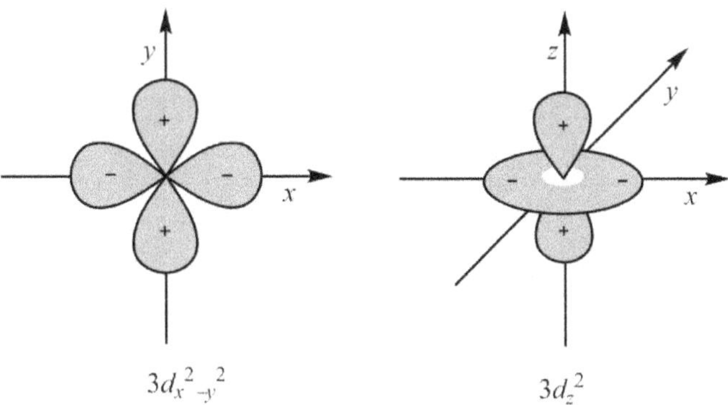

Figure 28. The shape of $3d$-orbitals.

Similarly, the shapes of some other $d$-orbitals are also given below to explain the concept more precisely. It is worthy to mention that the plots are easy to draw if we treat radial and angular parts consequently.

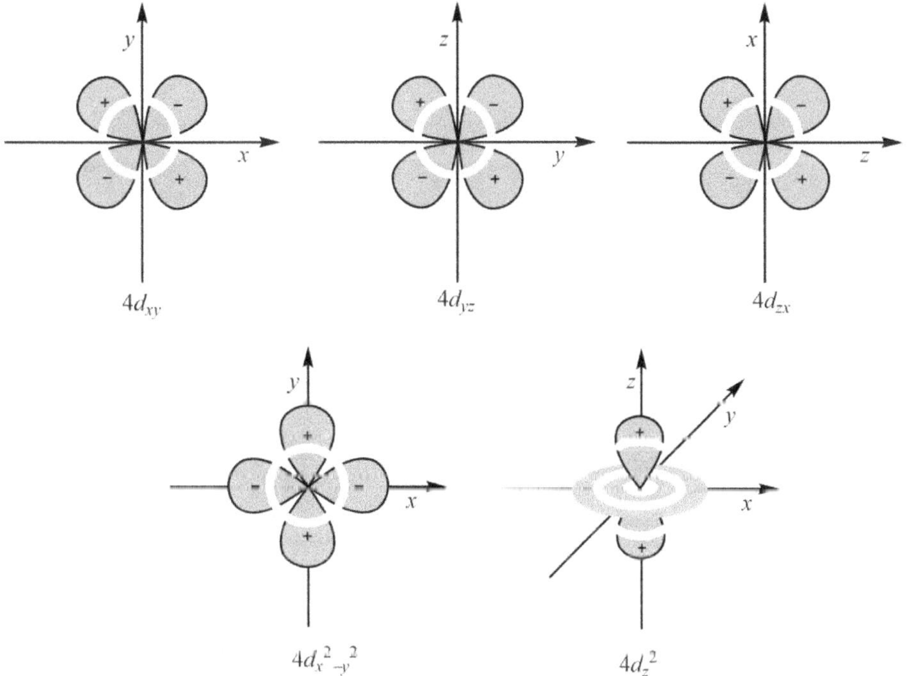

Figure 29. The shape of $4d$ orbitals.

It is also worthy to mention that the radii of maximum probability for $4d$ orbitals are larger than that of $3d$ orbitals. The same is true for $s$ and $p$ orbitals i.e. radius of maximum probability of $s$ and $p$ orbitals follow the order $3s > 2s > 1s$ and $4p > 3p > 2p$, respectively.

### ❖ Problems

Q 1. Derive and discuss the Schrodinger wave equation for a particle of mass $m$ trapped inside a cubical box with side $a$. Provided that the potential inside the box is zero while outside the box is infinite.

Q 2. Discuss the concept of the degeneracy of quantum mechanical states in a 3D box.

Q 3. What is the zero-point energy of a simple harmonic oscillator? How does it vary with force constant?

Q 4. Derive and discuss the Schrodinger wave equation for a diatomic rigid rotator. Also, draw the energy level diagram for the same.

Q 5. Define space quantization with special reference to diatomic rigid rotator.

Q 6. What is the difference between the radial and angular wave function for hydrogen atom? Write down both parts for $3d_z{}^2$ orbital.

Q 7. What are quantum numbers in the modern wave mechanical model of the atom? Also, discuss the main significance of the principal quantum number.

Q 8. Write down a short note on "probability distribution functions".

Q 9. What are "radial distribution functions"? Also, explain how you would determine the radius of maximum probability for $1s$ orbital of the hydrogen atom.

Q 10. What are the formulas to find the number of angular and radial nodes?

Q 11. Draw and discuss the shape of $4d_{xy}$ and $3d_z{}^2$ in detail

## ❖ Bibliography

[1] B. R. Puri, L. R. Sharma, M.S. Pathania, *Principles of Physical Chemistry*, Vishal Publications, Jalandhar, India, 2018.

[2] M. Reed, B. Simon *Functional Analysis*, Elsevier, Amsterdam, Netherlands, 2003.

[3] G. E. Bowman, Essential Quantum Mechanics, Oxford University Press, Oxford, UK, 2008.

[4] W. Heisenberg, *Über den anschaulichen Inhalt der quantentheoretischen Kinematik und Mechanik*, Zeitschrift für Physik, 1927, 172.

[5] E. Schrödinger, *Die gegenwärtige Situation in der Quantenmechanik*, Naturwissenschaften. 1935, 807.

[6] P. Alberto, C. Fiolhaisdag, V. Gil, *Relativistic particle in a box*, European Journal of Physics, 1996, 17.

[7] R. K. Prasad, *Quantum Chemistry*, New Age International Publishers, New Delhi, India, 2010.

[8] I. N. Levine, *Quantum Chemistry*, Pearson Prentice Hall, New Jersey, USA, 2009.

[9] D. A. McQuarrie, *Quantum Chemistry*, University Science Books, California, USA, 2008.

[10] E. Steiner, *The Chemistry Maths Book*, Oxford University Press, Oxford, UK, 2008.

[11] P. Atkins, J. Paula, *Physical Chemistry*, Oxford University Press, Oxford, UK, 2010.

# CHAPTER 6

# Thermodynamics – II

### ❖ Clausius-Clapeyron Equation

The Clausius-Clapeyron equation was initially proposed by a German physics Rudolf Clausius in 1834 and then further developed by French physicist Benoît Clapeyron in 1850. This equation is extremely useful in characterizing a discontinuous phase transition between two phases of a single constituent.

### ➢ *Derivation of Clausius-Clapeyron Equation*

In order to derive the Clausius-Clapeyron equation, consider a system at equilibrium i.e. the free energy change for the ongoing process is zero ($\Delta G = 0$). However, we know from the principles of thermodynamics that the variation of free energy with temperature and pressure can be formulated by the following differential equation

$$dG = VdP - SdT \tag{1}$$

For typical single-constituent equilibria, we have

$$Phase\ I \rightleftharpoons Phase\ II \tag{2}$$

Where phase I can be solid, liquid, or gas; whereas phase II can be liquid or vapor depending upon the nature of the transition whether it is melting, vaporization or sublimation, respectively.

Now consider a system at pressure $P$ and temperature $T$; and the free energies of phase I and phase II of the same system are $G_1$ and $G_2$, respectively. Moreover, if we change the temperature from $T$ to $T+dT$ and pressure from $P$ to $P+dP$; the free energy changes in phase I and phase II will be $dG_1$ and $dG_2$, respectively. Mathematically, we can formulate the situation as given below.

$$dG_1 = V_1 dP - S_1 dT \tag{3}$$

$$dG_2 = V_2 dP - S_2 dT \tag{4}$$

Where $V_1$ and $S_1$ represent the molar volume and entropy of phase I, while $V_2$ and $S_2$ represent the molar volume and entropy of phase II. However, if the phase I and phase II are in equilibrium with each other, we have

$$\Delta G = 0 \tag{5}$$

$$G_2 - G_1 = 0 \tag{6}$$

Now recalling the fact that $G_1$ and $G_2$ are simply the free energies of the phase I and phase II at temperature T and pressure P; we can say the same must also be true at temperature $T+dT$ and pressure $P+dP$ i.e.

$$(G_2 + dG_2) - (G_1 + dG_1) = 0 \tag{7}$$

$$(G_2 - G_1) + (dG_2 - dG_1) = 0 \tag{8}$$

Since $G_2 - G_1 = 0$ from equation (6), the above equation takes the form

$$dG_2 - dG_1 = 0 \tag{9}$$

$$dG_2 = dG_1 \tag{10}$$

After putting the values of $dG_1$ and $dG_2$ from equation (3, 4) in equation (10), we get

$$V_2 dP - S_2 dT = V_1 dP - S_1 dT \tag{11}$$

or

$$V_2 dP - V_1 dP = S_2 dT - S_1 dT \tag{12}$$

$$(V_2 - V_1)dP = (S_2 - S_1)dT \tag{13}$$

$$\Delta V. dP = \Delta S. dT \tag{14}$$

$$\frac{dP}{dT} = \frac{\Delta S}{\Delta V} \tag{15}$$

Now if the $\Delta H$ is the latent heat of phase transformation occurring at temperature, the entropy change is

or

$$\Delta S = \frac{\Delta H}{T} \tag{16}$$

After using the value of $\Delta S$ from equation (16) into equation (15), we get

$$\frac{dP}{dT} = \frac{\Delta H}{T\Delta V} \tag{17}$$

The equation was first developed by Calpeyron in 1834 and therefore is named after him Calpeyron equation. Now if phase I is solid while phase II is vapor i.e. solid $\rightleftharpoons$ melt equilibria, the equation (17) takes the form

$$\frac{dP}{dT} = \frac{\Delta_{fus}H}{T_f \Delta V} \tag{18}$$

Where $\Delta_{fus}H$ is the latent heat of fusion and $T_f$ is the melting point. The term $\Delta V = V_l - V_s$ i.e. the difference in the volume of melt and solid phase. Therefore, equation (18) can also be written as

$$\frac{dP}{dT} = \frac{\Delta_{fus}H}{T_f(V_l - V_s)} \tag{19}$$

For vaporisation equilibrium i.e. liquid $\rightleftharpoons$ vapour,

$$\frac{dP}{dT} = \frac{\Delta_{vap}H}{T_b(V_v - V_l)} \tag{20}$$

Where $\Delta_{vap}H$ is the latent heat of vaporization and $T_b$ is the boiling point. The term $\Delta V = V_v - V_l$ i.e. the difference in the volume of vapor and liquid phases. Now since the volume of the liquid is very small in comparison to the gaseous phase, the term $V_l$ in equation (20) can simply be neglected for simplicity i.e.

$$\frac{dP}{dT} = \frac{\Delta_{vap}H}{TV_v} \tag{21}$$

If the vapor act as an ideal gas i.e. $PV_v = RT$ or $V_v = RT/P$, the above equation can be written as

$$\frac{dP}{dT} = \frac{\Delta_{vap}H}{RT^2} \cdot P \tag{22}$$

or

$$\frac{1}{P} \cdot \frac{dP}{dT} = \frac{\Delta_{vap}H}{RT^2} \tag{23}$$

or

$$\frac{d \ln P}{dT} = \frac{\Delta_{vap}H}{RT^2} \tag{24}$$

The above equation is popularly known as the Clausius-Clapeyron equation.

One more popular form of Clausius-Clapeyron Equation is the integrated one which can easily be derived by using the rearranged form of equation (24) as given below.

$$d \ln P = \frac{\Delta_{vap}H}{RT^2} dT \tag{25}$$

If the temperature changes from $T_1$ to $T_2$ and pressure is varied from $P_1$ to $P_2$, then we can say that

$$\int_{P_1}^{P_2} d \ln P = \int_{T_1}^{T_2} \frac{\Delta_{vap}H}{RT^2} dT \tag{26}$$

or

$$\ln \frac{P_2}{P_1} = \frac{\Delta_{vap}H}{R} \int_{T_1}^{T_2} \frac{1}{T^2} dT \tag{27}$$

$$\ln \frac{P_2}{P_1} = \frac{\Delta_{vap}H}{R} \left[ \frac{1}{T_1} - \frac{1}{T_2} \right] \tag{28}$$

After converting to the common logarithm, we get

$$2.303 \, log \frac{P_2}{P_1} = \frac{\Delta_{vap}H}{R}\left[\frac{1}{T_1} - \frac{1}{T_2}\right] \tag{29}$$

or

$$log \frac{P_2}{P_1} = \frac{\Delta_{vap}H}{2.303 \, R}\left[\frac{1}{T_1} - \frac{1}{T_2}\right] \tag{30}$$

Which is the integrated form of Clausius-Clapeyron equation.

> ### Applications of Clausius-Clapeyron Equation

There are many applications of Clausius-Clapeyron equation ranging from chemistry to meteorology and climatology as well. Some of the most important applications in chemistry are given below.

*i*) For transitions between a gas and a condensed phase (liquid or solid), the Clausius-Clapeyron equation can be given as

$$ln \, P = -\frac{L}{R}\left(\frac{1}{T}\right) + C \tag{31}$$

Where $C$ represents a constant. The symbol $L$ represents the specific latent heat for a liquid-gas transition whereas specific latent heat of sublimation for a solid-gas transition. If the latent heat is known, then knowledge of one point on the coexistence curve determines the rest of the curve. Conversely, as the plot $ln P$ vs $1/T$ is linear with the zero intercept, the simple linear fitting is used to estimate the latent heat of phase change. The latent heat of vaporization of any liquid can also be determined if the vapor pressures of the same are known at two different temperatures.

*ii*) The boiling point of a liquid at any temperature can be determined if the boiling point at a particular temperature and latent heat of vaporization of the same are known.

*iii*) If the latent heat of vaporization and the vapor pressure of a liquid at a particular temperature are known, the vapor pressure of the same liquid at any other temperature can be known.

*iv*) One more application of this equation is to check if a phase transition will occur via compression or expansion in solid $\rightleftharpoons$ melt equilibria. To understand this, recall the

$$\frac{dP}{dT} = \frac{\Delta_{fus}H}{T_f \Delta V} \tag{32}$$

The sign for $dP/dT$ can be positive or negative for expansion and compression, respectively.

## ❖ Law of Mass Action and Its Thermodynamic Derivation

*According to the law of mass action, the rate of a chemical reaction is directly proportional to the product of the activities or simply the active masses of the reactants each term raised to its stoichiometric coefficients.*

To understand the law of mass action in mathematical language, consider a reaction in which two reactants $A$ and $B$ react to form the product $C$ and $D$ i.e.

$$aA + bB \rightarrow cC + dD \qquad (33)$$

Then the law of mass action says the rate of the above conversion should be

$$Rate \propto [A]^a[B]^b \qquad (34)$$

$$Rate = k[A]^a[B]^b \qquad (35)$$

Where $k$ is the constant of proportionality and is typically labeled as rate constant of the reaction.

However, the actual rate of the reaction may or may not be equal to what is suggested by the "law of mass action" because the actual rate law may have powers raised to the active masses different from their stoichiometric coefficients. Mathematically, the actual rate law for the reaction given by equation (33) is

$$Rate \propto [A]^\alpha[B]^\beta \qquad (36)$$

$$Rate = k[A]^\alpha[B]^\beta \qquad (37)$$

Now comparing equation (35) and equation (37); the law of mass action and actual rate law will give same results when $a = \alpha$ and $b = \beta$; whereas different results will be observed when $a \neq \alpha$ and $b \neq \beta$.

> #### *Modern Definition of the Law of Mass Action*

The law of mass action can be used to study the composition of a mixture in a reversible reaction under equilibrium conditions. To do so, consider a typical reversible reaction i.e.

$$aA + bB \rightleftharpoons cC + dD \qquad (38)$$

Now, from the law of mass action, we know that the rate of forward reaction ($R_f$) and rate backward reaction ($R_b$) will be

$$R_f = k_f[A]^a[B]^b \qquad (39)$$

$$R_b = k_b[C]^c[D]^d \qquad (40)$$

Where $k_f$ and $k_b$ are the rate constants for the forward and backward reactions, respectively. After equilibrium is reached, we have

$$R_f = R_b \qquad (41)$$



**D** DALAL INSTITUTE

$$k_f[A]^a[B]^b = k_b[C]^c[D]^d \tag{42}$$

or

$$\frac{k_f}{k_b} = \frac{[C]^c[D]^d}{[A]^a[B]^b} \tag{43}$$

Since the $k_f$ and $k_b$ are also constant at equilibrium, the ratio of the two is also a constant and is typically labeled as $K$ or the equilibrium constant. Therefore, equation (43) is modified as

$$K = \frac{k_f}{k_b} = \frac{[C]^c[D]^d}{[A]^a[B]^b} \tag{44}$$

All this leads to the modern definition of "law of mass action" that the ratio of the multiplication of molar concentrations of products raised to the power of their stoichiometric coefficients to the multiplication of the molar concentrations of the reactants raised to the power of their stoichiometric coefficients is constant at constant temperature and is called as "equilibrium constant". It is also worthy to mention that equation (44) is also known as the "law of chemical equilibrium".

> ➢ *Thermodynamic Derivation of the Law of Mass Action*

In order to derive the law of mass action thermodynamically, recall the general form of a typical reversible reaction under equilibrium conditions in which reactants and products are ideal gases i.e.

$$aA + bB \rightleftharpoons cC + dD \tag{45}$$

Now, as we know that the total free energy of the reactant ($G_R$) can be formulated as

$$G_R = a\mu_A + b\mu_B \tag{46}$$

Where $\mu_A$ and $\mu_B$ are the chemical potentials of reactant $A$ and $B$, respectively. Similarly, the total free energy of the products ($G_P$) can also be formulated i.e.

$$G_P = c\mu_C + d\mu_D \tag{47}$$

It is also important to mention that the temperature and pressure are kept constant. Moreover, the free energy change of the whole reaction can be obtained by subtracting equation (46) from equation (47) i.e.

$$\Delta G_{reaction} = G_P - G_R \tag{48}$$

$$\Delta G_{reaction} = (c\mu_C + d\mu_D) - (a\mu_A + b\mu_B) \tag{49}$$

Recalling the fact that the free energy change at equilibrium is zero, equation (49) is reduced to

$$(c\mu_C + d\mu_D) - (a\mu_A + b\mu_B) = 0 \tag{51}$$

Now recall the expression of the chemical potential of the $i$th species in gas phase i.e.

$$\mu_i = \mu_i^0 + RT \ln p_i \tag{52}$$

Where $p_i$ and $\mu_i^0$ are the partial pressure and standard chemical potential of $i$th species, respectively. Now using equation (52) in equation (51), we get

$$[c(\mu_C^0 + RT \ln p_C) + d(\mu_D^0 + RT \ln p_D)] - [a(\mu_A^0 + RT \ln p_A) + b(\mu_B^0 + RT \ln p_B)] = 0 \tag{53}$$

or

$$c\mu_C^0 + cRT \ln p_C + d\mu_D^0 + dRT \ln p_D - a\mu_A^0 - aRT \ln p_A - b\mu_B^0 - bRT \ln p_B = 0 \tag{54}$$

$$c\mu_C^0 + RT \ln p_C^c + d\mu_D^0 + RT \ln p_D^d - a\mu_A^0 - RT \ln p_A^a - b\mu_B^0 - RT \ln p_B^b = 0 \tag{55}$$

$$RT \ln p_C^c + RT \ln p_D^d - RT \ln p_A^a - RT \ln p_B^b = -c\mu_C^0 - d\mu_D^0 + a\mu_A^0 + b\mu_B^0$$

or

$$RT \ln \left(p_C^c p_D^d\right) - RT \ln \left(p_A^a p_B^b\right) = -[c\mu_C^0 + d\mu_D^0 - a\mu_A^0 - b\mu_B^0] \tag{56}$$

$$RT \ln \frac{\left(p_C^c p_D^d\right)}{\left(p_A^a p_B^b\right)} = -[G_P^o - G_R^o] \tag{57}$$

$$RT \ln \frac{\left(p_C^c p_D^d\right)}{\left(p_A^a p_B^b\right)} = -\Delta G_{reaction}^o \tag{58}$$

Where $\Delta G_{reaction}^o$ is the standard free energy change of the reaction can be simply abbreviated as $\Delta G^o$ only. Therefore, the equation (58) can be rearranged as given below.

$$\ln \frac{\left(p_C^c p_D^d\right)}{\left(p_A^a p_B^b\right)} = -\frac{\Delta G^o}{RT} \tag{59}$$

$$\frac{p_C^c p_D^d}{p_A^a p_B^b} = e^{-\frac{\Delta G^o}{RT}} \tag{61}$$

Now because $\Delta G^o$ is a function of temperature only and $R$ is a constant quantity, the right-hand side can be put equal to another constant, say '$K_p$'.

$$e^{-\frac{\Delta G^o}{RT}} = K_p \tag{62}$$

From equation (61) and equation (62), we have

$$K_p = \frac{p_C^c p_D^d}{p_A^a p_B^b} \tag{63}$$

Which is again the modern statement of "law of mass action" but in terms of partial pressures.

Other forms of equation (63) can also be written depending upon the reactants and products involved. If the chemical potentials of the reactants and products are in mole fractions ($x_i$) i.e.

$$\mu_i = \mu_i^0 + RT \ln x_i \tag{64}$$

Then equation (63) takes the form

$$K_x = \frac{x_C^c \, x_D^d}{x_A^a \, x_B^b} \tag{65}$$

Similarly, If the chemical potentials of the reactants and products are in molar concentrations ($c_i$) i.e.

$$\mu_i = \mu_i^0 + RT \ln c_i \tag{66}$$

Then equation (63) takes the form

$$K_c = \frac{[C]^c [D]^d}{[A]^a [B]^b} \tag{67}$$

Which is the popular form of "law of mass action".

## ❖ Third Law of Thermodynamics (Nernst Heat Theorem, Determination of Absolute Entropy, Unattainability of Absolute Zero) And Its Limitation

*The third law of thermodynamics states that the entropy of a system approaches a constant value as its temperature approaches absolute zero.*

In order to understand the abovementioned statement fully, we need to discuss some very important concepts and consequences of third law thermodynamics first such as "Nernst heat theorem" or "absolute entropies" etc.

### ➤ *Nernst Heat Theorem*

In 1906, Walther Nernst, a German chemist, studied the variation of enthalpy change and free energy change as a function of temperature. His results gained very much popularity as "Nernst heat theorem" in chemistry and built the foundation of many other discoveries. For detailed picture, recall the Gibbs-Helmholtz equation i.e.

$$\Delta G = \Delta H + T \left( \frac{\partial (\Delta G)}{\partial T} \right)_P \tag{68}$$

It is obvious from the above equation that the free energy change will become equal to the enthalpy change when the temperature is reduced to absolute zero i.e. $\Delta G = \Delta H$ at $T = 0$. Besides, Nernst also noted that the magnitude of $\partial(\Delta G)/\partial T$ declines gradually and approaches the zero with the decrease of temperature.

In other words, W. Nernst observed that as the temperature is decreased continuously, the enthalpy change was decreasing while the enthalpy change was increasing gradually with the same magnitude. Therefore, the change in slope in both curves must become zero near absolute zero i.e.

$$\lim_{T \to 0} \frac{\partial(\Delta G)}{\partial T} = \lim_{T \to 0} \frac{\partial(\Delta H)}{\partial T} = 0 \tag{69}$$

Which is the mathematical form of Nernst heat theorem. The pictorial representation of these observations is also given below for a more clear perspective.

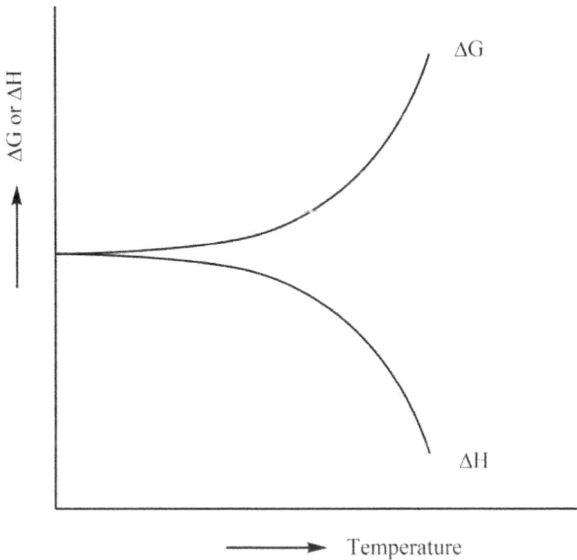


Figure 1. The graphical representation of Nernst heat theorem.

It is also worthy to mention that though we have shown $\Delta G$ greater than $\Delta H$ when the temperature is set greater than zero, the reverse may also be possible because $\partial(\Delta G)/\partial T$ can have positive as well as negative values.

Furthermore, we also know that variation of free energy change with the temperature at constant pressure is equal to the negative of entropy change i.e.

$$\left(\frac{\partial(\Delta G)}{\partial T}\right)_P = -\Delta S \tag{70}$$

Also, from the definition of change in the heat capacity, we have

$$\left(\frac{\partial(\Delta H)}{\partial T}\right)_P = \Delta C_P \tag{71}$$

After putting the values of equation (70, 71) in equation (69), we get

$$\lim_{T \to 0} (\Delta S) = 0 \quad and \quad \lim_{T \to 0} \Delta C_P = 0 \tag{72}$$

Now because, no gas or liquid at or in the vicinity of absolute zero, we can apply the Nernst heat theorem to solids only.

The right-hand side result given in equation (72) is extremely important as far as the third law of thermodynamics is concerned. Actually, we can say that the "third law of thermodynamics" follows from the results obtained by the Nernst heat theorem. In order to understand it more clearly, recall one of two results of Nernst heat theorem i.e.

$$\lim_{T \to 0} \Delta C_P = 0 \tag{73}$$

This means that as we approach absolute zero, the heat capacity of reactants becomes equal to the heat capacity of the product. In other words, we can say that the heat capacities of all substances are equal at absolute zero. Mathematically, it can be formulated as

$$\lim_{T \to 0} C_P(product) = \lim_{T \to 0} C_P(reactant) \tag{74}$$

Since the results of the quantum mechanics say that the heat capacity of tends to zero as the temperature approaches absolute zero, the above equation takes the form

$$\lim_{T \to 0} C_P = 0 \tag{75}$$

Similarly recalling the second result of Nernst heat theorem i.e.

$$\lim_{T \to 0} (\Delta S) = 0 \tag{76}$$

Which means that as we approach absolute zero, the entropy of reactants becomes equal to the entropy of the product. In other words, we can say that the entropy of all substances are equal at absolute zero. Mathematically, it can be formulated as

$$\lim_{T \to 0} S_{product} = \lim_{T \to 0} S_{reactant} \tag{77}$$

Using the same argument as in case of heat capacity, the above equation takes the form

$$\lim_{T \to 0} S = 0 \tag{78}$$

Therefore, the entropy of all crystalline substances can be taken as zero at absolute zero, which is the general statement of the third law of thermodynamics.

> ### *Determination of Absolute Entropy*

One of the most interesting and important applications of the third law of thermodynamics (or the Nernst heat theorem) is that it can be used to determine the absolute entropies of different substances at any temperature. The procedure employs the fact that we can calculate the entropy change easily and if the entropy of the initial state is zero then this difference will simply be equal to the absolute entropy i.e.

$$\Delta S = S_T - S_0 \tag{79}$$

Where $S_T$ and $S_0$ are the entropies at temperatures $T$ and 0K, respectively. Since we know from the third law of thermodynamics that $S_0 = 0$, the equation (79) takes the form

$$\Delta S = S_T - 0 = S_T \tag{80}$$

Before we proceed further, we must remember that a substance may change phase when it supplied with heat. Therefore, we need to discuss the absolute entropies in solid, liquid and gases separately.

**1. Absolute entropies in case of solids:** To calculate the absolute entropy of a solid at any temperature, we just need to find the total entropy change in shifting the absolute zero state to that temperature. The very small entropy change is given by

$$dS = \frac{\delta q}{T} \tag{81}$$

However, the general expression for heat capacity is

$$C_P = \frac{\delta q}{dT} \tag{82}$$

$$\delta q = C_P \, dT \tag{83}$$

After putting the value of $\delta q$ from equation (83) into equation (81), we have

$$dS = \frac{C_P \, dT}{T} \tag{84}$$

In order to find the total entropy change in the same phase (solid in this case) when the temperature is raised from 0 to $T$, we need to integrate the above equation over the range of interest i.e.

$$\int_0^T dS = \int_0^T \frac{C_P \, dT}{T} \tag{85}$$

$$S_T - S_0 = \int_0^T C_P \, d \ln T \tag{86}$$

Since $S_0 = 0$, the above equation becomes

$$S_T = \int_0^T C_P \, d \ln T \tag{87}$$

Thus, the entropy of any solid at temperature T can be obtained by heat capacity at many temperature points between $0K$ to $T$. The total integral of equation (87) can be obtained by measuring the area under the plot of $C_P$ vs $\ln T$.

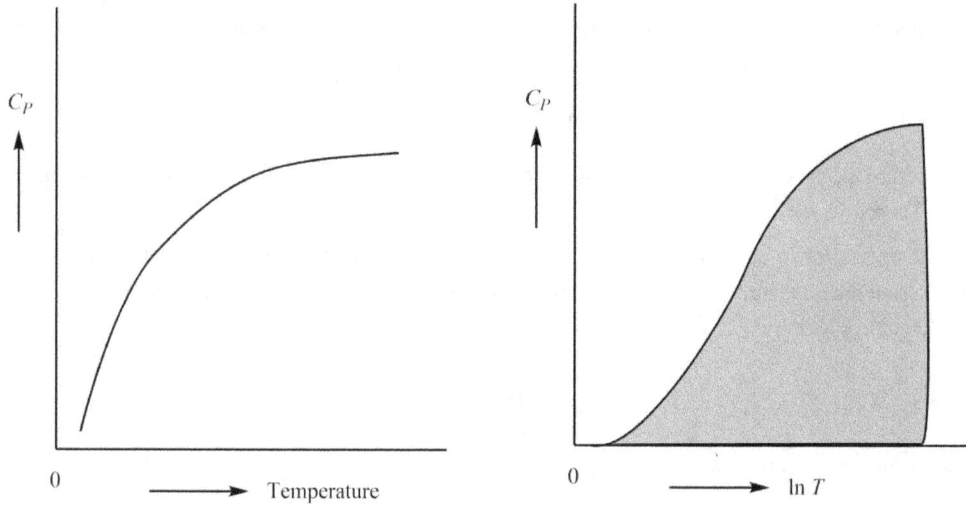


Figure 2. The plot of heat capacity vs $T$ and $\ln T$.

**2. Absolute entropies in case of liquids:** To calculate the absolute entropy of a liquid at any temperature, we just need to find the total entropy change in shifting the absolute zero state to that temperature. The whole process can be divided into three steps; one in heating up the solid phase form $0K$ to it melting or the fusion point, then phase change from solid to liquid at fusion temperature, and the last step in which liquid is heated up to temperature T.

*i) Entropy change from 0K to $T_f$:*

$$\Delta S_1 = \int_0^{T_f} C_P(s) \, d \ln T \tag{88}$$

*ii) Entropy change of fusion at $T_f$:*

$$\Delta S_2 = \frac{\Delta H_f}{T_f} \tag{89}$$

*iii) Entropy change from $T_f$ to $T$ :*

$$\Delta S_3 = \int_{T_f}^{T} C_P(l) \, d \ln T \tag{90}$$

The total entropy change for the process can be obtained by simply adding equations (88–90) i.e.

$$S_T - S_0 = \Delta S_1 + \Delta S_2 + \Delta S_3 \tag{91}$$

$$S_T = \int_{0}^{T_f} C_P(s) \, d \ln T + \frac{\Delta H_f}{T_f} + \int_{T_f}^{T} C_P(l) \, d \ln T \tag{92}$$

Where $C_P(s)$ and $C_P(l)$ are the heat capacities for solid and liquid phases, respectively. The symbol $\Delta H_f$ represents the latent heat of fusion.

**3. Absolute entropy in case of gases:** To calculate the absolute entropy of a gas at any temperature, we just need to find the total entropy change in shifting the absolute zero state to that temperature. The whole process can be divided into five steps; one in heating up the solid phase form $0K$ to it melting or the fusion point, then phase change from solid to liquid at fusion temperature, followed by the raising the temperature from $T_f$ to $T_b$. The fourth step includes the vaporization of liquid phase to gaseous phase at boiling point followed by the last step in which the temperature must be raised from $T_b$ to the required temperature $T$.

*i) Entropy change from $0K$ to $T_f$ :*

$$\Delta S_1 = \int_{0}^{T_f} C_P(s) \, d \ln T \tag{93}$$

*ii) Entropy change of fusion at $T_f$ :*

$$\Delta S_2 = \frac{\Delta H_f}{T_f} \tag{94}$$

*iii) Entropy change from $T_f$ to $T_b$ :*

$$\Delta S_3 = \int_{T_f}^{T_b} C_P(l) \, d \ln T \tag{95}$$

*iv) Entropy change of vaporization at $T_b$ :*

$$\Delta S_4 = \frac{\Delta H_{vap}}{T_b} \tag{96}$$

*v) Entropy change from $T_b$ to $T$ :*

$$\Delta S_5 = \int_{T_b}^{T} C_P(g)\, d\ln T \tag{97}$$

The total entropy change for the process can be obtained by simply adding equations (88–90) i.e.

$$S_T - S_0 = \Delta S_1 + \Delta S_2 + \Delta S_3 + \Delta S_4 + \Delta S_5 \tag{98}$$

$$S_T = \int_{0}^{T_f} C_P(s)\, d\ln T + \frac{\Delta H_f}{T_f} + \int_{T_f}^{T_b} C_P(l)\, d\ln T + \frac{\Delta H_{vap}}{T_b} + \int_{T_b}^{T} C_P(g)\, d\ln T \tag{99}$$

Where $C_P(s)$, $C_P(l)$ and $C_P(g)$ are the heat capacities for solid, liquid and gas phases, respectively. The symbol $\Delta H_{vap}$ represents the latent heat of vaporization.

> ### The unattainability of Absolute Zero

One more statement of the third law of thermodynamics is that the lowering of the temperature of a material body to the absolute zero is impossible in the finite number of steps. To understand this claim, recall the concept of Carnot refrigerator first. A Carnot refrigerator is basically just the reverse of the Carnot heat engine i.e. Carnot heat engine working in reverse cycle. Since a Carnot heat engine provides work through reversible isothermal-adiabatic compressions and expansions, a net amount of work must be done in Carnot refrigerator making it electricity consumer.

Let $q_1$ be the amount of heat absorbed by a Carnot refrigerator form a body at lower temperature $T_1$ and $q_2$ as the amount of heat rejected by the same refrigerator to a body at higher temperature $T_2$. The coefficient of performance ($\beta$) of the reversible Carnot refrigerator can be given by the following relation.

$$\beta = \frac{1}{q_2/q_1 - 1} \tag{100}$$

Since for a Carnot cycle, we know that

$$\frac{q_1}{T_1} = \frac{q_2}{T_2} \tag{101}$$

or

$$\frac{q_2}{q_1} = \frac{T_2}{T_1} \tag{102}$$

After using the value of $q_2/q_1$ from equation (102) in equation (100), we get

$$\beta = \frac{1}{T_2/T_1 - 1} \tag{103}$$

Furthermore, we also know that the coefficient of performance of an ideal Carnot refrigerator is simply the ratio of the cooling effect to the work done i.e.

$$\beta = \frac{q_1}{w} \tag{104}$$

Which means that how much heat is removed from the body at lower temperature per unit of work done. Now from equation (103) and equation (104), we have

$$\frac{q_1}{w} = \frac{1}{T_2/T_1 - 1} \tag{105}$$

$$\frac{q_1}{w} = \frac{T_1}{T_2 - T_1} \tag{106}$$

Taking reciprocal of the above result, we get

$$\frac{w}{q_1} = \frac{T_2 - T_1}{T_1} \tag{107}$$

Thus, it is obvious from the above equation that as the lower temperature $T_1$ approaches zero, more and more work will be needed to remove the same amount of heat ($w$ is inversely proportional to $T_1$), which is the unattainability of the absolute zero.

> ### *Limitation of Third Law of Thermodynamics*

Although the third law of thermodynamics is very useful in determining various thermodynamic properties of various substances, there are some limitations of the same. The reason being is the fact that entropies obtained using the third law of thermodynamics are thermal entropies and are somewhat smaller than the entropy values obtained using statistical mechanics. This deviation is typically in range of 3–5 $JK^{-1}\,mol^{-1}$.

From this, we may conclude that there is some entropy present even at the 0K temperature. This entropy is called as "residual entropy" and can be obtained using the following relation.

$$S = k \ln W \tag{108}$$

Where $k$ is Boltzmann constant and the $W$ is the thermodynamic probability which represents the number of equally probable orientations of the molecule under consideration. For instance, consider a sample of N number of carbon monoxide molecules. Since each molecule can have two orientations that are equally probable (CO CO OC CO OC OC), the thermodynamic probability will be $W = 2^N$. The residual entropy is

$$S = k \ln W = k \ln 2^N = kN \ln 2 \tag{109}$$

For one mole of sample $kN = R$, the above equation takes the form

$$S = R \ln 2 = 2.303R \log 2 = 5.85 \; JK^{-1}mol^{-1} \tag{110}$$

## ❖ Phase Diagram for Two Completely Miscible Components Systems

Before we discuss the phase diagram for two completely miscible components systems, it is better to recall the Gibbs phase rule which states that

$$F = C - P + 2 \tag{110}$$

Where $F$ is the number of degrees of freedom, $C$ represents the number of components and $P$ simply gives the total number of phases. For a two components system, the equation (110) reduces to

$$F = 2 - P + 2 = 4 - P \tag{111}$$

At this point, there are three major possibilities depending upon the number of phases involved which will be discussed one by one.

### ➢ *Types of Two-Component Systems Based on Number of Phases*

*i*) When we use $P = 1$ in equation (111), we get

$$F = 4 - 1 = 3 \tag{112}$$

This means that for a one-phase and two-component system, there are three degrees of freedom. This single-phase can be gaseous, solid solution, or two completely miscible liquids. Moreover, since the number of degrees of freedom is three, there are three conditions that can be varied without disturbing the total number of phases at equilibrium state.

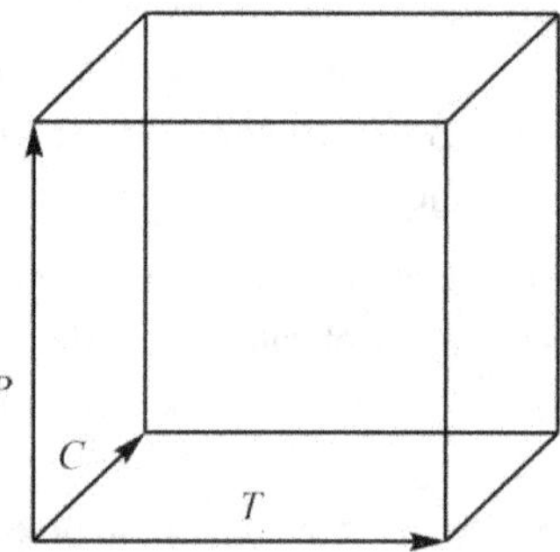


Figure 3. The graphical representation of a one-phase, two-component system.

Hence, the three degrees of freedom can be depicted along the three axes of a cube. However, it is more common to vary only two variables while keeping the third one as constant. For instance, in case of pressure-temperature diagrams, composition is kept constant; while in case of pressure-composition curves, the temperature is kept fixed at a particular value. Furthermore, in case of temperature-composition curves, the pressure of the system is always kept constant. Now although all the possibilities are there, it is more convenient to fix the pressure of the system at atmospheric pressure and vary the temperature and composition.

Therefore, we can say that instead of using three variables, it is more popular to keep one constant and vary the other two. This will also simplify the graphical representation of a simple two-component system.

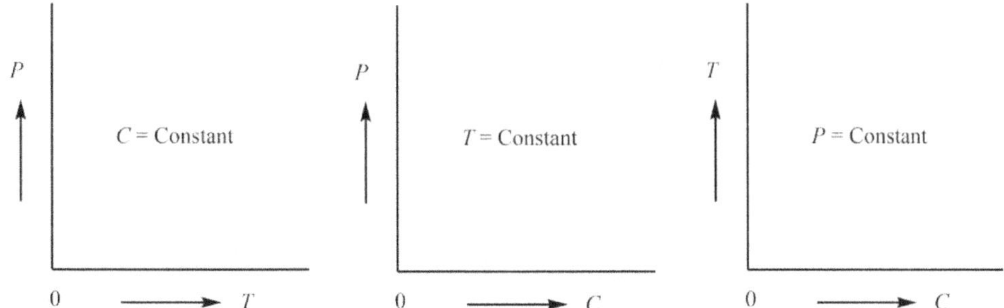


Figure 4. The two-dimensional variants of the one-phase-two-component system.

*ii*) When we use $P = 2$ in equation (111), we get

$$F = 4 - 2 = 2 \tag{113}$$

This means that for a two-phase and two-component system, there are two degrees of freedom. These phases can be liquid-vapour, liquid-liquid, or solid-liquid. Moreover, since the number of degrees of freedom is two, there are two variables that can be varied without disturbing total number of phases at equilibrium state.

*iii*) When we use $P = 3$ in equation (111), we get

$$F = 4 - 3 = 1 \tag{114}$$

Which means that for a three-phase and two-component system, there is only one degree of freedom. These phases can be liquid-liquid-vapour, solid-liquid-liquid, or solid-solid-liquid. Moreover, since the number of degrees of freedom is one, there is only one variable that can be varied without disturbing the total number of phases at equilibrium state.

Now although a complete phase diagram for two-component systems must be able to show the different phases up to number three, the situation is slightly simple for "condensed systems". More specifically, if all the phases in two-component systems are either solid or liquids only (solid-liquid equilibria), a minor pressure disturbance will have little to no effect on the system. Therefore, we can conclude that the number of degree of freedom in such "condensed systems" is actually reduced by one i.e.

$$F' = C - P + 1 \tag{115}$$

Which is the reduced or condensed phase rule where $F'$ is the total number of degrees of freedom excluding pressure. Now because the total number of phases that can exist simultaneously are reduced to 2 (system become univariant at $P = 2$), The complete phase diagram can be drawn on two dimensional paper with vertical and horizontal sides representing temperature and pressure, respectively.

# DALAL INSTITUTE

➢ *Types of Completely Miscible Two-Component Systems in Solid-Liquid Equilibria*

We know that in a solid-liquid equilibrium, the solid can be one of the constituents or solid solution which are completely miscible in the liquid phase.

Consider a liquid mixture of two components *A* and *B* at temperature *T*. Now if this liquid mixture is allowed to cool down below the freezing point of the mixture, the solid will start to separate out. At this point, assume that we have a number of such mixtures with the same components *A* and *B* but with different compositions (i.e. with different ratios of *A* and *B*). The cooling of all the mixtures is carried out in open vessels so that the pressure remains constant (atmospheric pressure). After that, the only thing we need to do is to plot these freezing point vs the composition. These completely miscible two-component systems can primarily be classified into three categories.

**1. Eutectic systems:** In these types of systems, the components do not react with each other but only the simple mixing takes place in the solution or in the molten state. The common examples of such systems are lead-silver system, bismuth-cadmium system, potassium iodide-water system.

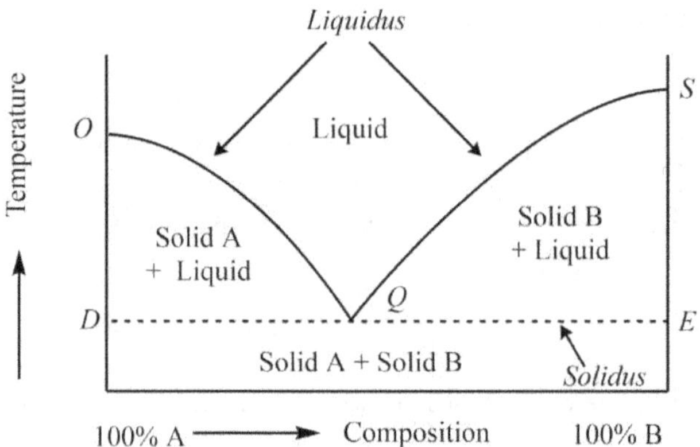


Figure 5. General phase diagram of eutectic systems.

**2. Systems forming solid compounds $A_xB_y$ with congruent and incongruent melting points:** In these types of systems, the components do react with each other and the formation of a compound takes place. The common example of such systems are the Mn-Zn system (forming $MgZn_2$) and $Na_2SO_4$-$H_2O$ system (forming $Na_2SO_4.10H_2O$). These types of systems can further be classified.

*i) Systems forming solid compounds with congruent melting point:* In these types of systems, the two components react to give a compound which is quite stable until its melting point is reached. When melted, the composition remains the same as that of the solid, and such compounds are said to have a congruent melting point. Some of the common examples of such systems are Mg-Zn and $FeCl_3$-$H_2O$ systems.

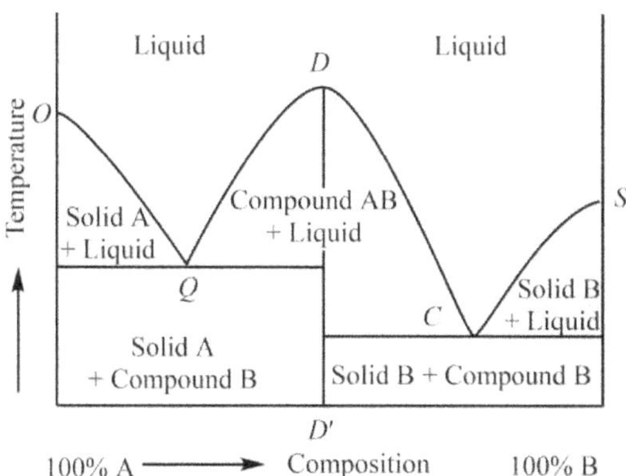


Figure 6. General phase diagram for systems forming solid compounds with congruent melting points.

*ii) Systems forming solid compounds with incongruent melting point:* In these types of systems, the two components react to give a compound which is not stable up to its melting point. When heated, the decomposition starts before the melting point is reached; and a new solid phase and a solution or melt with a different composition from the original solid are formed. Such compounds are said to undergo peritectic or transition reaction are labeled to have a congruent melting point. Some of the common examples of such systems are $Na_2SO_4$-$H_2O$ and $NaCl$-$H_2O$ systems. A typical transition can be represented as

$$C_1 \rightleftharpoons C_2 + \text{melt or solution} \tag{116}$$

Where $C_1$ is the compound formed by the reaction between participating components whereas $C_2$ represents the compound formed as a result of decomposition of $C_1$ below its fusion temperature.

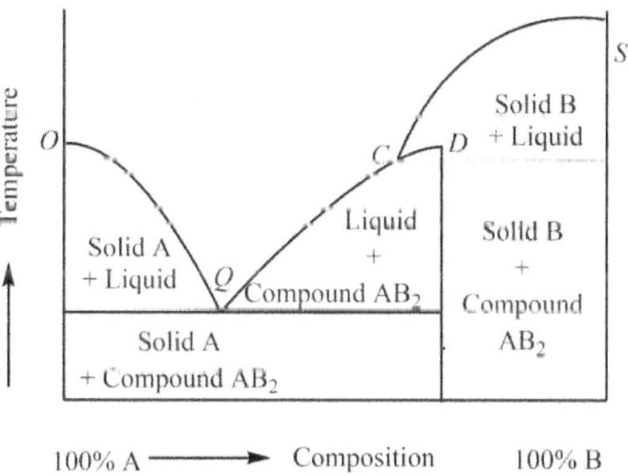


Figure 7. General phase diagram for systems forming solid compounds with incongruent melting points.

**3. Systems forming solid solutions:** In these types of systems, the components are completely miscible with each other in solid phase and completely homogeneous solid solutions are produced. The X-ray diffraction studies are typically employed to check that single crystalline phase is obtained rather than a mixture of two solid phases. The common example of such systems are Co-Ni system, Au-Ag system and AgCl-NaCl system.

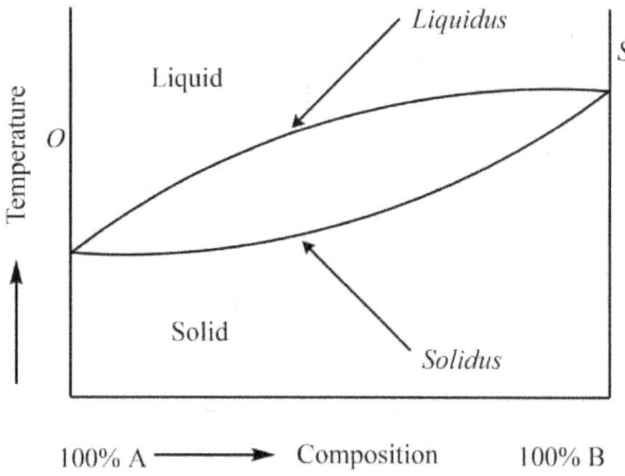


Figure 8. General phase diagram of eutectic systems.

> ### Experimental Methods for The Determination of Phase Diagram of Two-Component Systems

The most common approaches for the determination of phase diagram of two-component systems are "cooling curve" and "thaw melt" methods. These methods are quite popular due to their easiness and practicability to many systems.

**1. Cooling curve method:** In this approach, a liquid mixture of two components $A$ and $B$ at temperature $T$ is allowed to cool down below the freezing point of the mixture, the solid will start to separate out. A number of such mixtures with the same components $A$ and $B$ but with different compositions (i.e. with different ratios of $A$ and $B$) are then allowed to cool down in open vessels so that the pressure remains constant (atmospheric pressure). The temperature of the system is regularly recorded at different times to obtain temperature vs time plots. After that, from the breaks and arrests in those plots, important information such as freezing point or solidification time are obtained for different compositions.

Now in order to understand the concept more clearly, a variation of temperature with time for pure and a liquid mixture must be discussed. In the case of a liquid mixture, above point $a$, the system is completely liquid and it will cool down very rapidly in going from point $a$ to point $b$. Now because at point $a$, the crystallization of the second compound will also start releasing a large amount of heat, and therefore, the rate of cooling will be slightly slower until point $c$ is reached. Now since the system is getting solidified at eutectic temperature i.e. it is moving from $c$ to $d$; the temperature of the system will remain constant during this conversion. After point $d$, the system will cool down as a solid mixture with two degrees of freedom.

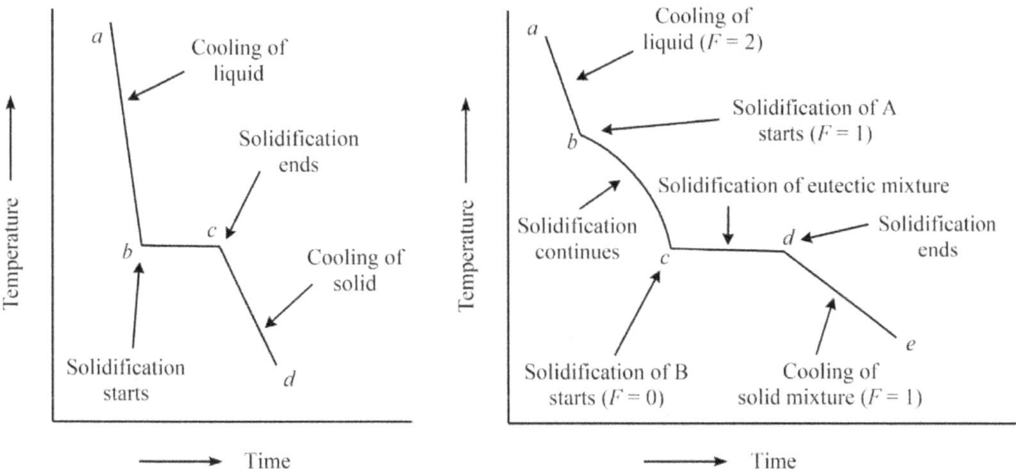



Figure 9. The typical cooling curves for pure a pure liquid (left) and a liquid mixture (right).

It is also worthy to mention that in some cases a phenomenon called "supercooling" is observed in which the break at point '$b$' is absent and cooling of liquid continues along $a$-$b$ with sudden rise after some time giving an abnormal break at $b'$ instead of $b$. The correct break is found by extrapolating from point $b'$ backward.

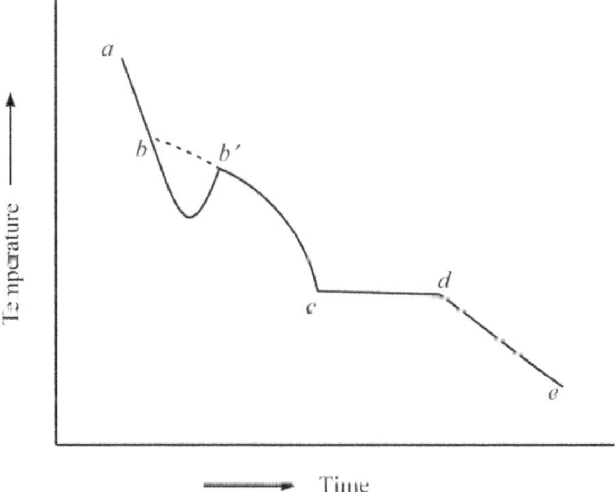



Figure 10. The typical cooling curves for a liquid mixture (right) with supercooling.

Now although the supercooling phenomenon does not create a major problem in most of the cases, it can be a real mess in some systems. Therefore, another method that can overcome these limitations must be discussed.

**2. Thaw-melt method:** This method is free from the limitation of the "supercooling phenomenon" posed by the cooling carve method. In this approach, a solid mixture of two components $A$ and $B$ at temperature $T$ is heated continuously. This is exactly opposite of what we follow in cooling curve method and we will note down melting points for different compositions. A number of such solid mixtures with same components $A$ and $B$ but with different compositions (i.e. with different ratios of $A$ and $B$) are then heated up in open vessels so that the pressure remains constant (atmospheric pressure). The temperature of system is regularly recorded at different times to obtain temperature vs time plots. After that, from the breaks and arrests in those plots, important information such as melting point or liquefaction time are obtained for different compositions.

Now in order to understand the concept more clearly, a variation of temperature with time for pure and a liquid mixture must be discussed. In the case of a solid mixture, below point $e$, the system is completely solid and it will heat up slowly in going from point $e$ to point $d$. Now since the system is getting liquefied at eutectic temperature i.e. it is moving from $d$ to $c$; the temperature of the system will remain constant during this conversion. Now because at point $c$, the melting of A compound will continue requiring a large amount of heat, and therefore, the rate of heating will be slightly slower until point $b$ is reached. After point $b$, the system will heat up as liquid mixture with two degrees of freedom very rapidly.

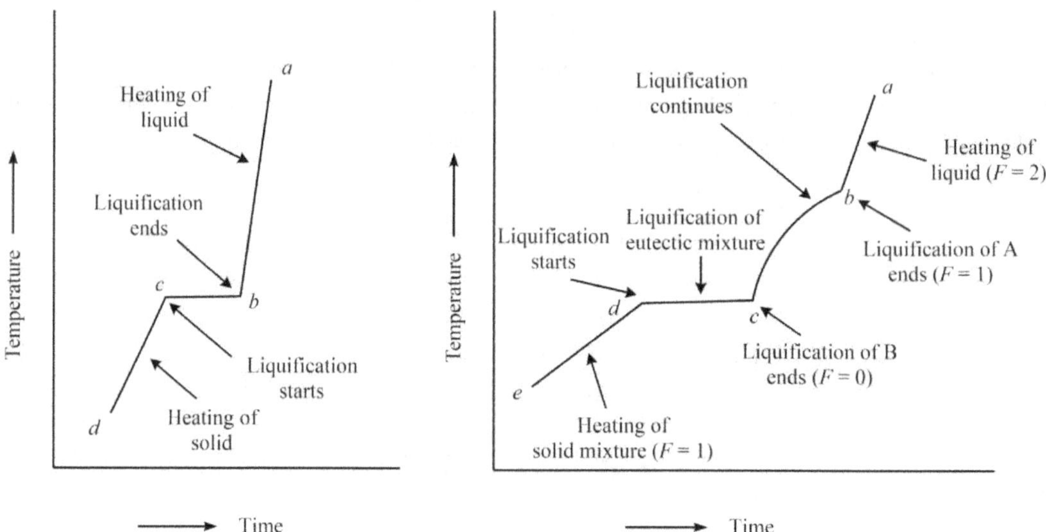


Figure 11. The typical Thaw-melting heating curves (temperature vs time) for a pure liquid (left) and a liquid mixture (right).

It is also worthy to mention that the temperature at which the melting of the solid mixture starts is typically called as "thaw point" and the temperate at which liquefaction ends is called as "melting point".

### ❖ Eutectic Systems (Calculation of Eutectic Point)

In order to understand the eutectic systems in a comprehensive manner, we need to recall the general phase diagram for eutectic systems i.e.

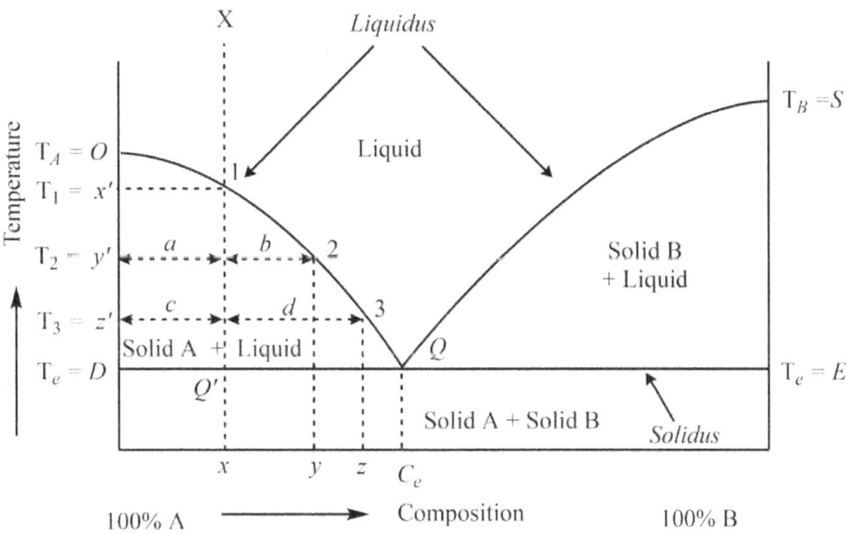


Figure 12. The general phase diagram of eutectic systems.

It is obvious from the phase diagram given above that the diagram has two curves $OQ$ and $QS$ enclosing the area $ODQ$ and $SQE$, respectively. The significance of different parts of the above-depicted diagram is discussed below.

**1. The $OQ$ and $SQ$ curves:** The point $O$ and point $S$ represent the freezing point the pure compound A and compound B, respectively. When component B is added to component A, its freezing point of decreases regularly along $OQ$. Likewise, when component A is added to component B, its freezing point decreases regularly along $SQ$. In conclusion, we can say that the curve $OQ$ and $SQ$ represent the temperature-conditions at which solid A and solid B are in equilibrium with the liquid mixture. Since there are only two phases involved, and both are condensed in nature (solid and liquid), we need to use condensed or reduced phase rule here i.e. after putting $C = 2$ and $P = 2$ in equation (114), we get

$$F' = 2 - 2 + 1 = 1 \tag{117}$$

Which means that the system is univariate. In other words, only one condition is needed to be defined to define the whole system; for instance, if we define a particular composition of A and B, the freezing point of the mixture is completely fixed. The curve $OQ$ and $SQ$ also called as "liquidus" because they separate the field of all liquid from that of liquid + solid crystals.

**2. Point $Q$ or the eutectic point:** At point '$Q$', all the three phases (solid A + solid B + liquid) are in equilibrium with each other; and therefore, zero degrees of freedom exists at this point i.e.

$$F' = 2 - 3 + 1 = 0 \qquad (118)$$

This means that we can neither change temperature nor composition without disturbing the mutual equilibrium of solid A, solid B and liquid. Moreover, below this temperature, the mixture freezes as a whole. This temperature is called as "eutectic point" and the composition corresponding to this point is called as eutectic composition.

**3. Area $ODQ$ and $SQE$:** In addition to the enclosed area $ODQ$ and $SQE$, there are also areas above $OQS$ line and below $DQE$ line as well.

*i) Area above $OQS$:* It is obvious from the phase diagram that the area lying above the line $OQS$ only has a liquid phase, which means

$$F' = 2 - 1 + 1 = 2 \qquad (119)$$

Two degrees of freedom means that we can change temperature well as composition without disturbing completely melt state as long as the temperature-composition coordinates lie above the line $OQS$. In other words, we need to define both, temperature as well as composition, to define the system completely in this area.

*ii) Area below $DQE$:* It is also obvious from the phase diagram that the area lying below the line $DQE$ has two solid phase, which means

$$F' = 2 - 2 + 1 = 1 \qquad (120)$$

The only one degree of freedom means that we can change temperature only without disturbing the state of the system. In other words, we only need to define the temperature to define the system completely in this area. The curve $DQE$ is also called as "solidus" because they separate the field of all solids from that of liquid + solid crystals.

*iii) Area enclosed by $ODQ$:* It is also obvious from the phase diagram that the area enclosed by $ODQ$ line has solid A + liquid phases, which means

$$F' = 2 - 2 + 1 = 1 \qquad (121)$$

The only one degree of freedom means that we can change temperature only without disturbing the state of the system. In other words, we only need to define the temperature to define the system completely in this area.

*iv) Area enclosed by $SQE$:* It is also obvious from the phase diagram that the area enclosed by $SQE$ line has solid B + liquid phases, which means

$$F' = 2 - 2 + 1 = 1 \qquad (122)$$

The only one degree of freedom means that we can change temperature only without disturbing the state of the system. In other words, we only need to define the temperature to define the system completely in this area.

It is also worthy to note that we can find out the equilibrium phase-composition at any temperature by simply plotting a horizontal line that cuts the line $OQ$ or $SQ$. For instance, let such a line $1x'$ that cuts $OQ$ at $x'$, or $2y'$ that cuts $OQ$ at $y'$. The importance of this line will be disused later in this section.

> ➤   *Cooling of a Liquid Mixture and the Calculation of Eutectic Point*

In order to understand the cooling profile of a liquid mixture and eutectic point, we must state the following rule first.

*During equilibrium melting or crystallization in a closed system, the initial composition of the system will be identical to the final composition of the system.*

Now consider the composition of the liquid mixture in which compound A is 80% while compound B is 20% (represented by the vertical line X). Therefore, it follows from the above-mentioned rule that after crystallization we must get the same composition i.e. 80% (solid A) + 20% (solid B).

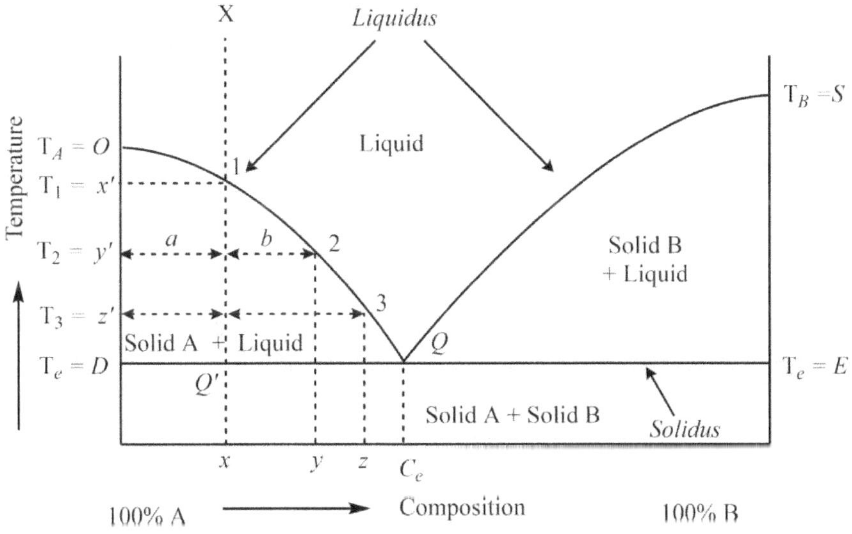


Figure 13. The cooling of a liquid mixture.

Above temperature $T_1$, the composition X lies in the field of "all liquid", and therefore, will exist as liquid completely. However, when the temperature is reduced to $T_1$ (corresponding to point 1) the crystal formation of A will be initiated. The further lowering of the temperature will make more crystals of A to form. All this will result in a continuous change of composition of the liquid mixture which is becoming more and more enriched in B. Therefore, as the temperature is lowered from $T_1$ to $T_2$, the composition of the liquid will change

from point $x$ to point $y$ (corresponding to point 2). Likewise, the further lowering of temperature from $T_2$ to $T_3$ and then to eutectic temperature ($T_E$) will change the composition from $y$ to $z$ (corresponding to point 3) and then to $C_e$ (corresponding to point Q), respectively.

As near point Q, the liquid will be rich in compound B relative to the initial composition, and the crystallization of compound B will also start at point Q; it is better to imagine the composition in solid mixture afterward. By recalling the fact that all the three phases coexist at point Q (solid A + solid B + liquid mixture), even an infinitesimal decrease in temperature will disturb the "eutectic point" and liquid B will also start converting into a solid phase. This will make the system move from point $Q$ to $Q'$, getting exactly the same composition $x$ i.e. 80% solid A + 20% solid B (represented by the vertical line X). A further lowering of temperature will make the system to move downward from point $Q'$ along the vertical line X again.

**1. Determination of phase composition:**

Now, it is clear from the phase diagram that above point 1 along line X, a 80% A and 20 % B will exist in the liquid state; whereas, below point $Q'$, a 80% A and 20 % B will exist in solid-state. However, between point 1 and $Q'$ (i.e. at temperatures between $T_1$ and $T_e$) two phases (liquid mixture of A and B + crystals of A) will be present in the system. If the crystallization process is at any point, the relative phase proportion can be obtained by simple lever rule. For this, left-hand side and right-hand side of line X are noted. For instance, at $T_2$ temperature, the amount of crystals of A and liquid could be found by just measuring the distances $a$ and $b$ i.e.

$$\% \text{ of crystals of } A = \frac{b}{a+b} \times 100 \tag{123}$$

$$\% \text{ of liquid mixture} = \frac{a}{a+b} \times 100 \tag{124}$$

Similarly, at $T_3$ temperature, the amount of crystals of A and liquid could be found by just measuring the distances $c$ and $d$ i.e.

$$\% \text{ of crystals of } A = \frac{d}{c+d} \times 100 \tag{125}$$

$$\% \text{ of liquid mixture} = \frac{c}{c+d} \times 100 \tag{126}$$

It is also worthy to note that the percentage composition of liquid at $T_3$ 65% compound A + 35% compound B. Now if $c = 2$ and $d = 3$ then putting in equation (125, 126), we get 60% of crystal A and 40% of the liquid mixture. This 60% solid is pure crystalline A while 40% liquid mixture is actually composed of 65% compound A (26) + 35% compound B (14).

At this point, three possibilities of calculating the eutectic point arise when the cooling curve under three different conditions. Now we need to study cooling curves at composition X, Y, pure A, and eutectic (let eutectic be 60% compound A and 40% compound B).

## 2. Calculation of eutectic point:

If the composition of the liquid mixture in which compound A is 80% while compound B is 20% (represented by the vertical line X). Therefore, it follows from the above-mentioned rule that after crystallization we must get the same composition i.e. 80% (solid A) + 20% (solid B). Now, if the system allowed to cool down and then the temperatures are plotted against time, we will get the cooling curves for the same systems. Since above point 1 the system is completely liquid, it will cool down very rapidly in going from point X to point 1. However, point 1, the crystalline of compound A will also start releasing a large amount of heat, and therefore, the rate of cooling will be slow until point Q is reached. Now since the system is getting solidified at eutectic temperature i.e. it is moving from $Q$ to $Q'$; the temperature of the system will remain constant during this conversion.

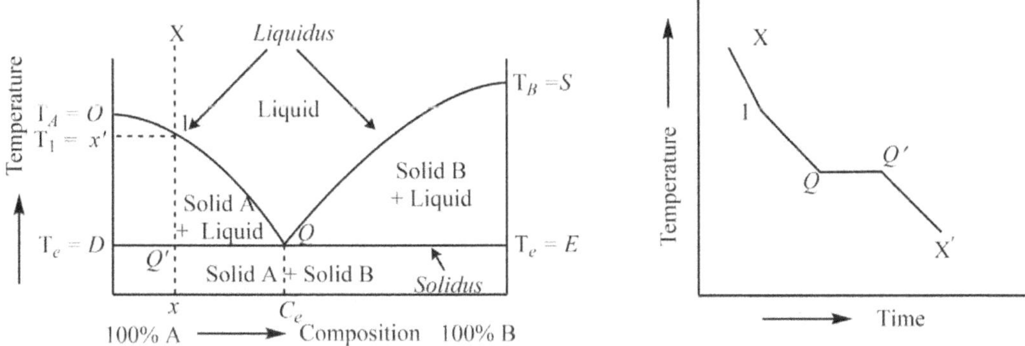


Figure 14. The cooling of a liquid mixture (compound A and compound B) when composition exists left to the eutectic composition.

After point $Q'$ both compounds will exist as solids, and therefore, the system will cool down below point $Q'$ along the vertical line X. The same type of curve will be obtained if we start from a composition left to the "eutectic composition". The only difference will be that the compound B will separate out first up to point Q, and then the solidification of compound A will start.

## 3. Some additional points:

Besides the features we already discussed, there are also some additional remarks which are based on the experimental results and can also be justified theoretically are given below.

*i)* Although the temperature corresponding to point '1' depends upon the initial composition, the temperature corresponding to Q to Q' always remains the same for compositions.

*ii)* If the composition of the mixture is more close to the eutectic composition, the portion 1–Q will be small whereas the portion $Q$–$Q'$. The reason for this type of behavior lies in the diagram itself. Assume a composition Y which is closer to the eutectic composition than composition X.

Now, if the system allowed to cool down and then the temperatures are plotted against time, we will get the cooling curves for the same systems. Since above point 2, the system is completely liquid, it will cool down very rapidly in going from point Y to point 2. After point 2, the crystallization of compound A will also start releasing a large amount of heat, and therefore, the rate of cooling will be slow until point Q is reached. Now, since the composition X has less A and more B than composition Y, the solidification of A (up to point Q) will take less time but solidification of B will take more time. This will make portion 2–Q smaller than 1–Q but will make portion Q–Q′ larger.

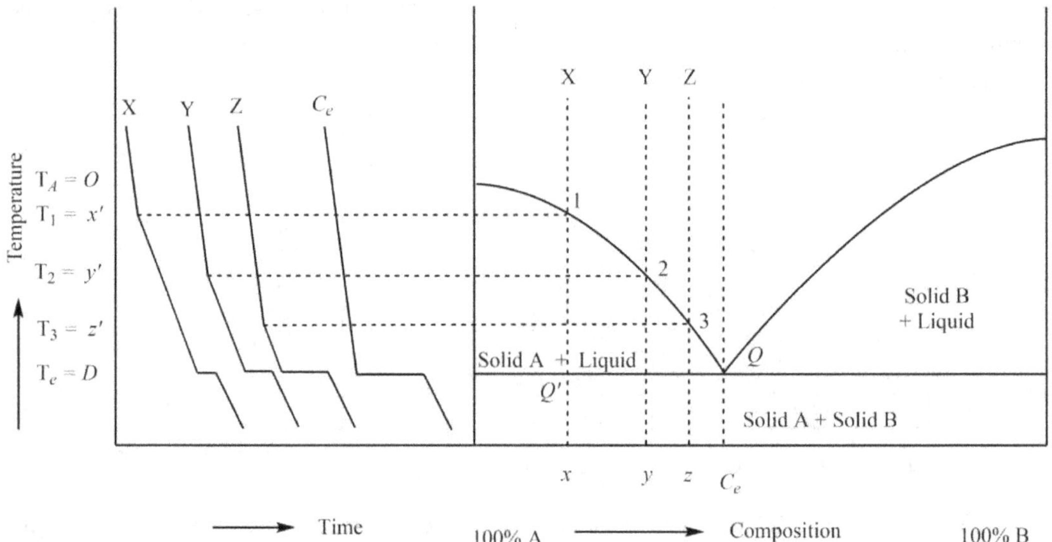



Figure 15. The cooling of a liquid mixture (compound A and Compound B) when compositions exist at eutectic point or left to the eutectic composition.

Now since the system is getting solidified at eutectic temperature i.e. it is moving from $Q$ to $Q'$; the temperature of the system will remain constant during this conversion.

iii) When the liquid mixture composition is the same as that of "eutectic composition" (shown by line Z), then the cooling will take place in all-liquid phase up to point $Q$ first, which means that there will be no break from point $Z$ to $Q$.

However, after reaching point $Q$, the solidification of both compounds (A as well as B) will start to take place. Now since point Q is at eutectic temperature, there will be no change in temperature until this process completes. This will create a long halt ($Q$–$Q'$) in the cooling curve since the whole liquid needs more time to solidify than some fraction left in solid-liquid equilibria.

> ### *Some Typical Eutectic Systems*

Some of the typical examples of two-components systems forming eutectic mixtures are given below for a more comprehensive analysis.

### 1. Bismuth-cadmium system (Bi-Cd):

The Bi-Cd is a typical case of solid-liquid equilibria in a two-component system that form a eutectic mixture. The phases involved in this case are solid Bi, solid Cd, liquid mixture of two (Cd + Bi) and gas-phase also. Now since a minor pressure disturbance will have little to no effect on the system and all the phases in two-component systems are either solid or liquids only (solid-liquid equilibria), Which is the reduced or condensed phase rule i.e.

$$F' = C - P + 1 \tag{127}$$

Now because the total number of phases that can exist simultaneously are reduced to 3 (system becomes non-variant at $P = 3$).

The complete phase diagram can be drawn on two dimensional paper with vertical and horizontal sides representing temperature and pressure, respectively. Consider a liquid mixture of Bi and Cd at temperature $T$. Now if this liquid mixture is allowed to cool down below the freezing point of the mixture, the solid will start to separate out. Prepare a number of such mixtures but with different compositions (i.e. with different ratios of Bi and Cd). The cooling of all the mixtures is carried out in open vessels so that the pressure remains constant (atmospheric pressure). After that, the plot freezing points vs the composition is

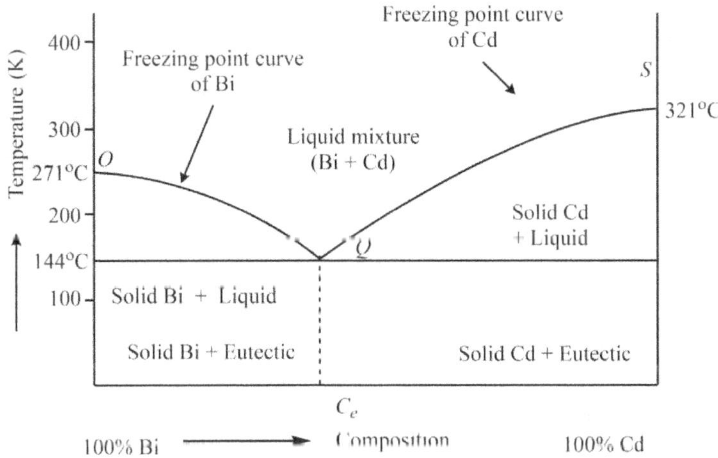


Figure 16. The phase diagram of Bi-Cd system.

The discussion on different parts (curve OQ, Curve SQ and the eutectic point) of the phase diagram of Bi-Cd system is given below.

*i) Curve OQ and SQ:* The point *O* and point *S* represent the freezing point the pure Bi (271 °C) and pure Cd (321 °C), respectively. When cadmium is added to bismuth, its freezing point of decreases regularly along *OQ*. Likewise, bismuth is added to cadmium, its freezing point of decreases regularly along *OQ*. In conclusion, we can say that the curve *OQ* and *SQ* represent the temperature-conditions at which solid Bi and solid Cd are in equilibrium with the liquid mixture. Since there are only two phases involved, i.e. after putting $C = 2$ and $P = 2$ in equation (127), we get

$$F' = 2 - 2 + 1 = 1 \tag{128}$$

Which means that the system is univariate. In other words, only one condition is needed to be defined to define the whole system; for instance, if we define a particular composition of Bi and Cd, the freezing point of the mixture is completely fixed. Since the points on *OQ* and *SQ* represent the initial freezing temperature whereas the points on solidus *DE* represent final freezing temperature; *OQ* and *SQ* can also be considered as the solubility curves of Bi in molten Cd and Cd in molten Bi, respectively.

*ii) Eutectic point (Q):* At point '*Q*', all the three phases (solid Bi + solid Cd + liquid) are in equilibrium with each other; and therefore, zero degrees of freedom exist at this point i.e.

$$F' = 2 - 3 + 1 = 0 \tag{129}$$

This means that we can neither change temperature nor composition without disturbing the mutual equilibrium of solid Bi, solid Cd and liquid. Moreover, below this temperature (144 °C), the mixture freezes as a whole. This temperature is called as "eutectic point" and the composition corresponding to this point is called as eutectic composition (40% Cd + 60% Bi).

**2. Lead-Silver system (Pb-Ag):**

The Pb-Ag is a typical case of solid-liquid equilibria in a two-component system that form eutectic mixture. The phases involved in this case are solid Ag, solid Pb, liquid mixture of two (Ag + Pb) and gas-phase also. Now since a minor pressure disturbance will have little to no effect on the system and all the phases in two-component systems are either solid or liquids only (solid-liquid equilibria), Which is the reduced or condensed phase rule i.e.

$$F' = C - P + 1 \tag{130}$$

Now because the total number of phases that can exist simultaneously are reduced to 3 (system becomes non-variant at $P = 3$).

The complete phase diagram can be drawn on two dimensional paper with vertical and horizontal sides representing temperature and pressure, respectively. Consider a liquid mixture of Ag and Pb at temperature *T*. Now if this liquid mixture is allowed to cool down below the freezing point of the mixture, the solid will start to separate out. Prepare a number of such mixtures but with different compositions (i.e. with different ratios of Ag and Pb). The cooling of all the mixtures is carried out in open vessels so that the pressure remains constant (atmospheric pressure). After that, the plot freezing points vs the composition is

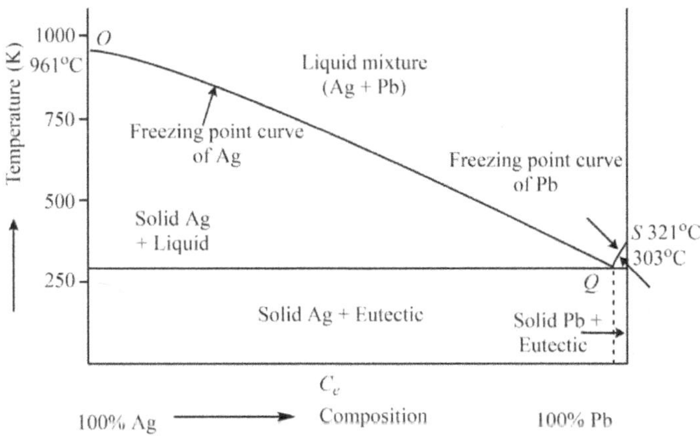


Figure 17. The phase diagram of Pb-Ag system.

The discussion on different parts (curve OQ, Curve SQ and the eutectic point) of the phase diagram of Pb-Ag system is given below.

*i) Curve OQ and SQ:* The point *O* and point *S* represent the freezing point the pure Ag (961°C) and pure Pb (327°C), respectively. When lead is added to silver, its freezing point of decreases regularly along *OQ*. Likewise, silver is added to lead, its freezing point of decreases regularly along *OQ*. In conclusion, we can say that the curve *OQ* and *SQ* represent the temperature-conditions at which solid Ag and solid Pb are in equilibrium with the liquid mixture. Since there are only two phases involved, i.e. after putting $C = 2$ and $P = 2$ in equation (130), we get

$$F' = 2 - 2 + 1 = 1 \tag{131}$$

Which means that the system is univariate. In other words, only one condition is needed to be defined to define the whole system; for instance, if we define a particular composition of Ag and Pb, the freezing point of the mixture is completely fixed. Since the points on *OQ* and *SQ* represent the initial freezing temperature whereas the points on solidus *DE* represent final freezing temperature; *OQ* and *SQ* can also be considered as the solubility curves of Ag in molten Pb and Pb in molten Ag, respectively.

*ii) Eutectic point (Q):* At point '*Q*', all the three phases (solid Ag + solid Pb + liquid) are in equilibrium with each other; and therefore, zero degree of freedom exists at this point i.e.

$$F' = 2 - 3 + 1 = 0 \tag{132}$$

This means that we can neither change temperature nor composition without disturbing the mutual equilibrium of solid Ag, solid Pb, and liquid. Moreover, below this temperature (303 °C), the mixture freezes as a whole. This temperature is called as "eutectic point" and the composition corresponding to this point is called as eutectic composition (2.6% Ag + 97.4% Bi).

### 3. Potassium iodide-water system (KI-H₂O):

The KI-H₂O is a typical case of solid-liquid equilibria in a two-component system that form eutectic mixture. The phases involved in this case are solid KI, ice, liquid mixture of two (KI + H₂O) and gas-phase also. Now since a minor pressure disturbance will have little to no effect on the system and all the phases in two-component systems are either solid or liquids only (solid-liquid equilibria), Which is the reduced or condensed phase rule i.e.

$$F' = C - P + 1 \qquad\qquad (133)$$

Now because the total number of phases that can exist simultaneously are reduced to 3 (system becomes non-variant at $P = 3$).

The complete phase diagram can be drawn on two dimensional paper with vertical and horizontal sides representing temperature and pressure, respectively. Consider a liquid mixture of KI and H₂O at temperature $T$. Now if this liquid mixture is allowed to cool down below the freezing point of the mixture, the solid will start to separate out. Prepare a number of such mixtures but with different compositions (i.e. with different ratios of KI and H₂O). The cooling of all the mixtures is carried out in open vessels so that the pressure remains constant (atmospheric pressure). After that, the plot freezing points vs the composition is

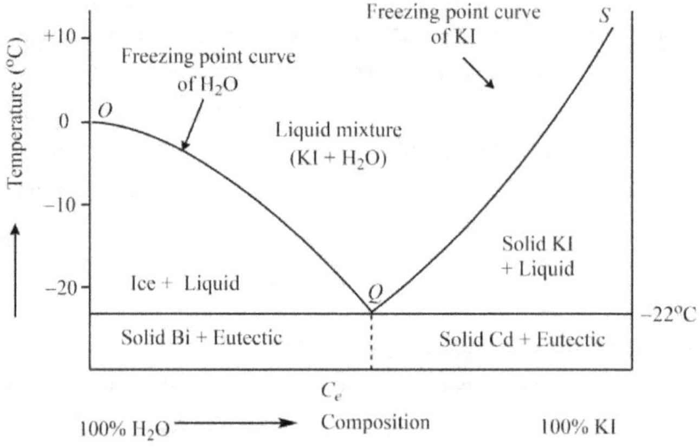


Figure 18. The phase diagram of KI-H₂O system.

The discussion on different parts (curve OQ, Curve SQ and the eutectic point) of phase diagram of KI-H₂O system is given below.

*i) Curve OQ and SQ:* The point $O$ represents the freezing point pure water (0°C). However, the curve SQ is incomplete because it is impossible to reach the melting point of KI in the presence of water. When KI is added to water, its freezing point of decreases regularly along $OQ$. Likewise, when water is added to KI, its freezing point of decreases regularly along $SQ$. In conclusion, we can say that the curve $OQ$ and $SQ$ represent the

temperature-conditions at which solid KI and solid $H_2O$ are in equilibrium with the liquid mixture. Since there are only two phases involved, i.e. after putting $C = 2$ and $P = 2$ in equation (133), we get

$$F' = 2 - 2 + 1 = 1 \qquad (134)$$

Which means that the system is univariate. Since the points on $OQ$ and $SQ$ represent the initial freezing temperature whereas the points on solidus $DE$ represent final freezing temperature; $OQ$ and $SQ$ can also be considered as the solubility curves of KI in molten $H_2O$ and ice in molten KI, respectively.

*ii) Eutectic point (Q):* At point '$Q$', all the three phases (solid KI + solid $H_2O$ + liquid) are in equilibrium with each other; and therefore, zero degree of freedom exists at this point i.e.

$$F' = 2 - 3 + 1 = 0 \qquad (135)$$

This means that we can neither change temperature nor composition without disturbing the mutual equilibrium of solid KI, ice and liquid. Moreover, below this temperature ($-22\,°C$), the mixture freezes as a whole. This temperature is called as "eutectic point" and the composition corresponding to this point is called as eutectic composition (52% KI + 48% $H_2O$).

## ❖ Systems Forming Solid Compounds $A_xB_y$ with Congruent and Incongruent Melting Points

In some two-component systems, the participants react together to form solid compounds $A_xB_y$. On the basis of the melting point of the compounds formed, these systems can be further divided.

### ➢ *Systems Forming Solid Compounds $A_xB_y$ with Congruent Melting Points*

In these systems, the solid compound melts sharply at temperature $T$ with the same composition as in the initial solid. These compound are said to possess a congruent melting point with the phase diagram as

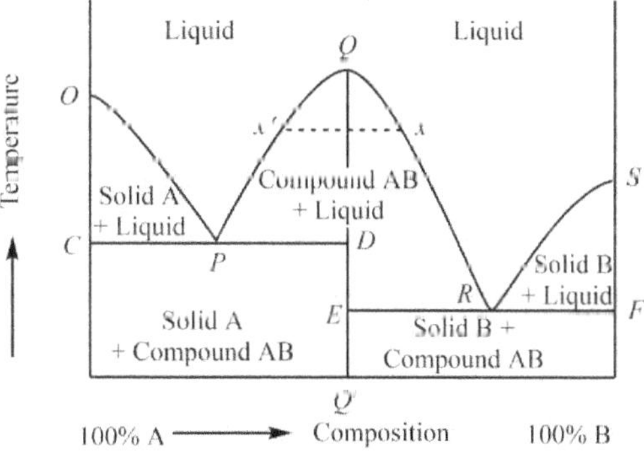


Figure 19. The general phase diagram of systems forming compounds with congruent melting points.

Consider two components $A$ and $B$ which also form a chemical compound AB by reacting with each other. Therefore, in the complete solid-state, there will be three phases named solid A, solid B and solid AB. Accordingly, there should also be three different freezing point curves i.e. $OP$, $PQR$ and $RS$. In addition to this liquidus, there will be two solidus $CE$ and $DE$ as well. The significance of different parts of the above-depicted diagram is discussed below.

**1. Curve *OP* and *RS*:** The point $O$ and point $S$ represent the freezing point the pure compound A and compound B, respectively. When compound AB is added to component A, its freezing point decreases regularly along $OP$. Likewise, when component AB is added to component B, its freezing point decreases regularly along $SR$. In conclusion, we can say that the curve $OP$ and $SR$ represent the temperature-conditions at which solid A and solid B are in equilibrium with the liquid mixture. Since there are only two phases involved, and both are condensed in nature (solid and liquid), we need to use condensed or reduced phase rule here i.e. after putting $C = 2$ and $P = 2$ in equation (114), we get

$$F' = 2 - 2 + 1 = 1 \tag{136}$$

Which means that the system is univariate. In other words, only one condition is needed to be defined to define the whole system; for instance, if we define a particular composition of A and AB (or B and AB), the freezing point of the mixture is completely fixed.

**2. Curve *PQR*:** When compound A is increased in compound AB, its freezing point of decreases regularly along $QP$. Likewise, when component B is added to compound AB, its freezing point decreases regularly along $QR$. In conclusion, we can say that the curve $PQR$ represents the temperature-conditions at which compound AB is in equilibrium with the liquid mixture. Since there are only two phases involved, and both are condensed in nature (solid and liquid), we need to use condensed or reduced phase rule here i.e. after putting $C = 2$ and $P = 2$ in equation (114), we get

$$F' = 2 - 2 + 1 = 1 \tag{137}$$

However, it is also worthy to mention that at the point Q also represents the congruent melting point of compound AB because liquid and solid phases have the same compositions. Consequently, we can also conclude that the system becomes one-component system at this point because the solid as well liquid phases contain only the compound AB alone. Furthermore, the congruent melting point of compound AB may lie above or below the congruent melting points of component A and component B.

**3. Eutectic points *P* and *R*:** In order to explain that the liquid phase can have two different compositions in equilibrium with the solid phase i.e. $x$ and $x'$, the curve $PQR$ is divided into two parts by the vertical line $QQ'$. This makes the concept very simple as the left and right parts can be treated as simple eutectic systems separately. The right half of the diagram is a eutectic system with components B and AB (solidus $CD$); whereas left hand side is a eutectic system with components A and AB (solidus is $EF$). The eutectic temperature and eutectic composition for the left-hand side portion are given by point $P$. Similarly, the eutectic temperature and eutectic composition for the right-hand side portion are given by point $R$.

> ➤ *Some Typical Examples of Systems Forming Compounds with Congruent Melting Points*

Some of the typical examples of two-components systems forming compounds with congruent melting points are given below for a more comprehensive analysis.

### 1. Magnesium-zinc system (Mg-Zn):

The Mg-Zn is a typical case of solid-liquid equilibria in a two-component system that form compounds with congruent melting points. The phases involved in this case are solid Mg, solid Zn, solid $MgZn_2$, liquid mixture of three (Mg + Zn + $MgZn_2$) and vapor phase too. Now since a minor pressure disturbance will have little to no effect on the system and all the phases in two-component systems are either solid or liquids only (solid-liquid equilibria), which is the reduced or condensed phase rule i.e.

$$F' = C - P + 1 \tag{138}$$

The complete phase diagram can be drawn on two dimensional paper with vertical and horizontal sides representing temperature and pressure, respectively. Consider a liquid mixture of Mg and Zn at temperature $T$. Now if this liquid mixture is allowed to cool down below the freezing point of the mixture, the solid will start to separate out. Prepare a number of such mixtures but with different compositions (i.e. with different ratios of Mg and Zn). The cooling of all the mixtures is carried out in open vessels so that the pressure remains constant (atmospheric pressure). After that, the plot freezing points vs the composition is

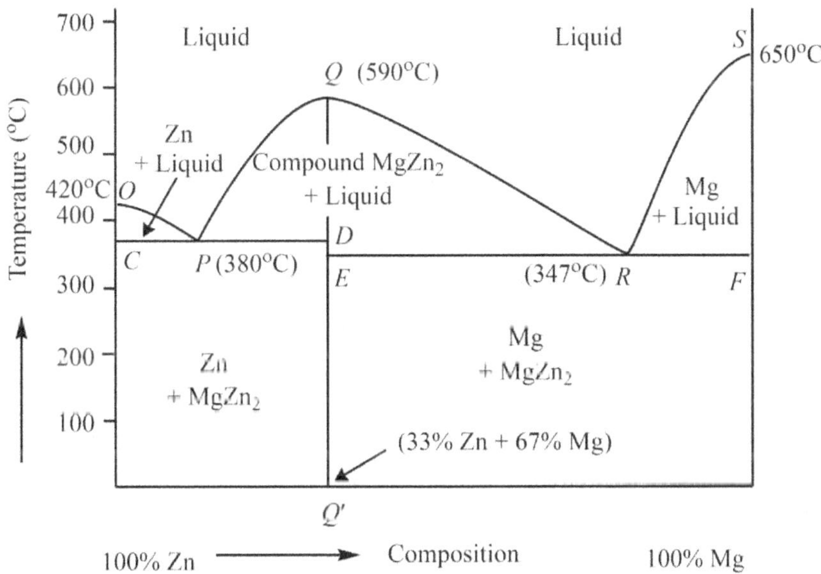


Figure 20. The phase diagram of Mg-Zn system.

The discussion on different parts (curve OQ, Curve SQ and the eutectic point) of the phase diagram of Bi-Cd system is given below.

*i) Curve OP and point P:* The point $O$ represents the freezing point of the pure Zn (420°C). When Mg is added to zinc, its freezing point of decreases regularly along $OP$ and Zn separates out simultaneously. Since there are only two phases involved, i.e. after putting $C = 2$ and $P = 2$ in equation (138), we get

$$F' = 2 - 2 + 1 = 1 \tag{139}$$

Which means that the system is univariate. However, after the point P is reached, the compound $MgZn_2$ is formed and also starts separating out as solid. We can say that there are three phases that coexist at point P i.e. solid Zn, solid $MgZn_2$ and melt which makes the system invariant (for $P = 3$, $F = 0$).

*ii) Curve PQ and point Q:* The point $Q$ represents the freezing point of the compound $MgZn_2$ (590°C). When Mg is added to zinc after point P, it combines with zinc to form MgZn2 which keeps on separating and its freezing point increases regularly until point Q is reached (33% magnesium). Since there are only two phases involved, i.e. after putting $C = 2$ and $P = 2$ in equation (138), we get

$$F' = 2 - 2 + 1 = 1 \tag{140}$$

Which means that the system is univariate. Now since the liquid and solid phases have the same composition at point Q, the corresponding temperature can be called a congruent melting point of compound $MgZn_2$. Furthermore, as the number of components becomes one at point Q i.e. solid $MgZn_2$ and melt $MgZn_2$, the system becomes invariant at Q (for $P = 3$, $F = 0$).

*iii) Curve QR and point R:* When Mg is further added to zinc after point Q, it goes into melt $MgZn_2$ separating out as solid, and therefore, the freezing point of $MgZn_2$ decreases regularly until point R is reached. Since there are only two phases involved along curve $QR$, i.e. after putting $C = 2$ and $P = 2$ in equation (138), we get the following

$$F' = 2 - 2 + 1 = 1 \tag{141}$$

Which means that the system is univariant. Moreover, we can also conclude here that there are three phases that coexist at point R i.e. solid Mg solid $MgZn_2$ and melt which makes the system invariant (for $P = 3$, $F = 0$).

*iv) Curve SR:* When Mg is further added to zinc after point R, it starts separating out as solid, and therefore, the freezing point of Mg increases regularly until point S is reached. Since there are only two phases involved along curve $SR$, i.e. after putting $C = 2$ and $P = 2$ in equation (138), we get

$$F' = 2 - 2 + 1 = 1 \tag{142}$$

Which means that the system remains univariant along curve $SR$. in reverse we can also say that the freezing point of Mg decreases regularly along curve $SR$ until point $R$ is reached i.e. solid Mg solid $MgZn_2$ and melt which makes the system invariant (for $P = 3$, $F = 0$).

### 2. Ferric chloride-water system (FeCl$_3$-H$_2$O):

The FeCl$_3$-H$_2$O is another typical case of solid-liquid equilibria in a two-component system that forms stable compounds with congruent melting point. The phases involved in this case are solid FeCl$_3$, ice, solid Fe$_2$Cl$_6$.12H$_2$O, Fe$_2$Cl$_6$.7H$_2$O, Fe$_2$Cl$_6$.5H$_2$O, Fe$_2$Cl$_6$.4H$_2$O, liquid mixture and vapor phase too. Now since a minor pressure disturbance will have little to no effect on the system and all the phases in two-component systems are either solid or liquids only (solid-liquid equilibria), the reduced phase rule can be used i.e.

$$F' = C - P + 1 \tag{143}$$

The complete phase diagram can be drawn on two dimensional paper with vertical and horizontal sides representing temperature and composition, respectively. The cooling of all the mixtures is carried out in open vessels so that the pressure remains constant (atmospheric pressure). After that, the plot of freezing points vs the composition is obtained.

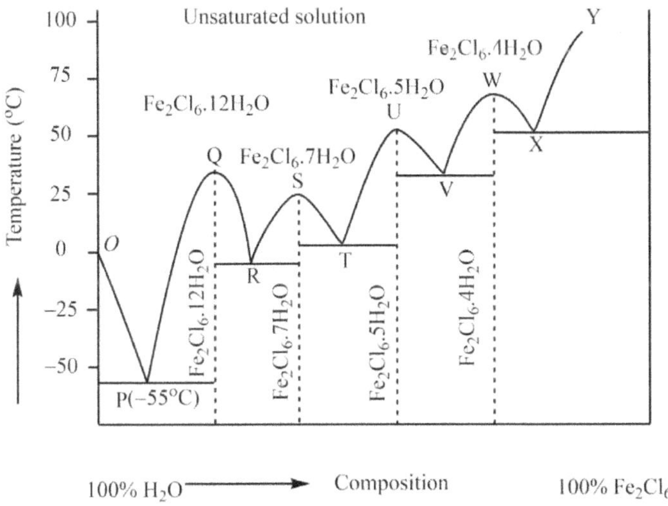


Figure 21. The phase diagram of FeCl$_3$-H$_2$O system.

*i) Curve OP and point P:* The point *O* represents the freezing point of the pure water (0°C). When FeCl$_3$ is added to water, the freezing point of water decreases regularly along *OP* and ice separates out simultaneously. Since there are only two phases involved along curve *OP* ($C - 2$ and $P - 2$), the equation (143) gives

$$F' = 2 - 2 + 1 = 1 \tag{144}$$

Which means that the system is univariant along curve *OP*. However, after reaching the point P, the liquid phase becomes saturated with compound Fe$_2$Cl$_6$.12H$_2$O which also starts separating out as solid afterward. We can say that there are three phases that coexist at point *P* i.e. solid ice, solid Fe$_2$Cl$_6$.12H$_2$O and solution which

makes the system invariant (for $P = 3$, $F = 0$). In other words, the point P is the "eutectic point for water and $Fe_2Cl_6.12H_2O$.

*ii) Curve PQ and point Q:* The point $Q$ represents the congruent melting point of the compound $Fe_2Cl_6.12H_2O$ (37°C). When $FeCl_3$ is added to the water after point P, it combines with water to form $Fe_2Cl_6.12H_2O$ which keeps on separating and its freezing point increases regularly until point Q is reached. Since there are only two phases involved along curve $PQ$, i.e. after putting $C = 2$ and $P = 2$ in equation (143), we get

$$F' = 2 - 2 + 1 = 1 \tag{145}$$

Which means that the system is univariant. Furthermore, as the number of components becomes one at point Q i.e. solid $Fe_2Cl_6.12H_2O$ and solution, the system becomes invariant at Q (for $P = 2$, $F = 0$).

*iii) Curve QR and point R:* When $FeCl_3$ is further added to the water after point Q, the freezing point of $Fe_2Cl_6.12H_2O$ decreases regularly until point R is reached. Since there are only two phases involved along curve $QR$, i.e. after putting $C = 2$ and $P = 2$ in equation (143), we get

$$F' = 2 - 2 + 1 = 1 \tag{146}$$

Which means that the system is univariant. After point R, the compound $Fe_2Cl_6.7H_2O$ is formed and also starts separating out as solid. However, we can say that there are three phases that coexist at point R i.e. solid $Fe_2Cl_6.12H_2O$, solid $Fe_2Cl_6.7H_2O$ and solution which makes the system invariant (for $P = 3$, $F = 0$).

*iv) Curve RS and point S:* The point $S$ represents the congruent melting point of the compound $Fe_2Cl_6.7H_2O$ (37°C). When $FeCl_3$ is added to the water after point R, it combines with water to form $Fe_2Cl_6.7H_2O$ which keeps on separating and its freezing point increases regularly until point S is reached. Since there are only two phases involved, i.e. after putting $C = 2$ and $P = 2$ in equation (143), we get

$$F' = 2 - 2 + 1 = 1 \tag{147}$$

Which means that the system is univariant. Furthermore, as the number of components becomes one at point S i.e. solid $Fe_2Cl_6.7H_2O$ and solution, the system becomes invariant at S (for $P = 2$, $F = 0$).

*v) Curve ST and point T:* When $FeCl_3$ is further added to the water after point S, the freezing point of $Fe_2Cl_6.7H_2O$ decreases regularly until point T is reached. Since there are only two phases involved along curve $ST$, i.e. after putting $C = 2$ and $P = 2$ in equation (143), we get

$$F' = 2 - 2 + 1 = 1 \tag{148}$$

Which means that the system is univariant. After point T, the compound $Fe_2Cl_6.5H_2O$ is formed and also starts separating out as solid. However, we can say that there are three phases that coexist at point T i.e. solid $Fe_2Cl_6.7H_2O$, solid $Fe_2Cl_6.5H_2O$ and solution which makes the system invariant (for $P = 3$, $F = 0$).

*vi) Curve TU and point U:* The point $U$ represents the congruent melting point of the compound $Fe_2Cl_6.5H_2O$ (37°C). When $FeCl_3$ is added to the water after point T, it combines with water to form $Fe_2Cl_6.5H_2O$ which

keeps on separating and its freezing point increases regularly until point U is reached. Since there are only two phases involved, i.e. after putting $C = 2$ and $P = 2$ in equation (138), we get

$$F' = 2 - 2 + 1 = 1 \tag{149}$$

Which means that the system is univariant. Furthermore, as the number of components becomes one at point U i.e. solid $Fe_2Cl_6.5H_2O$ and solution, the system becomes invariant at U (for $P = 2$, $F = 0$).

*vii) Curve UV and point V:* When $FeCl_3$ is further added to the water after point U, the freezing point of $Fe_2Cl_6.5H_2O$ decreases regularly until point V is reached. Since there are only two phases involved along curve *UV*, i.e. after putting $C = 2$ and $P = 2$ in equation (143), we get

$$F' = 2 - 2 + 1 = 1 \tag{150}$$

Which means that the system is univariant. After point V, the compound $Fe_2Cl_6.4H_2O$ is formed and also starts separating out as solid. However, we can say that there are three phases that coexist at point V i.e. solid $Fe_2Cl_6.5H_2O$, solid $Fe_2Cl_6.4H_2O$ and solution which makes the system invariant (for $P = 3$, $F = 0$).

*viii) Curve VW and point W:* The point $W$ represents the congruent melting point of the compound $Fe_2Cl_6.4H_2O$ (37°C). When $FeCl_3$ is added to the water after point V, it combines with water to form $Fe_2Cl_6.4H_2O$ which keeps on separating and its freezing point increases regularly until point W is reached. Since there are only two phases involved, i.e. after putting $C = 2$ and $P = 2$ in equation (143), we get

$$F' = 2 - 2 + 1 = 1 \tag{151}$$

Which means that the system is univariant. Furthermore, as the number of components becomes one at point W i.e. solid $Fe_2Cl_6.4H_2O$ and solution, the system becomes invariant at Q (for $P = 2$, $F = 0$).

*ix) Curve WX and point X:* When $FeCl_3$ is further added to the water after point W, the freezing point of $Fe_2Cl_6.4H_2O$ decreases regularly until point $X$ is reached. Since there are only two phases involved along curve *WX*, i.e. after putting $C = 2$ and $P = 2$ in equation (143), we get

$$F' = 2 - 2 + 1 = 1 \tag{152}$$

Which means that the system is univariant. After point X, anhydrous $Fe_2Cl_6$ is formed and also starts separating out as solid. However, we can say that there are three phases that coexist at point R i.e. solid $Fe_2Cl_6.4H_2O$, anhydrous $Fe_2Cl_6$ and solution which makes the system invariant (for $P = 3$, $F = 0$).

*x) Curve XY:* The point $Y$ represents the freezing point of the compound $Fe_2Cl_6$. When $FeCl_3$ is added to the water after point X, it combines with water to yield anhydrous $Fe_2Cl_6$ which keeps on separating and its freezing point increases regularly along XY is reached. Since there are only two phases involved along curve *XY*, i.e. after putting $C - 2$ and $P = 2$ in equation (143), we get

$$F' = 2 - 2 + 1 = 1 \tag{153}$$

Which means that the system is univariant.

➢ *Systems Forming Solid Compounds $A_xB_y$ with Incongruent Melting Points*

In these types of systems, the two components react to give a compound which is not stable up to its melting point. When heated, the decomposition starts before the melting point is reached; and a new solid phase and a solution or melt with a different composition from the original solid are formed. Such compounds are said to undergo peritectic or transition reaction are labeled to have congruent melting point. A typical transition can be represented as

$$C_1 \rightleftharpoons C_2 + \text{melt or solution} \tag{154}$$

Where $C_1$ is the compound formed by the reaction between participating components whereas $C_2$ represents the compound formed as a result of decomposition of $C_1$ below its fusion temperature.

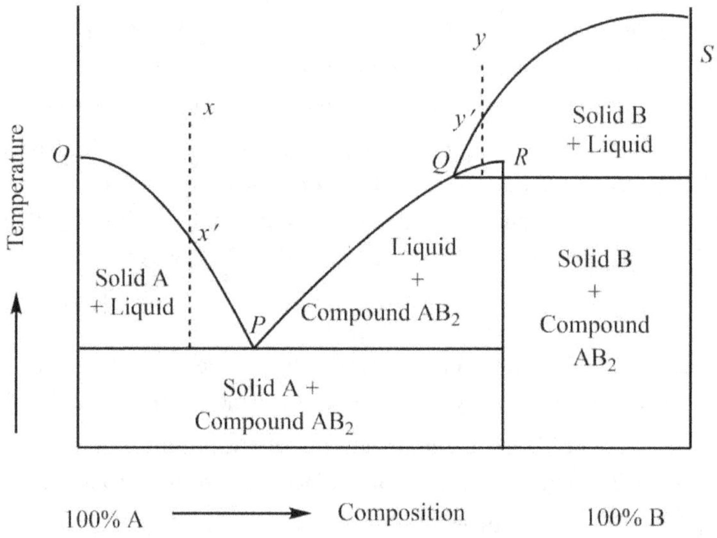


Figure 22. The general phase diagram of systems forming compounds with incongruent melting points.

Consider two components $A$ and $B$ which also form a chemical compound $AB_2$ by reacting with each other. The cooling of all the mixture-compositions is carried out in open vessels so that the pressure remains constant (atmospheric pressure). After that, the plot of freezing points vs the composition is obtained. The significance of different parts of the above-depicted diagram is discussed below.

**1. Point $O$, $S$, $Q$ and $R$:** The point $O$ and point $S$ represent the freezing point the pure compound A and compound B, respectively. The point $Q$, however, is quite strange because it represents the transition temperature (incongruent melting point of $AB_2$) where compound $AB_2$ decomposes into compound B. If does not decompose at this temperature, its congruent melting point would be $R$. In other words, we can say that the point $R$ represents the hypothetical congruent melting point of compound $AB_2$.

**2. Curve *OP*, *SQ* and *QP*:** When compound B is added to component A, its freezing point decreases regularly along *OP*. In other words, *OP* is the fusion curve of compound A along which solid A is in equilibrium with melt or solution. The line *SQ* is the fusion curve of compound B along which solid B is in equilibrium with melt or solution. Similarly, *QP* is the fusion curve of compound $AB_2$ along which solid $AB_2$ is in equilibrium with melt or solution.

When a liquid mixture with composition X is allowed to cool down, it will do so by keeping its composition the same until point 1 is reached where the compound A will just start to separate out as solid. Further cooling will lead to a change in composition along the line $1P$. When point $P$ is attained, the formation of compound AB will be started; and since three phases coexist at this point (solid A, solid AB and liquid), the reduced phase rule gives

$$F' = C - P + 1 \tag{155}$$

$$F' = 2 - 3 + 1 = 0 \tag{156}$$

Which means that there will be no degree of freedom at point $P$ ($P$ is non-variant). On the other hand, if liquid mixture with composition Y is allowed to cool down, it will do so by keeping its composition the same until point 2 is reached where the compound B will just start to separate out as solid. Further cooling will lead to a change in composition along the line $2Q$. When point $Q$ is attained, the following meritectic reaction will take place

$$Solid\ B + Solution \rightleftharpoons Solid\ AB_2 \tag{157}$$

Therefore, the formation of compound $AB_2$ will be started; and since three phases coexist at this point (solid B, solid AB and liquid), the point Q also become invariant. Furthermore, it is also worthy to mention that the transformation of compound B to compound $AB_2$ occurs at a constant temperature, and therefore, the point Q is also called a peritectic point.

> ➢ *Some Typical Examples of Systems Forming Compounds with Incongruent Melting Points*

Some of the typical examples of two-components systems forming compounds with incongruent melting points are given below for a more comprehensive analysis.

**1. Sodium chloride-water system (NaCl-H₂O):**

The NaCl-H₂O is a typical case of solid-liquid equilibria in a two-component system that forms compounds with an incongruent melting point. The phases involved in this case are solid NaCl, solid NaCl.2H₂O, ice, liquid mixture, and vapour phase too. Now since a minor pressure disturbance will have little to no effect on the system and all the phases in two-component systems are either solid or liquids only (solid-liquid equilibria), which is the reduced or condensed phase rule i.e.

$$F' = C - P + 1 \tag{158}$$

The complete phase diagram can be drawn on two dimensional paper with vertical and horizontal sides representing temperature and pressure, respectively. Consider a liquid mixture of water and NaCl at

temperature $T$. Now if this liquid mixture is allowed to cool down below the freezing point of the mixture, the solid will start to separate out. Prepare a number of such mixtures but with different compositions (i.e. with different ratios of $H_2O$ and $NaCl$). The cooling of all the mixtures is carried out in open vessels so that the pressure remains constant (atmospheric pressure). After that, the plot freezing points vs the composition is

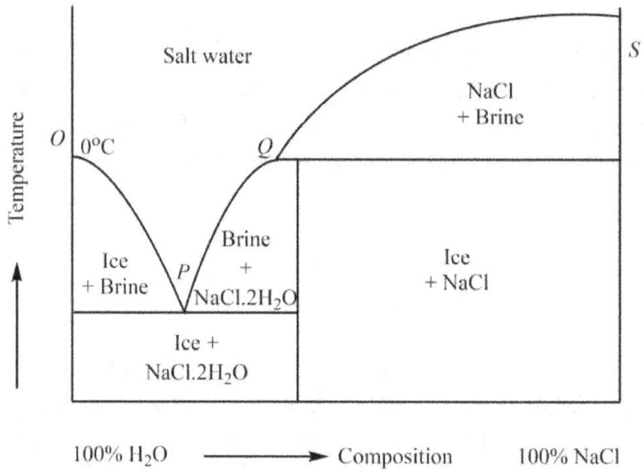


Figure 23. The phase diagram of NaCl-H₂O system.

*i) Point O, S and Q:* The point $O$ and point $S$ represent the freezing point the pure water and pure NaCl, respectively. The point $Q$, however, is quite strange because it represents the transition temperature (incongruent melting point of NaCl.2H₂O) where compound NaCl.2H₂O decomposes into pure NaCl.

*ii) Curve OP, SQ and QP:* When NaCl is added to water, its freezing point decreases regularly along *OP*. In other words, *OP* is the fusion curve of water along which the ice is in equilibrium with the solution. The line *SQ* is the fusion curve of pure NaCl along which solid NaCl is in equilibrium with the brine solution. Similarly, *QP* is the fusion curve of compound NaCl.2H₂O along which solid NaCl.2H₂O is in equilibrium with solution.

When NaCl is added to pure water, ice will just start to separate out as solid along path *OP*. This will lead to a change in composition along the line *OP*. When point $P$ is attained, the formation of compound NaCl.2H₂O will start; and since three phases coexist at $P$ (ice, NaCl.2H₂O and liquid), the reduced phase rule

$$F' = 2 - 3 + 1 = 0 \tag{159}$$

Which means that there will be no degree of freedom at point $P$ ($P$ is non-variant). On the other hand, the curve *SQ* is the fusion curve of NaCl. The further cooling will lead to a change in composition along the line *SQ*. When point $Q$ is attained, the following meritectic reaction will take place

$$Solid\ NaCl + Solution \rightleftharpoons Solid\ NaCl.2H_2O \tag{160}$$

Therefore, the formation of compound NaCl.2H$_2$O will be started; and since three phases coexist at this point (solid NaCl, solid NaCl.2H$_2$O and liquid), the point $Q$ also become invariant. Furthermore, it is also worthy to mention that the transformation of compound NaCl to compound NaCl.2H$_2$O occurs at a constant temperature, and therefore, the point $Q$ is also called as peritectic point.

**2. Sodium sulphate-water system (Na$_2$SO$_4$-H$_2$O):**

The Na$_2$SO$_4$-H$_2$O is another typical case of solid-liquid equilibria in a two-component system that forms compounds with an incongruent melting point. The phases involved in this case are solid Na$_2$SO$_4$, solid Na$_2$SO$_4$.10H$_2$O, ice, liquid mixture, and vapor phase too. Now since a minor pressure disturbance will have little to no effect on the system and all the phases in two-component systems are either solid or liquids only (solid-liquid equilibria), which is the reduced or condensed phase rule i.e.

$$F' = C - P + 1 \tag{161}$$

The complete phase diagram can be drawn on two dimensional paper with vertical and horizontal sides representing temperature and pressure, respectively. Consider a liquid mixture of water and Na$_2$SO$_4$ at temperature $T$. Now if this liquid mixture is allowed to cool down below the freezing point of the mixture, the solid will start to separate out. Prepare a number of such mixtures but with different compositions (i.e. with different ratios of H$_2$O and Na$_2$SO$_4$). The cooling of all the mixtures is carried out in open vessels so that the pressure remains constant (atmospheric pressure). After that, the plot freezing points vs the composition is

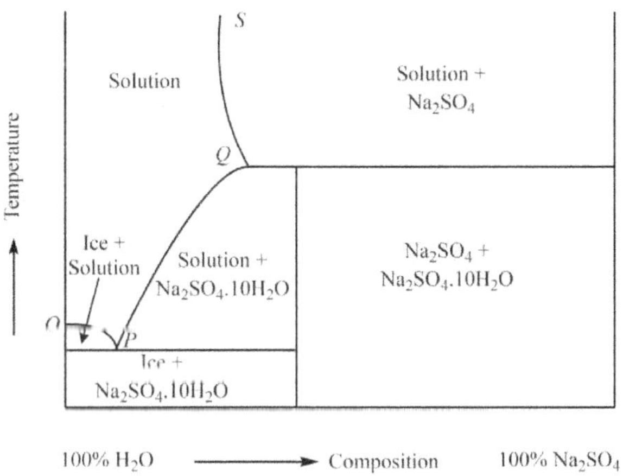


Figure 24. The phase diagram of Na$_2$SO$_4$-H$_2$O system.

*i) Point O and Q:* The point $O$ represents the freezing point the pure water. The point $Q$, however, is quite strange because it represents the transition temperature (incongruent melting point of Na$_2$SO$_4$.10H$_2$O) where compound Na$_2$SO$_4$.10H$_2$O decomposes into pure Na$_2$SO$_4$.

*ii) Curve OP, SQ and QP:* When $Na_2SO_4$ is added to water, its freezing point decreases regularly along *OP*. In other words, *OP* is the fusion curve of compound water along which ice is in equilibrium with the solution. The line *SQ* is the fusion curve of pure $Na_2SO_4$ along which $Na_2SO_4$ is in equilibrium with the solution. Similarly, *QP* is the fusion curve of compound $Na_2SO_4.10H_2O$ along which solid $Na_2SO_4.10H_2O$ is in equilibrium with the solution.

When $Na_2SO_4$ is added to water, ice will start to separate out along *OP*. This will lead to a change in composition along the line *OP*. When point *P* is attained, the formation of compound $Na_2SO_4.10H_2O$ will be started; and since three phases coexist at *P* (ice, $Na_2SO_4$-$H_2O$ and solution), the reduced phase rule gives

$$F' = 2 - 3 + 1 = 0 \tag{162}$$

Which means that there will be no degree of freedom at point *P* (*P* is non-variant). On the other hand, the curve *SQ* is fusion curve of $Na_2SO_4$. The further cooling will lead to a change in composition along the line *SQ*. When point *Q* is attained, the following meritectic reaction will take place

$$Solid\ Na_2SO_4 + Solution \rightleftharpoons Solid\ Na_2SO_4.10H_2O \tag{163}$$

Therefore, the formation of compound $Na_2SO_4.10H_2O$ will be started; and since three phases coexist at this point (solid $Na_2SO_4$, solid $Na_2SO_4.10H_2O$ and liquid), the point Q also become invariant. Furthermore, it is also worthy to mention that the transformation of compound $Na_2SO_4$ to compound $Na_2SO_4.10H_2O$ occurs at a constant temperature, and therefore, the point Q is also called a peritectic point.

## ❖ Phase Diagram and Thermodynamic Treatment of Solid Solutions

In these types of systems, the components are completely miscible with each other in solid phase and completely homogeneous solid solutions are produced. The X-ray diffraction studies are typically employed to check that single crystalline phase is obtained rather than a mixture of two solid phases. In order to draw and understand the phase diagrams of solid solutions, we need to discuss the same in a comprehensive thermodynamic framework first.

### ➤ *General Thermodynamic Treatment of Solid Solutions*

The general thermodynamic treatment of solid solutions includes the shift in the solvent's freezing point during the crystallization and the nature of the cooling curve as well. The necessary discussion on both concepts is given below.

**1. The shifting of solvent's freezing point during crystallization:** The thermodynamic expression for the shift of solvent's freezing point when the solid begins to solidify during cooling can be obtained by assuming the solid solution as an ideal solution. Therefore, the chemical potential for *i*th constituent ($\mu_i$) can be given by the following relation.

$$\mu_i = \mu_i^* + RT \ln x_i \tag{164}$$

Where $x_i$ represents the fractional composition of the constituent whereas $\mu_i^*$ is the chemical potential of the pure solid. After setting the primary condition for the phase equilibria between solid and liquid, we have

$$\mu_1(s) = \mu_1(l) \tag{165}$$

Considering both as ideal, we can write above equation using equation (164) as

$$\mu_1^*(s) + RT \ln x_1(s) = \mu_1^*(l) + RT \ln x_1(l) \tag{166}$$

After rearranging, we get

$$RT \ln x_1(s) - RT \ln x_1(l) = \mu_1^*(l) - \mu_1^*(s) \tag{167}$$

$$RT[\ln x_1(s) - \ln x_1(l)] = \mu_1^*(l) - \mu_1^*(s) \tag{168}$$

$$R \ln \frac{x_1(s)}{x_1(l)} = \frac{\mu_1^*(l) - \mu_1^*(s)}{T} \tag{169}$$

Now since $\mu_1^*(l) - \mu_1^*(s)$ represents the molar free energy of the fusion for the pure solvent at $p$ pressure and T temperature, the equation can also be written as

$$R \ln \frac{x_1(s)}{x_1(l)} = \frac{\Delta_{fus}\,\mu_1^*}{T} \tag{170}$$

Now putting the value of molar free energy of fusion ($\Delta_{fus}\,\mu_1^* = \Delta_{fus}H_{1,m}^* - T\Delta_{fus}S_{1,m}^*$) in the above equation, we get

$$R \ln \frac{x_1(s)}{x_1(l)} = \frac{\Delta_{fus}H_{1,m}^* - T\Delta_{fus}S_{1,m}^*}{T} \tag{171}$$

$$R \ln \frac{x_1(s)}{x_1(l)} = \frac{\Delta_{fus}H_{1,m}^*}{T} - \Delta_{fus}S_{1,m}^* \tag{172}$$

Recalling the entropy of fusion at for the pure solvent at a temperature $T_1^*$ (melting point) i.e.

$$\Delta_{fus}S_{1,m}^* = \frac{\Delta_{fus}H_{1,m}^*}{T_1^*} \tag{173}$$

Using the above result in equation (172), we get

$$R \ln \frac{x_1(s)}{x_1(l)} = \frac{\Delta_{fus}H_{1,m}^*}{T} - \frac{\Delta_{fus}H_{1,m}^*}{T_1^*} \tag{174}$$

$$R \ln \frac{x_1(s)}{x_1(l)} = \Delta_{fus}H_{1,m}^* \left[\frac{1}{T} - \frac{1}{T_1^*}\right] \tag{175}$$

or

$$R \ln \frac{x_1(s)}{x_1(l)} = \Delta_{fus}H_{1,m}^* \left[\frac{T_1^* - T}{TT_1^*}\right] \tag{176}$$

After putting $T - T_1^* = \Delta T_f$, the above equation takes the form

$$R \ln \frac{x_1(s)}{x_1(l)} = \Delta_{fus}H_{1,m}^* \left[\frac{-\Delta T_f}{TT_1^*}\right] \tag{177}$$

If the solution is very dilute, the melting point $T_1^*$ will be quite close to temperature $T$; therefore, $TT_1^*$ can be replaced by $T_1^{*2}$. Therefore, the equation (177) takes the form

$$R[\ln x_1(s) - \ln x_1(l)] = \Delta_{fus}H_{1,m}^* \left[\frac{-\Delta T_f}{T_1^{*2}}\right] \tag{178}$$

Furthermore, for very dilute solutions, we have

$$\ln x_1(s) = \ln [1 - x_2(s)] \approx -x_2(s) \tag{179}$$

$$\ln x_1(l) = \ln [1 - x_2(l)] \approx -x_2(l) \tag{180}$$

Using equation (179, 180) in equation (178), we get

$$R[-x_2(s) + x_2(l)] = \Delta_{fus}H_{1,m}^* \left[\frac{-\Delta T_f}{T_1^{*2}}\right] \tag{181}$$

or

$$-\Delta T_f = \frac{R\, T_1^{*2}}{\Delta_{fus}H_{1,m}^*} [x_2(l) - x_2(s)] \tag{182}$$

$$-\Delta T_f = \frac{R\, T_1^{*2}}{\Delta_{fus}H_{1,m}^*} x_2(l) \left[1 - \frac{x_2(s)}{x_2(l)}\right] \tag{183}$$

Recalling the expression for molality i.e.

$$x_2(l) = \frac{n_2(l)}{n_1(l) + n_2(l)} \approx \frac{n_2(l)}{n_1(l)} = \frac{n_2(l)}{m_1(l)/M_1} = \left[\frac{n_2(l)}{m_1(l)}\right] M_1 = mM_1 \tag{184}$$

Using the above result in equation (183), we have

$$-\Delta T_f = \frac{R\, T_1^{*2} M_1}{\Delta_{fus}H_{1,m}^*} m \left[1 - \frac{x_2(s)}{x_2(l)}\right] \tag{185}$$

or

$$\Delta T_f = -K_f m[1 - K] \tag{186}$$

Where $K$ is the distribution coefficient given by $x_2(s)/x_2(l)$. In other words, the distribution coefficient the ratio of the fractional amount of the solute in the solid phase to the fractional amount of the solute in the solution phase.

Therefore, we can conclude that the value of $\Delta T_f$ can be negative or positive depending upon the magnitude of $K$.

*i) When $\Delta T_f$ is negative:* The magnitude of $\Delta T_f$ will be negative if $K < 1$; which means that $T < T_1^*$. The physical significance is that depression in the freezing point of the solvent will be observed.

*ii) When $\Delta T_f$ is positive:* The magnitude of $\Delta T_f$ will be positive if $K > 1$; which means that $T > T_1^*$. The physical significance is that a less elevation in the freezing point of the solvent will be observed.

The two above-mentioned conditions can also be can be summarized in one statement that if the addition of a component induces a decline in the freezing point of the solvent, it fractional amount must be greater in liquid phase than in the solids phase, and vice-versa.

**2. The nature of cooling curve of the solid solution:** Now because the two components of a solid solution are completely miscible with each other, the maximum number of phases which can coexist is only two (solid solution + liquid). Therefore, recalling reduced phase

$$F' = C - P + 1 \tag{187}$$

After putting $C = 2$ and $P = 2$, we get

$$F' = 2 - 2 + 1 = 1 \tag{188}$$

Which means that there is only one degree of freedoms i.e. univariant system. In other words, only one variable is needed to be defined to define the system completely which can either be temperature or the composition. Therefore, we can conclude that the solid-liquid equilibria can exist at different conditions of temperatures. Also, the composition of the two solutions will be fixed at a given temperature.

Furthermore, there will be two breaks in the cooling curve; one at the start of the freezing of the solid solution, and another at the end of the freezing of solid solution. However, if the composition of the solid solution is exactly same to the solid solution, it will become a one-component system, and therefore, after putting $C = 1$ and $P = 2$ in equation (187), we get

$$F' = 1 - 2 + 1 = 0 \tag{189}$$

Which means that there is no degree of freedoms i.e. non-variant system. In other words, we will not be able to change temperature or composition without disturbing the state of equilibrium. Furthermore, the melting or freezing process at this point will occur at constant temperature, which in turn, would result in arrests in corresponding cooling curve.

> ### *General Discussion on the Phase Diagrams of Solid Solutions*

Depending upon the value of $K_{A\ in\ B}$ and $K_{B\ in\ A}$, the phase diagrams of very dilute solid solutions of component A in B or B in A can primary be classified into three categories. A general discussion on these three types of solid solutions is given below.

### 1. Ascending solid solutions:

In these types of solutions, $K_{B\ in\ A} > 1$ (B is solute and A is solvent i.e. $x_B(s)/x_B(l)$) and $K_{A\ in\ B} < 1$ (A is solute and B is solvent i.e. $x_A(s)/x_A(l)$). Furthermore, the solid solutions with $K_{B\ in\ A} < 1$ and $K_{A\ in\ B} > 1$ also fall into this category. The freezing points of such solutions exist in-between the freezing points of pure solvents. Now, in order to draw the phase diagram of such solid solutions, we need to understand nature of the solidus and liquidus in the thermodynamic framework first. To do so, rearrange equation (175).

$$R \ln \frac{x_1(l)}{x_1(s)} = -\Delta_{fus}H^*_{1,m}\left[\frac{1}{T} - \frac{1}{T^*_1}\right] \tag{190}$$

or

$$\frac{x_1(l)}{x_1(s)} = e^{-a} \tag{191}$$

Where the parameter $a$ is defined as

$$a = \frac{\Delta_{fus}H^*_{1,m}}{R}\left[\frac{1}{T} - \frac{1}{T^*_1}\right] \tag{192}$$

Likewise, the second exponent can be written as

$$\frac{x_2(l)}{x_2(s)} = e^{-b} \tag{193}$$

with

$$b = \frac{\Delta_{fus}H^*_{1,m}}{R}\left[\frac{1}{T} - \frac{1}{T^*_2}\right] \tag{194}$$

For solid and liquid phases, we have

$$x_1(l) = 1 - x_2(l) \tag{195}$$

and

$$x_1(s) = 1 - x_2(s) \tag{196}$$

Now putting the value of equation (196) in equation (191), we get

$$\frac{x_1(l)}{1 - x_2(s)} = e^{-a} \tag{197}$$

Now putting the value of $x_2(s)$ from equation (193), we get

$$\frac{x_1(l)}{1 - x_2(l)/e^{-b}} = e^{-a} \tag{198}$$

After putting the value of $x_2(l)$ from equation (195), we get

$$\frac{x_1(l)}{1 - [1 - x_1(l)]/e^{-b}} = e^{-a} \tag{199}$$

The solution of the above equation for $x_1(l)$ gives

$$x_1(l) = \frac{e^{-a}(e^{-b} - 1)}{e^{-b} - e^{-a}} \tag{200}$$

Using the above result in equation (191), we get

$$x_1(s) = \frac{e^{-b} - 1}{e^{-b} - e^{-a}} \tag{201}$$

Similarly, we can solve for $x_2(l)$ and $x_2(l)$ as

$$x_2(l) = 1 - x_1(l) = \frac{e^{-b}(e^{-a} - 1)}{e^{-a} - e^{-b}} \tag{202}$$

and

$$x_2(s) = 1 - x_1(s) = \frac{e^{-a} - 1}{e^{-a} - e^{-b}} \tag{203}$$

It is obvious from the equation (200-203) that none of them is linear; and the expressions for $x_1(l)$ and $x_1(s)$ are different from each other excepting at $T_1^*$ and $T_2^*$ only. At any temperature $T$ that lies in the range of $T_1^*$ and $T_2^*$, $a$ is negative ($e^{-a} > 1$) and $b$ is positive ($e^{-b} < 1$). Therefore, at this point, we can conclude from equation (200) that

$$x_1(l) = \frac{e^{-a}(e^{-b} - 1)}{e^{-b} - e^{-a}} = \frac{Negative}{Positive} = Positive\ value \tag{204}$$

Similarly, the equations (201-203) can also be proved to have positive values. The physical significance of these results can be summarized by the statement that the physically meaningful compositions are obtained if, and only if, the temperature of the system ($T$) lies in-between the $T_1^*$ and $T_2^*$.

*i) If $T_1^* \leq T \leq T_2^*$:* The equations (191, 193) result in the following conclusions.

$$\frac{x_1(l)}{x_1(s)} > 1 \tag{205}$$

$$\frac{x_2(l)}{x_2(s)} < 1 \tag{206}$$

*ii) If* $T_2^* \leq T \leq T_1^*$*:* The equations (191, 193) result in the following conclusions.

$$\frac{x_1(l)}{x_1(s)} < 1 \tag{207}$$

$$\frac{x_2(l)}{x_2(s)} > 1 \tag{208}$$

Hence, we can conclude that if the liquid is richer that solid, a depression in the freezing point will be observed; whereas if solid is richer than liquid, an elevation in the freezing point will be observed. The general phase diagram is for such systems is given below.

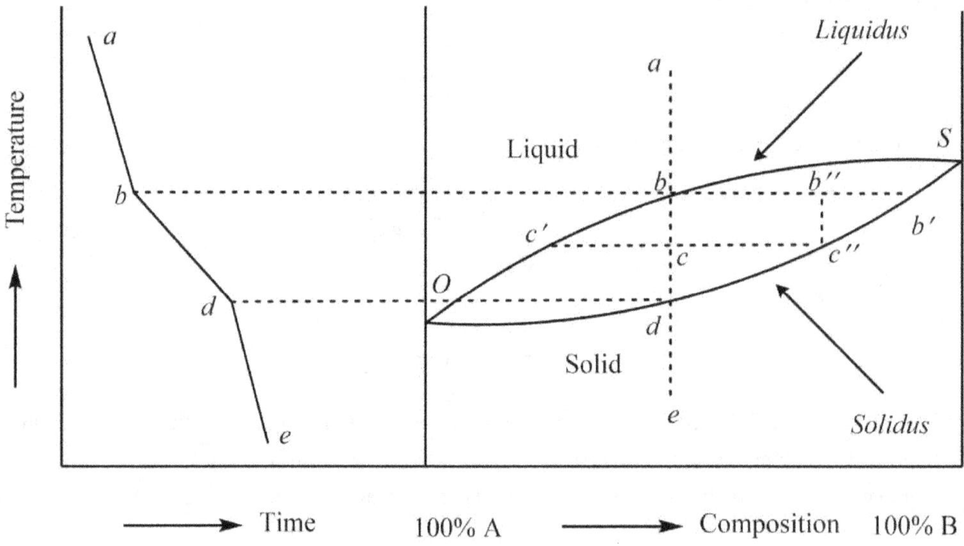


Figure 25. The phase diagram of ascending solid solutions.

The overall phase diagram can be easily understood by analyzing the cooling curve given in the left-hand side of the diagram. Consider a liquid composition represented by the vertical line *abcde* which is allowed to cool down. The system will maintain its liquid state until point *b* is attained where the crystals of the solid solution will start to form. The point corresponding to the crystal formation can be obtained by the tie line *b-b''*. Now because the solid solution is richer in *B* while the liquid phase is less rich in B, the compositional point of the liquid phase will move towards left. Continuing the process of crystallization, the temperature of the system will decrease and the composition of the liquid phase will move along $bc'd'O$. When the overall system is at point $c$, we have a solid solution with composition $c''$ in equilibrium with the liquid solution with composition $c'$. These two points are obtained by the tie line through point $c$ easily; while the relative amounts in solid and liquid phase can be calculated via lever rule. Furthermore, it is also very obvious from the phase diagram that as the system goes from point $b$ to $d$, the left part of the tie lines is increasing whereas right hand side decreases continuously. This means that in going from point $b$ to point $d$, the amount of solid solution increases continuously. When the temperature is lowered to point $d$, almost all of the liquid is transformed into solid which has the same composition as of the starting liquid solution.

The point O and S represent the freezing points of pure A and pure B, respectively. The curve $ObS$ is the freezing point curve of the liquid solution whereas the curve $OdS$ represents the fusion point curve of the solid solution. In the area above the curve $ObS$, the system is completely liquid; while it is completely solid below the curve $OdS$. In the area between the curve $ObS$ and $OdS$, the liquid solution is in equilibrium with the solid solution.

## 2. Minimum-type solid solutions:

In these types of solutions, $K_{B\ in\ A} < 1$ (B is solute and A is solvent i.e. $x_B(s)/x_B(l)$) and $K_{A\ in\ B} < 1$ (A is solute and B is solvent i.e. $x_A(s)/x_A(l)$). The freezing points of such solutions exist below the freezing points of pure solvents. In other words, the freezing points are depressed in these cases.

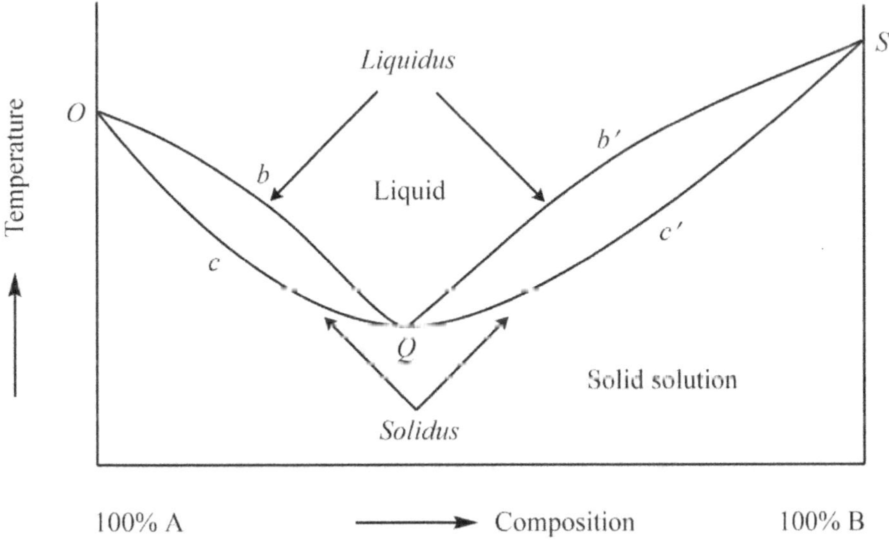


Figure 26. The phase diagram of minimum-type solid solutions.

The point O and S represent the freezing points of pure A and pure B, respectively. The point $Q$ is the minimum freezing point of the liquid solution where solid and liquid solutions have the same composition. The system becomes non-variant at point $Q$ and freezing takes place at a constant temperature with an arrest in cooling curve.

The curve $ObQb'S$ is the freezing point curve of the liquid solution whereas the curve $OcQc'S$ represents the fusion point curve of the solid solution. In the area above the curve $ObQb'S$, the system is completely liquid; while it is completely solid below the curve $OcQc'S$. In the area $ObQcO$, the liquid solution (composition on $ObQ$) is in equilibrium with the solid solution (composition on $OcQ$). In the area $Sb'Qc'S$, the liquid solution (composition on $Sb'Q$) is in equilibrium with the solid solution (composition on $Sc'Q$).

### 3. Maximum-type solid solutions:

In these types of solutions, $K_{B\ in\ A} > 1$ (B is solute and A is solvent i.e. $x_B(s)/x_B(l)$) and $K_{A\ in\ B} > 1$ (A is solute and B is solvent i.e. $x_A(s)/x_A(l)$). The freezing points of such solutions exist above the freezing points of pure solvents. In other words, the freezing points are elevated in these cases.

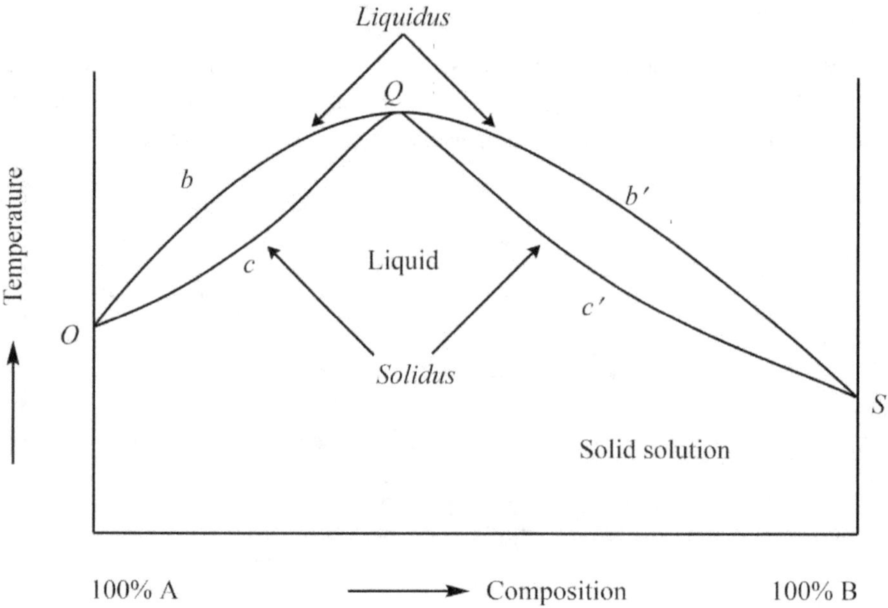


Figure 27. The phase diagram of maximum-type solid solutions.

The point $O$ and $S$ represent the freezing points of pure A and pure B, respectively. The point $Q$ is the maximum freezing point of the liquid solution where solid and liquid solutions have the same composition. The system becomes non-variant at point $Q$ and freezing takes place at a constant temperature with an arrest in cooling curve.

The curve $ObQb'S$ is the freezing point curve of the liquid solution whereas the curve $OcQc'S$ represents the fusion point curve of the solid solution. In the area above the curve $ObQb'S$, the system is completely liquid; while it is completely solid below the curve $OcQc'S$. In the area $ObQcO$, the liquid solution (composition on $ObQ$) is in equilibrium with the solid solution (composition on $OcQ$). In the area $Sb'Qc'S$, the liquid solution (composition on $Sb'Q$) is in equilibrium with the solid solution (composition on $Sc'Q$).

### ❖ Problems

Q 1. Derive and discuss the Clausius-Clapeyron equation.

Q 2. What is the law of mass actions? How is it different from actual rate law?

Q 3. Derive the law of mass action in the thermodynamic framework.

Q 4. State and discuss the Nernst heat theorem.

Q 5. What is the third law of thermodynamics? How it can be used to obtain the absolute entropies?

Q 6. What do you mean by reduced phase rule? Discuss the same for two-component systems.

Q 7. Draw and discuss the phase diagram of eutectic systems with reference to the cooling curve method.

Q 8. Define congruent and incongruent melting points.

Q 9. Draw and discuss the phase diagram of lead-silver system.

Q 10. Draw and discuss the phase diagram of $NaCl$-$H_2O$ system.

Q 11. What are solid solutions? Discuss the thermodynamic treatment for the same with special reference to ascending-type solid solutions.

## ❖ Bibliography

[1] B. R. Puri, L. R. Sharma, M. S. Pathania, *Principles of Physical Chemistry*, Vishal Publications, Jalandhar, India, 2008.

[2] P. Atkins, J. Paula, *Physical Chemistry*, Oxford University Press, Oxford, UK, 2010.

[3] E. Steiner, *The Chemistry Maths Book*, Oxford University Press, Oxford, UK, 2008.

[4] K. L. Kapoor, *A Textbook of Physical Chemistry Volume 3*, Macmillan Publishers, New Delhi, India, 2012.

[5] G. Raj, *Thermodynamics*, Krishna Prakashan, Meerut, India, 2004.

[6] S. Carnot, R. Fox, *Reflections on the Motive Power of Fire: a Critical Edition with the Surviving Scientific Manuscripts*, Manchester University Press, New York, USA, 1986.

[7] P. A. Rock, *Chemical Thermodynamics*, University Science Books, California, USA, 1983.

[8] E. B. Smith, *Basic Chemical Thermodynamics*, Imperial College Press, London, UK, 2014.



**DALAL INSTITUTE**

# CHAPTER 7

## Chemical Dynamics – II

### ❖ Chain Reactions: Hydrogen-Bromine Reaction, Pyrolysis of Acetaldehyde, Decomposition of Ethane

A German chemist, Max Bodenstein, proposed the idea of chemical chain reactions in 1913. He suggested that if two molecules react with each other, some unstable molecules may also be formed along-products that can further react with the reactant molecules with a much higher probability than the initial reactants. Another German chemist, Walther Nernst, explained the quantum yield phenomena in 1918 by suggesting that the photochemical reaction between hydrogen and chlorine is actually a chain reaction. He proposed that only one photon of light is actually accountable for the formation of 106 molecules of the final product. W. Nernst proposed that the incident photon breaks a chlorine molecule into two individual Cl atoms, each of which initiates a long series of stepwise reactions giving a large amount of hydrochloric acid.

A Danish scientist, Christian Christiansen; alongside a Dutch chemist, Hendrik Anthony Kramers; observed the polymer-synthesis in 1923 and conclude that a chain reaction doesn't need a photon always but can also be started by the violent collision of two molecules. They also concluded that if two or more unstable molecules are produced during this reaction, the reaction chain could branch itself to grow enormously resulting in an explosion as well. These ideas were the very initial explanations for the mechanism responsible for the chemical explosions. A more sophisticated theory was proposed by Soviet physicist Nikolay Semyonov in 1934 to explain the quantitative aspects. N. Semyonov received the Nobel Prize in 1956 for his work (along with Sir Cyril Norman Hinshelwood for his independent developments).

**Steps Involved in a Typical Chain Reaction:** The primary steps involved in a typical chain reaction are discussed below.

*i) Initiation:* This step includes the formation of chain carriers or simply the active particles usually free radicals in a photochemically or thermally induced chemical change.

*ii) Propagation:* This step may include many elementary reactions in a cycle in which chain carriers react to forms another chain carrier that continues the chain by entering the next elementary reaction. In other words, we can label these chain carriers as a catalyst for the overall propagation.

*iii) Termination:* This step includes the elementary chemical change in which the chain carriers lose their activity by combining with each other.

It is also worthy to note that the average number of times the propagation cycle is repeated is equal to the ratio of the overall reaction rate to the rate of initiation, and is called as "chain length". Furthermore, some chain reactions follow very complex rate laws with mixed or fractional order kinetics. In this section, we will discuss nature and kinetics some of the most popular chain reactions such as ethane's decomposition.

---

> *General Kinetics of Chain Reactions*

On the basis of the chain carriers produced in each propagation step, the chain reactions can primarily be classified into two categories; non-branched or the stationary reactions and branched or the non-stationary reactions. A typical chain reaction can be written as given below.

$$\text{Initiation:} \qquad A \xrightarrow{\phantom{aa}k_1\phantom{aa}} R^* \tag{1}$$

$$\text{Propagation:} \qquad R^* + A \xrightarrow{\phantom{aa}k_2\phantom{aa}} P + nR^* \tag{2}$$

$$\text{Termination:} \qquad R^* \xrightarrow{\phantom{aa}k_3\phantom{aa}} destruction \tag{3}$$

Where $P$, $R^*$ and $A$ represent the product, radical (or chain carrier) and reactant molecules, respectively. The symbol represents a number that equals unity for stationary chain reactions and greater than one for non-stationary or the branched-chain reactions. Furthermore, the destruction of the radical in the termination step can occur either via its collision with another radical (in gas phase) or by striking the walls of the container.

The steady-state approximation can be employed to determine the concentration of intermediate or the radical involved in the propagation step. The general procedure for which is to put the overall rate of formation equals to zero. In other words, the rate of formation of $R^*$ must be equal to the rate of decomposition of the same i.e.

$$Rate\ of\ formation\ of\ R^* = Rate\ of\ disappearance\ of\ R^* \tag{4}$$

$$k_1[A] = K_3[R^*] - k_2(n-1)[R^*][A] \tag{5}$$

$$k_1[A] - k_3[R^*] + k_2(n-1)[R^*][A] = 0 = \frac{d[R^*]}{dt} \tag{6}$$

or

$$-k_3[R^*] + k_2(n-1)[R^*][A] = -k_1[A] \tag{7}$$

$$[R^*] = \frac{k_1[A]}{k_2(1-n)[A] + k_3} \tag{8}$$

Now because the destruction of the radical in the termination step can occur either via its collision with another radical (in the gas phase) or via striking the walls of the container, the rate constant $k_3$ can be replaced by the sum of the rate constants of two i.e. $k_3 = k_w + k_g$. After using the value of $k_3$ in equation (7), we have

$$[R^*] = \frac{k_1[A]}{k_2(1-n)[A] + k_w + k_g} \tag{9}$$

Now we are ready to apply the concept on different types of chain reactions.

➢ *Hydrogen-Bromine Reaction*

The hydrogen-bromine or the $H_2$-$Br_2$ reaction is a typical case of stationary type chain reactions ($n = 1$) for which the overall reaction can be written as given below.

$$H_2 + Br_2 \longrightarrow 2HBr \tag{10}$$

Furthermore, the elementary steps for the same can be proposed as

*Initiation:* $\qquad\qquad Br_2 \xrightarrow{\ \ k_1\ \ } 2Br \tag{11}$

*Propagation:* $\qquad\quad Br + H_2 \xrightarrow{\ \ k_2\ \ } HBr + H \tag{12}$

$\qquad\qquad\qquad\qquad H + Br_2 \xrightarrow{\ \ k_3\ \ } HBr + Br \tag{13}$

*Inhibition:* $\qquad\quad H + HBr \xrightarrow{\ \ k_4\ \ } H_2 + Br \tag{14}$

*Termination:* $\qquad\quad Br + Br \xrightarrow{\ \ k_5\ \ } Br_2 \tag{15}$

The net rate of formation of HBr must be equal to the sum of the rate of formation and the rate of disappearance of the same i.e.

$$\frac{d[\text{HBr}]}{dt} = k_2[Br][H_2] + k_3[H][Br_2] - k_4[H][HBr] \tag{16}$$

Now, in order to obtain the overall rate expression, we need to apply the steady-state approximation on the H and Br first i.e.

$$\frac{d[\text{H}]}{dt} = 0 = k_2[Br][H_2] - k_3[H][Br_2] - k_4[H][HBr] \tag{17}$$

Similarly,

$$\frac{d[\text{Br}]}{dt} = 0 = 2k_1[Br_2] - k_2[Br][H_2] + k_3[H][Br_2] + k_4[H][HBr] - 2k_5[Br]^2 \tag{18}$$

Taking negative both side of equation (17), we have

$$-k_2[Br][H_2] + k_3[H][Br_2] + k_4[H][HBr] = 0 \tag{19}$$

Using the above result in equation (18), we get

$$2k_1[Br_2] + 0 - 2k_5[Br]^2 = 0 \tag{20}$$

$$2k_5[Br]^2 = 2k_1[Br_2] \tag{21}$$

$$[Br] = \left(\frac{k_1}{k_5}\right)^{1/2} [Br_2]^{1/2} \tag{22}$$

Similarly, rearranging equation (17) again

$$k_3[H][Br_2] + k_4[H][HBr] = k_2[Br][H_2] \tag{23}$$

$$[H] = \frac{k_2[Br][H_2]}{k_3[Br_2] + k_4[HBr]} \tag{24}$$

Now using the value of [Br] from equation (22), the above equation takes the form

$$[H] = \frac{k_2(k_1/k_5)^{1/2}[Br_2]^{1/2}[H_2]}{k_3[Br_2] + k_4[HBr]} \tag{25}$$

Now rearranging equation (17) again in different mode i.e.

$$k_2[Br][H_2] - k_4[H][HBr] = k_3[H][Br_2] \tag{26}$$

Using the above result in equation (16), we have

$$\frac{d[HBr]}{dt} = k_3[H][Br_2] + k_3[H][Br_2] \tag{27}$$

$$\frac{d[HBr]}{dt} = 2k_3[H][Br_2] \tag{28}$$

After putting the value of [H] from equation (24), the equation (28) takes the form

$$\frac{d[HBr]}{dt} = \frac{2k_3k_2(k_1/k_5)^{1/2}[Br_2]^{3/2}[H_2]}{k_3[Br_2] + k_4[HBr]} \tag{29}$$

Taking $k_3[Br_2]$ as common in the denominator and then canceling out the same form numerator, we get

$$\frac{d[HBr]}{dt} = \frac{2k_2(k_1/k_5)^{1/2}[Br_2]^{1/2}[H_2]}{1 + (k_4/k_3)[HBr]/[Br_2]} \tag{30}$$

Now consider two new constants as

$$k' = 2k_2(k_1/k_5)^{1/2} \quad and \quad k'' = k_4/k_3 \tag{31}$$

Using the above results in equation (30), we get

$$\frac{d[HBr]}{dt} = \frac{k'[Br_2]^{1/2}[H_2]}{1 + k''[HBr]/[Br_2]} \tag{32}$$

The initial reaction rate expression can be obtained by neglecting $[HBr]$ i.e. $1 + k''[HBr]/[Br_2] \approx 1$ as

$$\left[\frac{d[\text{HBr}]}{dt}\right]_0 = k'[Br_2]^{1/2}[H_2]_0 \tag{33}$$

Hence, the order of the hydrogen-bromine reaction in the initial stage will be 1.5 only i.e. first-order w.r.t. hydrogen and half w.r.t. bromine.

> ➤ *Pyrolysis of Acetaldehyde*

The pyrolysis of acetaldehyde is another typical case of stationary type chain reactions ($n = 1$) for which the overall reaction can be written as given below.

$$CH_3CHO \longrightarrow CH_4 + CO \tag{34}$$

Furthermore, the elementary steps for the same can be proposed as

*Initiation:* $\qquad CH_3CHO \xrightarrow{\quad k_1 \quad} {}^{\cdot}CH_3 + {}^{\cdot}CHO \tag{35}$

*Propagation:* $\qquad {}^{\cdot}CH_3 + CH_3CHO \xrightarrow{\quad k_2 \quad} CH_4 + {}^{\cdot}CH_2CHO \tag{36}$

$$\qquad\qquad {}^{\cdot}CH_2CHO \xrightarrow{\quad k_3 \quad} {}^{\cdot}CH_3 + CO \tag{37}$$

*Termination:* $\qquad {}^{\cdot}CH_3 + {}^{\cdot}CH_3 \xrightarrow{\quad k_4 \quad} CH_3CH_3 \tag{38}$

The net rate of formation of $CH_4$ must be equal to the sum of the rate of formation and the rate of disappearance of the same i.e.

$$\frac{d[CH_4]}{dt} = k_2[{}^{\cdot}CH_3][CH_3CHO] \tag{39}$$

Now, in order to obtain the overall rate expression, we need to apply the steady-state approximation on the $[{}^{\cdot}CH_3]$ and $[{}^{\cdot}CH_2CHO]$ first i.e.

$$\frac{d[{}^{\cdot}CH_3]}{dt} = 0 = k_1[CH_3CHO] - k_2[{}^{\cdot}CH_3][CH_3CHO] + k_3[{}^{\cdot}CH_2CHO] - 2k_4[{}^{\cdot}CH_3]^2 \tag{40}$$

Similarly,

$$\frac{d[{}^{\cdot}CH_2CHO]}{dt} = 0 = k_2[{}^{\cdot}CH_3][CH_3CHO] - k_3[{}^{\cdot}CH_2CHO] \tag{41}$$

Taking negative both side of equation (41), we have

$$-k_2[{}^{\cdot}CH_3][CH_3CHO] + k_3[{}^{\cdot}CH_2CHO] = 0 \tag{42}$$

Using the above result in equation (40), we get

$$k_1[CH_3CHO] + 0 - 2k_4[\dot{C}H_3]^2 = 0 \tag{43}$$

$$[\dot{C}H_3] = \left(\frac{k_1}{2k_4}\right)^{1/2} [CH_3CHO]^{1/2} \tag{44}$$

After putting the value of $[\dot{C}H_3]$ from equation (44) in equation (39), we have

$$\frac{d[CH_4]}{dt} = k_2 \left(\frac{k_1}{2k_4}\right)^{1/2} [CH_3CHO]^{1/2}\,[CH_3CHO] \tag{45}$$

$$\frac{d[CH_4]}{dt} = k_2 \left(\frac{k_1}{2k_4}\right)^{1/2} [CH_3CHO]^{3/2} \tag{46}$$

Now consider a new constant as

$$k = k_2 \left(\frac{k_1}{2k_4}\right)^{1/2} \tag{47}$$

Using in equation (46), we get

$$\frac{d[CH_4]}{dt} = k[CH_3CHO]^{3/2} \tag{48}$$

The kinetic chain length for the same can be obtained by dividing the rate of formation of the product by rate of initiation step i.e.

$$Kinetic\ chain\ length = \frac{R_p}{R_i} \tag{49}$$

Where $R_p$ and $R_i$ are the rate of propagation and rate of initiation respectively. Now since the rate of propagation is simply equal to the overall rate law i.e. equation (48), the rate of initiation can be given as

$$R_i = k_1[CH_3CHO] \tag{50}$$

After using the values of $R_p$ and $R_i$ from equation (48, 50) in equation (49), we get the expression for kinetic chain length as

$$Kinetic\ chain\ length = \frac{k[CH_3CHO]^{3/2}}{k_1[CH_3CHO]} \tag{51}$$

or

$$Kinetic\ chain\ length = \frac{k}{k_1}[CH_3CHO]^{1/2} \tag{52}$$

> ### Decomposition of Ethane

The decomposition of ethane is another typical case of stationary type chain reactions ($n = 1$) for which the overall reaction can be written as given below.

$$CH_3CH_3 \longrightarrow CH_2 = CH_2 + H_2 \tag{53}$$

Furthermore, the elementary steps for the same can be proposed as

*Initiation:* $\qquad CH_3CH_3 \xrightarrow{\ \ k_1\ \ } {}^\cdot CH_3 + {}^\cdot CH_3 \tag{54}$

*Propagation:* $\qquad {}^\cdot CH_3 + CH_3CH_3 \xrightarrow{\ \ k_2\ \ } CH_4 + {}^\cdot CH_2CH_3 \tag{55}$

$$ {}^\cdot CH_2CH_3 \xrightarrow{\ \ k_3\ \ } CH_2 = CH_2 + {}^\cdot H \tag{56}$$

$$ CH_3CH_3 + {}^\cdot H \xrightarrow{\ \ k_4\ \ } {}^\cdot CH_2CH_3 + H_2 \tag{57}$$

*Termination:* $\qquad {}^\cdot CH_2CH_3 + {}^\cdot H \xrightarrow{\ \ k_5\ \ } CH_3CH_3 \tag{58}$

The net rate of decomposition of ethane must be equal to the rate of formation ethylene i.e.

$$-\frac{d[C_2H_6]}{dt} = \frac{d[C_2H_4]}{dt} = k_3[{}^\cdot CH_2CH_3] \tag{59}$$

Now, in order to obtain the overall rate expression, we need to apply the steady-state approximation on the $[{}^\cdot H]$, $[{}^\cdot CH_3]$ and $[{}^\cdot CH_2CH_3]$ first i.e.

$$\frac{d[{}^\cdot CH_3]}{dt} = 0 = 2k_1[CH_3CH_3] - k_2[{}^\cdot CH_3][CH_3CH_3] \tag{60}$$

Similarly,

$$\frac{d[{}^\cdot CH_2CH_3]}{dt} = 0 = k_2[{}^\cdot CH_3][CH_3CH_3] - k_3[{}^\cdot CH_2CH_3] + k_4[{}^\cdot H][CH_3CH_3] - k_5[{}^\cdot H][{}^\cdot CH_2CH_3] \tag{61}$$

Similarly,

$$\frac{d[{}^\cdot H]}{dt} = 0 = k_3[{}^\cdot CH_2CH_3] - k_4[{}^\cdot H][CH_3CH_3] - k_5[{}^\cdot H][{}^\cdot CH_2CH_3] \tag{62}$$

Rearranging equation (60), we get

$$2k_1 - k_2[{}^\cdot CH_3] = 0 \tag{63}$$

$$[{}^\cdot CH_3] = \frac{2k_1}{k_2} \tag{64}$$

Rearranging equation (62), we get

$$k_4[\text{H}][\text{CH}_3\text{CH}_3] + k_5[\text{H}][\text{CH}_2\text{CH}_3] = k_3[\text{CH}_2\text{CH}_3] \tag{65}$$

$$[\text{H}] = \frac{k_3[\text{CH}_2\text{CH}_3]}{k_4[\text{CH}_3\text{CH}_3] + k_5[\text{CH}_2\text{CH}_3]} \tag{66}$$

After putting the value of $[\text{CH}_3]$ and $[\text{H}]$ from equation (64, 66) in equation (61), we have

$$k_2\left(\frac{2k_1}{k_2}\right)[\text{CH}_3\text{CH}_3] - k_3[\text{CH}_2\text{CH}_3] + k_4\left(\frac{k_3[\text{CH}_2\text{CH}_3]}{k_4[\text{CH}_3\text{CH}_3] + k_5[\text{CH}_2\text{CH}_3]}\right)[\text{CH}_3\text{CH}_3] \tag{67}$$

$$- k_5\left(\frac{k_3[\text{CH}_2\text{CH}_3]}{k_4[\text{CH}_3\text{CH}_3] + k_5[\text{CH}_2\text{CH}_3]}\right)[\text{CH}_2\text{CH}_3] = 0$$

or

$$k_3k_5[\text{CH}_2\text{CH}_3]^2 - k_1k_5[\text{CH}_2\text{CH}_3][\text{CH}_3\text{CH}_3] - k_1k_4[\text{CH}_3\text{CH}_3]^2 = 0 \tag{68}$$

The equation (68) is quadric in nature and can be solved to give

$$[\text{CH}_2\text{CH}_3] = \frac{k_1k_5 + \left(k_1^2k_5^2 + 4k_1k_5k_3k_4\right)^{1/2}}{2k_5k_3}[\text{CH}_3\text{CH}_3] \tag{69}$$

Now putting the result in equation (59), we get

$$-\frac{d[\text{C}_2\text{H}_6]}{dt} = \frac{d[\text{C}_2\text{H}_4]}{dt} = k_3\frac{k_1k_5 + \left(k_1^2k_5^2 + 4k_1k_5k_3k_4\right)^{1/2}}{2k_5k_3}[\text{CH}_3\text{CH}_3] \tag{70}$$

$$-\frac{d[\text{C}_2\text{H}_6]}{dt} = \frac{k_1k_5 + \left(k_1^2k_5^2 + 4k_1k_5k_3k_4\right)^{1/2}}{2k_5}[\text{CH}_3\text{CH}_3] \tag{71}$$

At this stage, defining a new constant as

$$k = \frac{k_1k_5 + \left(k_1^2k_5^2 + 4k_1k_5k_3k_4\right)^{1/2}}{2k_5} \tag{72}$$

Now owing to the very small rate of initiation step, all the terms $k_1k_5$ and $k_1^2k_5^2$ can be neglected i.e.

$$k = \frac{(4k_1k_5k_3k_4)^{1/2}}{2k_5} = \left(\frac{k_1k_3k_4}{k_5}\right)^{1/2} \tag{73}$$

the equation (71) takes the form

$$-\frac{d[\text{C}_2\text{H}_6]}{dt} = k[\text{CH}_3\text{CH}_3] \tag{74}$$

The kinetic chain length for the same can be obtained by dividing the rate of formation of the product by rate of initiation step i.e.

$$Kinetic\ chain\ length = \frac{R_p}{R_i} \tag{75}$$

Where $R_p$ and $R_i$ are the rate of propagation and rate of initiation respectively. Now since the rate of propagation is simply equal to the overall rate law i.e. equation (74), the rate of initiation can be given as

$$R_i = k_1[CH_3CH_3] \tag{76}$$

After using the values of $R_p$ and $R_i$ from equation (74, 76) in equation (75), we get the expression for kinetic chain length as

$$Kinetic\ chain\ length = \frac{k[CH_3CH_3]}{k_1[CH_3CH_3]} \tag{77}$$

or

$$Kinetic\ chain\ length = \frac{k}{k_1} = \frac{1}{k_1}\left(\frac{k_1 k_3 k_4}{k_5}\right)^{1/2} = \left(\frac{k_3 k_4}{k_1 k_5}\right)^{1/2} \tag{78}$$

## ❖ Photochemical Reactions (Hydrogen-Bromine & Hydrogen-Chlorine Reactions)

There are many chemical reactions which occur also when the reactants are exposed to light. These reactions are called as photochemical reactions. Now before we discuss the nature and types of photochemical reactions, we need to discuss two laws first; the first is Grotthuss-Draper law while the second one is Stark-Einstein law.

The first law of photochemistry was given by Theodor Grotthuss and John W. Draper, and therefore, got its unique name.

*The first law of photochemistry states that when the light is allowed to strike the reactions mixture, it can be partially transmitted, reflected and absorbed; and it is the absorbed portion of the incident light which is responsible to carry out any chemical change.*

The second law of photochemistry was given by Johannes Stark and Albert Einstein, and therefore, is popularly known as Stark-Einstein law.

*The second law of photochemistry states that one photon of light must be absorbed for one molecule to get activated in a photochemical reaction by a chemical system.*

The quantum yield of a reaction is simply the ratio of number of molecules reacting in a given time to the number of photons absorbed in the same time i.e.

$$Quantum\ yield(\phi) = \frac{Number\ of\ moles\ of\ reactant\ reacting\ per\ second}{Number\ of\ einsteins\ absorbed\ per\ second} \tag{79}$$

Where one einstein represents one mole of photons ($6.022 \times 10^{23}$). The quantum yield is of two types in photochemical reactions; one for the primary process and the second one for the secondary process. During the course of the primary process, light is absorbed by the reactant and the quantum yield of this part is always unity. Therefore, we can say that the number of radiation-absorbing molecules consumed per unit time in the primary process is equal to intensity of absorbed radiation i.e. $I_{ab}$ (number of photons striking the sample per unit time). For instance, consider the photo-decomposition

$$Br_2 + h\nu \xrightarrow{\quad k \quad} 2Br \tag{80}$$

Then, the quantum yield will be

$$\phi = \frac{Number\ of\ moles\ of\ Br_2\ decomposed\ per\ unit\ time}{Number\ of\ einsteins\ absorbed\ per\ unit\ time} \tag{81}$$

or

$$\phi = \frac{-d[Br_2]/dt}{I_{ab}} \tag{82}$$

Since $\phi = 1$ for primary process, we have

$$1 = \frac{-d[Br_2]/dt}{I_{ab}} \tag{83}$$

or

$$-\frac{d[Br_2]}{dt} = I_{ab} \tag{84}$$

$$-\frac{d[Br_2]}{dt} = k[Br_2] = I_{ab} \tag{85}$$

If rate of formation of Br is asked, then

$$-\frac{d[Br_2]}{dt} = +\frac{1}{2}\frac{d[Br]}{dt} = k[Br_2] = I_{ab} \tag{86}$$

$$\frac{d[Br]}{dt} = 2k[Br_2] = 2I_{ab} \tag{87}$$

Two of the most common examples of photochemical reactions are hydrogen-bromine and hydrogen-chlorine reactions whose kinetics and nature will be discussed in this section.

➢ *Kinetics of Photochemical Reaction Between Hydrogen and Bromine*

The hydrogen-bromine or the $H_2$-$Br_2$ reaction is a typical case of photochemical reactions for which the overall reaction can be written as given below.

$$H_2 + Br_2 \xrightarrow{\;h\nu\;} 2HBr \tag{88}$$

Since it is a chain reaction, the elementary steps for the same can be proposed as

$$\textit{Initiation:} \qquad Br_2 + h\nu \xrightarrow{\;k_1\;} 2Br \tag{89}$$

$$\textit{Propagation:} \qquad Br + H_2 \xrightarrow{\;k_2\;} HBr + H \tag{91}$$

$$H + Br_2 \xrightarrow{\;k_3\;} HBr + Br \tag{92}$$

$$\textit{Inhibition:} \qquad H + HBr \xrightarrow{\;k_4\;} H_2 + Br \tag{93}$$

$$\textit{Termination:} \qquad Br + Br \xrightarrow{\;k_5\;} Br_2 \tag{94}$$

The net rate of formation of HBr must be equal to the sum of the rate of formation and the rate of disappearance of the same i.e.

$$\frac{d[HBr]}{dt} = k_2[Br][H_2] + k_3[H][Br_2] - k_4[H][HBr] \tag{95}$$

Now, in order to obtain the overall rate expression, we need to apply the steady-state approximation on the H and Br first i.e.

$$\frac{d[H]}{dt} = k_2[Br][H_2] - k_3[H][Br_2] - k_4[H][HBr] = 0 \tag{96}$$

Similarly,

$$\frac{d[Br]}{dt} = 2k_1[Br_2] - k_2[Br][H_2] + k_3[H][Br_2] + k_4[H][HBr] - 2k_5[Br]^2 = 0 \tag{97}$$

Taking negative both side of equation (96), we have

$$-k_2[Br][H_2] + k_3[H][Br_2] + k_4[H][HBr] = 0 \tag{98}$$

Using the above result in equation (97), we get

$$2k_1[Br_2] + 0 - 2k_5[Br]^2 = 0 \tag{99}$$

$$2k_5[Br]^2 = 2k_1[Br_2] \tag{100}$$

$$[Br] = \left(\frac{k_1}{k_5}\right)^{1/2} [Br_2]^{1/2} \tag{101}$$

Similarly, rearranging equation (96) again

$$k_3[H][Br_2] + k_4[H][HBr] = k_2[Br][H_2] \tag{102}$$

$$[H] = \frac{k_2[Br][H_2]}{k_3[Br_2] + k_4[HBr]} \tag{103}$$

Now using the value of [Br] from equation (101), the above equation takes the form

$$[H] = \frac{k_2(k_1/k_5)^{1/2}[Br_2]^{1/2}[H_2]}{k_3[Br_2] + k_4[HBr]} \tag{104}$$

Now rearranging equation (96) again in different mode i.e.

$$k_2[Br][H_2] - k_4[H][HBr] = k_3[H][Br_2] \tag{105}$$

Using the above result in equation (95), we have

$$\frac{d[HBr]}{dt} = k_3[H][Br_2] + k_3[H][Br_2] \tag{106}$$

$$d[HBr]/dt = 2k_3[H][Br_2] \tag{107}$$

After putting the value of [H] from equation (104), the equation (107) takes the form

$$\frac{d[HBr]}{dt} = \frac{2k_3k_2(k_1/k_5)^{1/2}[Br_2]^{3/2}[H_2]}{k_3[Br_2] + k_4[HBr]} \tag{108}$$

Taking $k_3[Br_2]$ as common in the denominator and then canceling out the same form numerator, we get

$$\frac{d[HBr]}{dt} = \frac{2k_2(k_1/k_5)^{1/2}[Br_2]^{1/2}[H_2]}{1 + (k_4/k_3)[HBr]/[Br_2]} \tag{109}$$

Now since $2k_1[Br_2] = 2I_{ab}$, then

$$k_1^{1/2}[Br_2]^{1/2} = (I_{ab})^{1/2} \tag{110}$$

Using the above result in equation (109), we get

$$\frac{d[HBr]}{dt} = \frac{2k_2(I_{ab}/k_5)^{1/2}[H_2]}{1 + (k_4/k_3)[HBr]/[Br_2]} \tag{111}$$

Hence, the rate of hydrogen-bromine reaction is directly proportional to the square root of the intensity of absorbed radiation.

> ➤ *Kinetics of Photochemical Reaction Between Hydrogen and Chlorine*

The hydrogen-bromine or the $H_2$-$Cl_2$ reaction is a typical case of photochemical reactions for which the overall reaction can be written as given below.

$$H_2 + Cl_2 \xrightarrow{\phantom{xx}h\nu\phantom{xx}} 2HCl \tag{112}$$

Since it is a chain reaction, the elementary steps for the same can be proposed as

$$\textit{Initiation:} \qquad Cl_2 + h\nu \xrightarrow{\phantom{xx}k_1\phantom{xx}} 2Cl \tag{113}$$

$$\textit{Propagation:} \qquad Cl + H_2 \xrightarrow{\phantom{xx}k_2\phantom{xx}} HCl + H \tag{114}$$

$$H + Cl_2 \xrightarrow{\phantom{xx}k_3\phantom{xx}} HCl + Cl \tag{115}$$

$$\textit{Termination:} \qquad 2Cl(\text{at the walls}) \xrightarrow{\phantom{xx}k_4\phantom{xx}} Cl_2 \tag{116}$$

The net rate of formation of HCl must be equal to the sum of the rate of formation and the rate of disappearance of the same i.e.

$$\frac{d[\text{HCl}]}{dt} = k_2[Cl][H_2] + k_3[H][Cl_2] \tag{117}$$

Now, in order to obtain the overall rate expression, we need to apply the steady-state approximation on the H and Cl first i.e.

$$\frac{d[\text{H}]}{dt} = k_2[Cl][H_2] - k_3[H][Cl_2] = 0 \tag{118}$$

Similarly,

$$\frac{d[\text{Cl}]}{dt} = 2k_1[Cl_2] - k_2[\text{Cl}][H_2] + k_3[H][Cl_2] - 2k_4[Cl] = 0 \tag{119}$$

Taking negative both side of equation (118), we have

$$-k_2[Cl][H_2] + k_3[H][Cl_2] = 0 \tag{120}$$

Using the above result in equation (119), we get

$$2k_1[Cl_2] + 0 - 2k_4[Cl] = 0 \tag{121}$$

or

$$2k_1[Cl_2] = 2k_4[Cl] \tag{122}$$

or

$$[Cl] = \frac{k_1}{k_4}[Cl_2] \tag{123}$$

Similarly, rearranging equation (118) again

$$k_2[Cl][H_2] - k_3[H][Cl_2] = 0 \tag{124}$$

or

$$[H] = \frac{k_2[Cl][H_2]}{k_3[Cl_2]} \tag{125}$$

Now using the value of [Cl] from equation (123), the above equation takes the form

$$[H] = \frac{k_2[H_2]}{k_3[Cl_2]}\frac{k_1}{k_4}[Cl_2] = \frac{k_1 k_2[H_2]}{k_3 k_4} \tag{126}$$

Using values of $[Cl]$ and $[H]$ from equation (123, 126) in equation (117), we have

$$\frac{d[\text{HCl}]}{dt} = k_2\frac{k_1}{k_4}[Cl_2][H_2] + k_3\frac{k_1 k_2[H_2]}{k_3 k_4}[Cl_2] \tag{127}$$

or

$$\frac{d[\text{HCl}]}{dt} = \frac{k_1 k_2[Cl_2][H_2]}{k_4} + \frac{k_1 k_2[H_2][Cl_2]}{k_4} \tag{128}$$

Now since $2k_1[Cl_2] = 2I_{ab}$, then

$$k_1[Cl_2] = I_{ab} \tag{129}$$

Using the above result in equation (129), we get

$$\frac{d[\text{HCl}]}{dt} = \frac{k_2 I_{ab}[H_2]}{k_4} + \frac{k_2 I_{ab}[H_2]}{k_4} \tag{130}$$

or

$$\frac{d[\text{HCl}]}{dt} = \frac{2k_2 I_{ab}[H_2]}{k_4} \tag{131}$$

Hence, we can conclude that the rate of hydrogen-bromine reaction is directly proportional to the intensity of absorbed radiation. It is also worthy to note that the quantum yield of the hydrogen-chlorine reaction is much higher than that of hydrogen-bromine reaction which may simply be attributed to the exothermic nature of the second step in $H_2$-$Cl_2$ reaction which makes it spontaneous in nature.

Copyright © *Mandeep Dalal*

## ❖ General Treatment of Chain Reactions (Ortho-Para Hydrogen Conversion and Hydrogen-Bromine Reactions)

In this section, we will discuss the general treatment of some common chain reactions like ortho-para hydrogen conversion and hydrogen-bromine reactions. To do so, we may also recall some concepts discussed earlier in this chapter.

### ➤ *Ortho-Para Hydrogen Conversion*

The ortho-hydrogen can convert into para-hydrogen and can easily be measured in the temperature range of 700–800°C. The conversion is completely homogeneous in nature and the order of the conversion is 1.5 as total. The widely accepted mechanism is given below.

$$p\text{-}H_2 \rightleftharpoons 2H \tag{132}$$

$$H + p\text{-}H_2 \xrightarrow{\;\;k_2\;\;} H + o\text{-}H_2 \tag{133}$$

The equilibrium constant for the above-mentioned molecular-atomic equilibria given by equation (132) can be written as

$$K = \frac{[H]^2}{[p\text{-}H_2]} \tag{134}$$

$$[H]^2 = K[p\text{-}H_2] \tag{135}$$

$$[H] = K^{1/2} [p\text{-}H_2]^{\frac{1}{2}} \tag{136}$$

The net rate of conversion of $p\text{-}H_2$ must be equal to the sum of the rate of formation and the rate of disappearance of the same i.e.

$$\frac{d[p\text{-}H_2]}{dt} = k_2 [p\text{-}H_2][H] \tag{137}$$

After using the value of $[H]$ from equation (136) in equation (137), we get

$$\frac{d[p\text{-}H_2]}{dt} = k_2 [p\text{-}H_2] K^{1/2} [p\text{-}H_2]^{\frac{1}{2}} \tag{137}$$

or

$$\frac{d[p\text{-}H_2]}{dt} = k_2 K^{1/2} [p\text{-}H_2]^{3/2} \tag{137}$$

The overall activation energy of the conversion is $E = E_2 + D/2$ where $E_2$ is the activation energy of the second step while D is the dissociation energy of dihydrogen molecule.

---

> ### *Hydrogen-Bromine Reactions*

The hydrogen-bromine or the $H_2$-$Br_2$ reaction is a typical case of stationary type chain reactions ($n = 1$) for which the overall reaction can be written as given below.

$$H_2 + Br_2 \longrightarrow 2HBr \tag{138}$$

Furthermore, the elementary steps for the same can be proposed as

$$\textit{Initiation:} \qquad Br_2 \xrightarrow{\;\;k_1\;\;} 2Br \tag{139}$$

$$\textit{Propagation:} \qquad Br + H_2 \xrightarrow{\;\;k_2\;\;} HBr + H \tag{140}$$

$$H + Br_2 \xrightarrow{\;\;k_3\;\;} HBr + Br \tag{141}$$

$$\textit{Inhibition:} \qquad H + HBr \xrightarrow{\;\;k_4\;\;} H_2 + Br \tag{142}$$

$$\textit{Termination:} \qquad Br + Br \xrightarrow{\;\;k_5\;\;} Br_2 \tag{143}$$

The net rate of formation of HBr must be equal to the sum of the rate of formation and the rate of disappearance of the same i.e.

$$\frac{d[\text{HBr}]}{dt} = k_2[Br][H_2] + k_3[H][Br_2] - k_4[H][HBr] \tag{144}$$

Now, in order to obtain the overall rate expression, we need to apply the steady-state approximation on the H and Br first i.e.

$$\frac{d[\text{H}]}{dt} = 0 = k_2[Br][H_2] - k_3[H][Br_2] - k_4[H][HBr] \tag{145}$$

Similarly,

$$\frac{d[\text{Br}]}{dt} = 0 = 2k_1[Br_2] - k_2[Br][H_2] + k_3[H][Br_2] + k_4[H][HBr] - 2k_5[Br]^2 \tag{146}$$

Taking negative both side of equation (145), we have

$$-k_2[Br][H_2] + k_3[H][Br_2] + k_4[H][HBr] = 0 \tag{147}$$

Using the above result in equation (146), we get

$$2k_1[Br_2] + 0 - 2k_5[Br]^2 = 0 \tag{148}$$

$$2k_5[Br]^2 = 2k_1[Br_2] \tag{149}$$

$$[Br] = \left(\frac{k_1}{k_5}\right)^{1/2} [Br_2]^{1/2} \tag{150}$$

Similarly, rearranging equation (145) again

$$k_3[H][Br_2] + k_4[H][HBr] = k_2[Br][H_2] \tag{151}$$

$$[H] = \frac{k_2[Br][H_2]}{k_3[Br_2] + k_4[HBr]} \tag{152}$$

Now using the value of [Br] from equation (150), the above equation takes the form

$$[H] = \frac{k_2(k_1/k_5)^{1/2}[Br_2]^{1/2}[H_2]}{k_3[Br_2] + k_4[HBr]} \tag{153}$$

Now rearranging equation (145) again in different mode i.e.

$$k_2[Br][H_2] - k_4[H][HBr] = k_3[H][Br_2] \tag{154}$$

Using the above result in equation (144), we have

$$\frac{d[HBr]}{dt} = k_3[H][Br_2] + k_3[H][Br_2] \tag{155}$$

$$\frac{d[HBr]}{dt} = 2k_3[H][Br_2] \tag{156}$$

After putting the value of $[H]$ from equation (152), the equation (156) takes the form

$$\frac{d[HBr]}{dt} = \frac{2k_3k_2(k_1/k_5)^{1/2}[Br_2]^{3/2}[H_2]}{k_3[Br_2] + k_4[HBr]} \tag{157}$$

Taking $k_3[Br_2]$ as common in the denominator and then canceling out the same form numerator, we get

$$\frac{d[HBr]}{dt} = \frac{2k_2(k_1/k_5)^{1/2}[Br_2]^{1/2}[H_2]}{1 + (k_4/k_3)[HBr]/[Br_2]} \tag{158}$$

Now consider two new constants as

$$k' = 2k_2(k_1/k_5)^{1/2} \quad and \quad k'' = k_4/k_3 \tag{159}$$

Using in equation (158), we get

$$\frac{d[HBr]}{dt} = \frac{k'[Br_2]^{1/2}[H_2]}{1 + k''[HBr]/[Br_2]} \tag{160}$$

The initial reaction rate expression can be obtained by neglecting $[HBr]$ i.e. $1 + k''[HBr]/[Br_2] \approx 1$ as

$$\left[\frac{d[HBr]}{dt}\right]_0 = k'[Br_2]^{1/2}[H_2]_0 \tag{161}$$

Hence, the order of the hydrogen-bromine reaction in the initial stage will be 1.5 only i.e. first-order w.r.t. hydrogen and half w.r.t. bromine.

However, if the initiation occurs by the photon of the incident light, the expression for the overall rate can slightly be modified. Recall the initiation step again via photochemical decomposition i.e.

$$Br_2 + h\nu \xrightarrow{\ \ k\ \ } 2Br \tag{162}$$

Then, the quantum yield for this primary change will be

$$\phi = \frac{-d[Br_2]/dt}{I_{ab}} \tag{163}$$

Where $I_{ab}$ is the intensity of absorbed radiation. Since $\phi = 1$ for primary process, we have

$$1 = \frac{-d[Br_2]/dt}{I_{ab}} \tag{164}$$

$$-\frac{d[Br_2]}{dt} = I_{ab} \tag{165}$$

$$-\frac{d[Br_2]}{dt} = k[Br_2] = I_{ab} \tag{166}$$

For the rate of formation of Br, we have

$$-\frac{d[Br_2]}{dt} = +\frac{1}{2}\frac{d[Br]}{dt} = k[Br_2] = I_{ab} \tag{167}$$

$$\frac{d[Br]}{dt} = 2k[Br_2] = 2I_{ab} \tag{168}$$

Now since $2k_1[Br_2] = 2I_{ab}$, then

$$k_1^{1/2}[Br_2]^{1/2} = (I_{ab})^{1/2} \tag{169}$$

Using the above result in equation (158), we get

$$\frac{d[HBr]}{dt} = \frac{2k_2(I_{ab}/k_5)^{1/2}[H_2]}{1 + (k_4/k_3)[HBr]/[Br_2]} \tag{170}$$

Hence, the rate of hydrogen-bromine reaction is directly proportional to the square root of the intensity of absorbed radiation.

### ❖ Apparent Activation Energy of Chain Reactions

In order to understand the activation energy profile of chain reactions, consider a typical chain reaction such as the synthesis of hydrogen bromide from its elements.

$$H_2 + Br_2 \longrightarrow 2HBr \tag{171}$$

Furthermore, the elementary steps for the same can be proposed as

$$\textit{Initiation:} \qquad Br_2 + M \xrightarrow{\ k_1\ } 2Br \tag{172}$$

$$\textit{Propagation:} \qquad Br + H_2 \xrightarrow{\ k_2\ } HBr + H \tag{173}$$

$$H + Br_2 \xrightarrow{\ k_3\ } HBr + Br \tag{174}$$

$$\textit{Inhibition:} \qquad H + HBr \xrightarrow{\ k_4\ } H_2 + Br \tag{175}$$

$$\textit{Termination:} \qquad 2Br + M \xrightarrow{\ k_5\ } M^* + Br_2 \tag{176}$$

It is obvious from the stepwise mechanism that the reaction is actually initiated by the formation of a free radical when one $Br_2$ molecule collides with some other molecule M (most another $Br_2$). All this just requires a sufficiently high temperature of the gas. Another mode of formation of Br free radicals is the photochemical activation in which the sample is exposed to light. In second and third reactions, the consumption of a free radical generates another, and therefore, propagates the chain. The same happens in the fourth step, but the molecule of the product is destroyed, and thus inhibited the whole process. Now if these first four steps occur, then the chain would continue for infinite. However, during the course of the 5th step, two Br combine together, and thus terminating the chain. The molecule $M^*$ in the last step represents a thermally-excited which dissipates its energy to other molecules quickly.

The overall rate laws for such chain reactions are quite complex and usually have non-integral orders. The net rate of formation of HBr can be given as

$$\frac{d[\text{HBr}]}{dt} = \frac{2k_2(k_1/k_5)^{1/2}[Br_2]^{1/2}[H_2]}{1 + (k_4/k_3)[\text{HBr}]/[Br_2]} \tag{177}$$

Now consider two new constants as

$$k' = 2k_2(k_1/k_5)^{1/2} \qquad and \qquad k'' = k_4/k_3 \tag{178}$$

Using in equation (177), we get

$$\frac{d[\text{HBr}]}{dt} = \frac{k'[Br_2]^{1/2}[H_2]}{1 + k''[\text{HBr}]/[Br_2]} \tag{179}$$

The initial reaction rate expression can be obtained by neglecting $[HBr]$ i.e. $1 + k''[HBr]/[Br_2] \approx 1$ as

$$\left[\frac{d[HBr]}{dt}\right]_0 = k'[Br_2]^{1/2}[H_2]_0 \tag{180}$$

Hence, the order of the hydrogen-bromine reaction in the initial stage will be 1.5 only i.e. first-order w.r.t. hydrogen and half w.r.t. bromine. Expanding equation (180, we get

$$\left[\frac{d[HBr]}{dt}\right]_0 = 2k_2\left(\frac{k_1}{k_5}\right)^{1/2}[Br_2]^{1/2}[H_2]_0 \tag{181}$$

Hence the overall rate constant can be written as

$$k_{overall} = 2k_2\left(\frac{k_1}{k_5}\right)^{1/2} \tag{182}$$

Now assuming $E_1$, $E_2$ and $E_5$ are the activation energies for first (initiation), second (propagation) and fifth (termination) steps; we can write general expressions for individual rate constants as

$$k_1 = Ae^{-E_1/RT} \tag{183}$$

$$k_2 = Ae^{-E_2/RT} \tag{184}$$

$$k_5 = Ae^{-E_5/RT} \tag{185}$$

After putting values of $k_1$, $k_2$ and $k_5$ from equation (183-185) in equation (182), we get

$$k_{overall} = 2\,Ae^{-E_2/RT}\left(\frac{Ae^{-E_1/RT}}{Ae^{-E_5/RT}}\right)^{1/2} = 2\,Ae^{-E_2/RT}\left(e^{-E_1+E_5/RT}\right)^{1/2} \tag{186}$$

$$k_{overall} = 2\,Ae^{\frac{-E_2}{RT}}\,e^{\frac{-E_1+E_5}{2RT}} = 2\,Ae^{\frac{-2E_2-E_1+E_5}{2RT}} = 2\,Ae^{-\frac{1}{2}\frac{(2E_2+E_1-E_5)}{RT}} \tag{187}$$

Comparing the above result with Arrhenius rate constant $k = Ae^{-E_a/RT}$, we get

$$E_a = \frac{1}{2}(2E_2 + E_1 - E_5) = E_2 - \frac{1}{2}(E_1 - E_5) \tag{188}$$

Recalling that $E_1$, $E_2$ and $E_5$ are the activation energies for initiation, propagation and termination steps; the equation (188) can be written in more general form i.e.

$$E_a = E_p - \frac{1}{n}(E_i - E_t) \tag{189}$$

Where $n$ is the order of termination step w.r.t. chain carrier.

## ❖ Chain Length

*The chain length for any chemical chain reaction may simply be defined as the average number of times that the closed cycle of chain propagation steps is repeated.*

In other words, the chain length of any chain reaction is equal to the ration of the rate of the overall reaction to the rate of the initiation step in which the active particles are generated. Mathematically,

$$Kinetic\ chain\ length = \frac{R_p}{R_i} \tag{190}$$

Where $R_p$ and $R_i$ are the rate of propagation and rate of initiation respectively. The concept can be well understood by taking the example of thermal decomposition of acetaldehyde and ethane.

### ➤ *Chain length in Case of Thermal Decomposition of Acetaldehyde*

The thermal decomposition of acetaldehyde is a typical case of stationary type chain reactions for which the overall reaction can be written as

$$CH_3CHO \longrightarrow CH_4 + CO \tag{191}$$

Furthermore, the elementary steps for the same can be proposed as

$$Initiation: \qquad CH_3CHO \xrightarrow{\ \ k_1\ \ } {}^\bullet CH_3 + {}^\bullet CHO \tag{192}$$

$$Propagation: \qquad {}^\bullet CH_3 + CH_3CHO \xrightarrow{\ \ k_2\ \ } CH_4 + {}^\bullet CH_2CHO \tag{193}$$

$$\qquad\qquad {}^\bullet CH_2CHO \xrightarrow{\ \ k_3\ \ } {}^\bullet CH_3 + CO \tag{194}$$

$$Termination: \qquad {}^\bullet CH_3 + {}^\bullet CH_3 \xrightarrow{\ \ k_4\ \ } CH_3CH_3 \tag{195}$$

The net rate of formation of $CH_4$ must be equal to the sum of the rate of formation and the rate of disappearance of the same i.e.

$$\frac{d[CH_4]}{dt} = k[CH_3CHO]^{3/2} \tag{196}$$

Using equation (192), the rate of initiation can be given as

$$R_i = k_1[CH_3CHO] \tag{197}$$

Using the values of $R_p$ and $R_i$ from equation (196, 197) in equation (190), we get the kinetic chain length as

$$Kinetic\ chain\ length = \frac{k[CH_3CHO]^{3/2}}{k_1[CH_3CHO]} = \frac{k}{k_1}[CH_3CHO]^{1/2} \tag{198}$$

> ## Chain Length in Case of Dehydrogenation of Ethane

The dehydrogenation of ethane is another typical case of stationary type chain reactions for which the overall reaction can be written as given below.

$$CH_3CH_3 \longrightarrow CH_2 = CH_2 + H_2 \tag{199}$$

Furthermore, the elementary steps for the same can be proposed as

*Initiation:* $\qquad CH_3CH_3 \xrightarrow{\quad k_1 \quad} \dot{C}H_3 + \dot{C}H_3 \tag{200}$

*Propagation:* $\qquad \dot{C}H_3 + CH_3CH_3 \xrightarrow{\quad k_2 \quad} CH_4 + \dot{C}H_2CH_3 \tag{201}$

$$\dot{C}H_2CH_3 \xrightarrow{\quad k_3 \quad} CH_2 = CH_2 + \dot{H} \tag{202}$$

$$CH_3CH_3 + \dot{H} \xrightarrow{\quad k_4 \quad} \dot{C}H_2CH_3 + H_2 \tag{203}$$

*Termination:* $\qquad \dot{C}H_2CH_3 + \dot{H} \xrightarrow{\quad k_5 \quad} CH_3CH_3 \tag{204}$

The net rate of decomposition of ethane will be

$$-\frac{d[C_2H_6]}{dt} = \left(\frac{k_1 k_3 k_4}{k_5}\right)^{1/2} [CH_3CH_3] \tag{205}$$

Since the rate of propagation is simply equal to the overall rate law i.e. equation (205), the rate of initiation is

$$R_i = k_1[CH_3CH_3] \tag{206}$$

After using the values of $R_p$ and $R_i$ from equation (205, 206) in equation (190), we get the expression for kinetic chain length as

$$\textit{Kinetic chain length} = \frac{\left(\frac{k_1 k_3 k_4}{k_5}\right)^{\frac{1}{2}} [CH_3CH_3]}{k_1[CH_3CH_3]} \tag{207}$$

or

$$\textit{Kinetic chain length} = \frac{1}{k_1}\left(\frac{k_1 k_3 k_4}{k_5}\right)^{1/2} \frac{[CH_3CH_3]}{[CH_3CH_3]} \tag{208}$$

$$\textit{Kinetic chain length} = \left(\frac{k_3 k_4}{k_1 k_5}\right)^{1/2} \tag{209}$$

## ❖ Rice-Herzfeld Mechanism of Organic Molecules Decomposition (Acetaldehyde)

In 1934, Frank O. Rice and Karl F. Herzfeld performed an extensive study on the chain reactions and developed special mechanics to account the observed rate laws. In the honor of researchers, this mechanism is popularly known as Rice-Herzfeld mechanism. The conceptual foundation of this mechanism can easily be understood by taking the example of an organic compound, acetaldehyde.

The pyrolysis of acetaldehyde is a typical case of stationary type chain reactions ($n = 1$) for which the overall reaction can be written as given below.

$$CH_3CHO \longrightarrow CH_4 + CO \tag{210}$$

Furthermore, the elementary steps for the same can be proposed as

*Initiation:* $\qquad CH_3CHO \xrightarrow{\quad k_1 \quad} {}^{\cdot}CH_3 + {}^{\cdot}CHO$ $\tag{211}$

*Propagation:* $\qquad {}^{\cdot}CH_3 + CH_3CHO \xrightarrow{\quad k_2 \quad} CH_4 + {}^{\cdot}CH_2CHO$ $\tag{212}$

$\qquad {}^{\cdot}CH_2CHO \xrightarrow{\quad k_3 \quad} {}^{\cdot}CH_3 + CO$ $\tag{213}$

*Termination:* $\qquad {}^{\cdot}CH_3 + {}^{\cdot}CH_3 \xrightarrow{\quad k_4 \quad} CH_3CH_3$ $\tag{214}$

The net rate of formation of $CH_4$ must be equal to the sum of the rate of formation and the rate of disappearance of the same i.e.

$$\frac{d[CH_4]}{dt} = k_2[{}^{\cdot}CH_3][CH_3CHO] \tag{215}$$

Now, in order to obtain the overall rate expression, we need to apply the steady-state approximation on the $[{}^{\cdot}CH_3]$ and $[{}^{\cdot}CH_2CHO]$ first i.e.

$$\frac{d[{}^{\cdot}CH_3]}{dt} = 0 = k_1[CH_3CHO] - k_2[{}^{\cdot}CH_3][CH_3CHO] + k_3[{}^{\cdot}CH_2CHO] - 2k_4[{}^{\cdot}CH_3]^2 \tag{216}$$

Similarly,

$$\frac{d[{}^{\cdot}CH_2CHO]}{dt} = 0 = k_2[{}^{\cdot}CH_3][CH_3CHO] - k_3[{}^{\cdot}CH_2CHO] \tag{217}$$

Taking negative both side of equation (217), we have

$$-k_2[{}^{\cdot}CH_3][CH_3CHO] + k_3[{}^{\cdot}CH_2CHO] = 0 \tag{218}$$

Using the above result in equation (216), we get

---

$$k_1[CH_3CHO] + 0 - 2k_4[^{\cdot}CH_3]^2 = 0 \tag{219}$$

$$[^{\cdot}CH_3] = \left(\frac{k_1}{2k_4}\right)^{1/2} [CH_3CHO]^{1/2} \tag{220}$$

After putting the value of $[^{\cdot}CH_3]$ from equation (220) in equation (215), we have

$$\frac{d[CH_4]}{dt} = k_2 \left(\frac{k_1}{2k_4}\right)^{1/2} [CH_3CHO]^{1/2} [CH_3CHO] \tag{221}$$

$$\frac{d[CH_4]}{dt} = k_2 \left(\frac{k_1}{2k_4}\right)^{1/2} [CH_3CHO]^{3/2} \tag{222}$$

Now consider a new constant as

$$k = k_2 \left(\frac{k_1}{2k_4}\right)^{1/2} \tag{223}$$

Using in equation (222), we get

$$\frac{d[CH_4]}{dt} = k[CH_3CHO]^{3/2} \tag{224}$$

The kinetic chain length for the same can be obtained by dividing the rate of formation of the product by rate of initiation step i.e.

$$Kinetic\ chain\ length = \frac{R_p}{R_i} \tag{225}$$

Where $R_p$ and $R_i$ are the rate of propagation and rate of initiation respectively. Now since the rate of propagation is simply equal to the overall rate law i.e. equation (224), the rate of initiation can be given as

$$R_i = k_1[CH_3CHO] \tag{226}$$

After using the values of $R_p$ and $R_i$ from equation (224, 226) in equation (225), we get the expression for kinetic chain length as

$$Kinetic\ chain\ length = \frac{k[CH_3CHO]^{3/2}}{k_1[CH_3CHO]} \tag{227}$$

or

$$Kinetic\ chain\ length = \frac{k}{k_1}[CH_3CHO]^{1/2} \tag{228}$$

### ❖ Branching Chain Reactions and Explosions ($H_2$-$O_2$ Reaction)

On the basis of the chain carriers produced in each propagation step, the chain reactions can primarily be classified into two categories; non-branched or the stationary reactions and branched or the non-stationary reactions. A typical chain reaction can be written as given below.

$$Initiation: \qquad A \xrightarrow{\quad k_1 \quad} R^* \tag{229}$$

$$Propagation: \qquad R^* + A \xrightarrow{\quad k_2 \quad} P + nR^* \tag{230}$$

$$Termination: \qquad R^* \xrightarrow{\quad k_3 \quad} destruction \tag{231}$$

Where $P$, $R^*$ and $A$ represent the product, radical (or chain carrier) and reactant molecules, respectively. The symbol represents a number that equals unity for stationary chain reactions and greater than one for non-stationary or the branched-chain reactions. Furthermore, the destruction of the radical in the termination step can occur either via its collision with another radical (in gas phase) or by striking the walls of the container.

The steady-state approximation can be employed to determine the concentration of intermediate or the radical involved in the propagation step. The general procedure for which is to put the overall rate of formation equals to zero. In other words, the rate of formation of $R^*$ must be equal to the rate of decomposition of the same i.e.

$$Rate \ of \ formation \ of \ R^* = Rate \ of \ disappearance \ of \ R^* \tag{232}$$

$$k_1[A] = k_3[R^*] - k_2(n-1)[R^*][A] \tag{233}$$

$$k_1[A] - k_3[R^*] + k_2(n-1)[R^*][A] = 0 = \frac{d[R^*]}{dt} \tag{234}$$

or

$$-k_3[R^*] + k_2(n-1)[R^*][A] = -k_1[A] \tag{235}$$

$$[R^*] = \frac{k_1[A]}{k_2(1-n)[A] + k_3} \tag{236}$$

Now because the destruction of the radical in the termination step can occur either via its collision with another radical (in gas phase) or via striking the walls of the container, the rate constant $k_3$ can be replaced by the sum of the rate constants of two i.e. $k_3 = k_w + k_g$. After using the value of $k_3$ in equation (236), we have

$$[R^*] = \frac{k_1[A]}{k_2(1-n)[A] + k_w + k_g} \tag{237}$$

or

---

$$[R^*] = \frac{k_1[A]}{-k_2(n-1)[A] + k_w + k_g} \tag{238}$$

For non-stationary or branched chain reactions, more and more radicals are generated in each successive step ($n > 1$). In other words, for every radical consumed in a chain propagation step, more than one chain carriers or radicals are generated. Therefore, the possibility of explosion arises when

$$k_2(n-1)[A] = k_w + k_g \tag{239}$$

The situation can be explained in terms of equation (238) because the abovementioned condition will make the denominator zero, and therefore, making radical concentration to approach infinite. Now because the rate is usually proportional to radical's concentration, a very high rate may lead to an explosion.

> ➤ *Conditions for Different Types of Explosion limits*

However, it is also worthy to mention that the occurrence of explosion depends upon the experimental temperature and pressure. Typically, three explosion limits are observed as the pressure of the reacting system is raised. To understand the different explosion limits, the behavior of the denominator in equation (238) must be analysed with pressure.

**1. The first explosion limit:** When the pressure is very low, the movement of chain carriers towards the wall of the container is very fast resulting in a very large rate of radicals' destruction at walls i.e. high $k_w$. Conversely, the probability of radicals colliding with each other at very low pressure resulting in a very low value of $k_g$. Therefore, we can conclude that the denominator in equation (238) has a sufficiently large positive value at low pressure giving smooth progression of the reaction without any explosion. However, with the rise in pressure, $k_w$ declines very rapidly than the increase in $k_g$. When a certain pressure value is achieved, the explosion condition is satisfied i.e.

$$k_2(n-1)[A] = k_w + k_g \tag{240}$$

Which is the first explosion limit.

**2. The second explosion limit:** The first explosion limit exists over a wide range of pressure. However, if the pressure is raised continuously, the movement of chain carriers towards the wall of the container is more and more hindered resulting in a very small rate of radicals' destruction at walls i.e. low $k_w$. Conversely, the probability of radicals colliding with each other further increases with pressure resulting in the very large value of $k_g$. Eventually, we can conclude that the denominator in equation (238) again becomes sufficiently positive giving a steady progression of the reaction. Which is the second explosion limit.

**3. The third explosion limit:** After the second explosion limit, the steady reaction-rate continues over a range of pressure. However, if the pressure is raised continuously, the heat produced in various propagating steps would not be able to leave the system at a rate equal to the rate at which it is produced. Therefore, this thermal effect will keep supporting the rate, eventually leading to a thermally-induced explosion. Which is the third explosion limit.

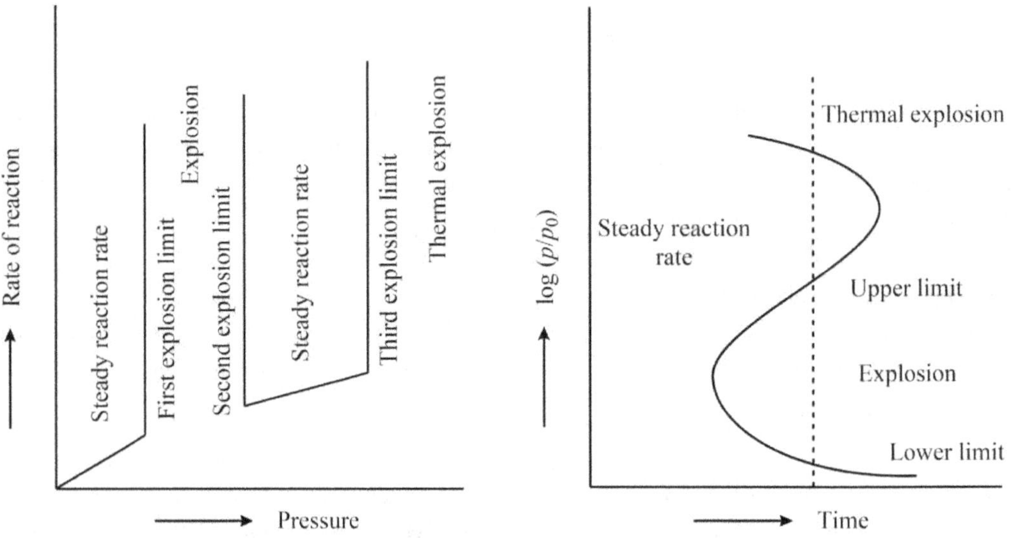



Figure 1. The variation of reaction with pressure in branching chain reactions (left) and variation of relative pressure with time on a logarithmic scale (right).

> ## Reaction Profile of H₂-O₂ Explosion

The reaction between $H_2$ and $O_2$ is a typical example in which all three explosion limits are observed. The net reaction is

$$2H_2 + O_2 \longrightarrow 2H_2O \tag{241}$$

Furthermore, the elementary steps for the same can be proposed as

*Initiation:* $\qquad H_2 + O_2 \xrightarrow{\ \ k_1\ \ } HO^{\cdot}_2 + H^{\cdot}$ $\qquad\qquad$ (242)

*Propagation:* $\qquad H_2 + HO^{\cdot}_2 \xrightarrow{\ \ k_2\ \ } HO^{\cdot} + H_2O$ $\qquad\qquad$ (243)

$\qquad\qquad\qquad H_2 + HO^{\cdot} \xrightarrow{\ \ k_3\ \ } H^{\cdot} + H_2O$ $\qquad\qquad$ (244)

$\qquad\qquad\qquad H^{\cdot} + O_2 \xrightarrow{\ \ k_4\ \ } HO^{\cdot} + O^{\cdot}$ $\qquad\qquad$ (245)

$\qquad\qquad\qquad O^{\cdot} + H_2 \xrightarrow{\ \ k_5\ \ } HO^{\cdot} + H^{\cdot}$ $\qquad\qquad$ (246)

The step $3^{rd}$ and $4^{th}$ produce more radicals than they consume, producing an explosion.

### ❖ Kinetics of (One Intermediate) Enzymatic Reaction: Michaelis-Menten Treatment

In 1901, a French Chemist, Victor Henri, proposed that enzyme reactions are actually initiated by a bond formed between the substrate and the enzyme. The concept was further developed by German researches Leonor Michaelis and Canadian scientist Maud Menten, who examined the kinetics of an enzymatic reaction mechanism of invertase. The final model of Michaelis and Menten's treatment was published in 1913 and gained immediate popularity among the scientific community.

$$E + S \underset{k_{-1}}{\overset{k_1}{\rightleftharpoons}} ES \overset{k_2}{\longrightarrow} P \tag{247}$$

Where $E$ is the enzyme that binds to a substrate S, and forms a complex $ES$. This enzyme-substrate complex then turns into product $P$. The overall rate of product formation should be

$$+\frac{d\lfloor P\rfloor}{dt} = r = k_2[ES] \tag{248}$$

Since $[ES]$ is intermediate, the steady-state approximation can be applied i.e.

$$Rate\ of\ formation\ of\ [ES] = Rate\ of\ decomposition\ of\ [ES] \tag{249}$$

$$k_1[E][S] = k_{-1}[ES] + k_2[ES] \tag{250}$$

$$k_1[E][S] = (k_{-1} + k_2)[ES] \tag{251}$$

$$[ES] = \frac{k_1[E][S]}{k_{-1} + k_2} \tag{252}$$

Now, defining a new parameter named as Michaelis- Menten constant $(K_m)$ with the following expression,

$$K_m = \frac{k_{-1} + k_2}{k_1} \tag{253}$$

equation (252) takes the form

$$[ES] = \frac{[E][S]}{K_m} \tag{254}$$

Now, if we put the $[ES]$ concentration given above into equation (248), we will get the rate of enzyme-catalyzed reaction in terms of substrate concentration and enzyme left unused i.e. $[E]$. This is again a problem that can be solved if we obtain the rate expression in terms of total enzyme concentration i.e. $[E_0]$; which would make the calculation of catalytic efficiency much easier. In order to do so, recall the total enzyme concentration mathematically i.e.

$$[E_0] = [E] + [ES] \tag{255}$$

Where $[E]$ is the enzyme concentration left unused whereas $[ES]$ represents the enzyme concentration that is bound with the substrate. Rearranging equation (253), we get

$$[E] = [E_0] - [ES] \tag{256}$$

After using the value of $[E]$ from equation (256) in equation (254), we have

$$[ES] = \frac{\{[E_0] - [ES]\}[S]}{K_m} \tag{257}$$

$$[ES] = \frac{[E_0][S] - [ES][S]}{K_m} \tag{258}$$

Rearranging for $[ES]$ again

$$K_m[ES] = [E_0][S] - [ES][S] \tag{259}$$

$$K_m[ES] + [ES][S] = [E_0][S] \tag{260}$$

$$[ES] = \frac{[E_0][S]}{K_m + [S]} \tag{261}$$

After putting the value of $[ES]$ from above equation into equation (248), we get

$$r = \frac{k_2[E_0][S]}{K_m + [S]} \tag{262}$$

If the rate of the enzyme-catalyzed reaction is recorded in the very initial stage, the substrate concentration can also be replaced by the initial substrate concentration i.e. $[S_0]$. Therefore, the initial enzyme-catalyzed rate $(r_0)$ should be

$$r_0 = \frac{k_2[E_0][S_0]}{K_m + [S_0]} \tag{263}$$

Which is the well-known Michaelis-Menten equation.

Although the equation (263) is the complete form of the Michaelis-Menten equation that has measurable quantities like $[E_0]$ and $[S_0]$, the more popular form of the same includes the maximum initial rate. At very large initial substrate concentration, the $K_m$ can simply be neglected in comparison to $[S_0]$ in the denominator, which makes the equation (263) to take the form

$$r_{max} = \frac{k_2[E_0][S_0]}{[S_0]} = k_2[E_0] \tag{264}$$

After using the $k_2[E_0] = r_{max}$ in equation (263), we have

$$r_0 = \frac{r_{max}[S_0]}{K_m + [S_0]} \tag{265}$$

Which is the most popular form of the Michaelis-Menten equation. The plot of the initial reaction rate for the enzyme-catalyzed reaction is given below.

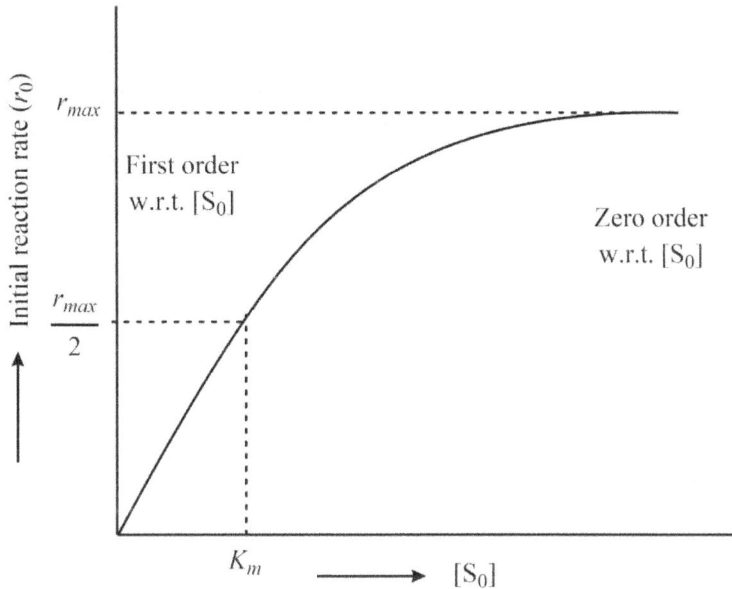


Figure 2. The variation of initial reaction rate as a function of substrate concentration for an enzyme-catalyzed reaction.

> ***Order of Enzyme Catalysed Reactions***

The reaction kinetics of the enzyme-catalyzed reactions can be well understood by looking at the general form of Michaelis-Menten equation.

*i) At high* [S$_0$]: When initial substrate concentration is very high, $K_m$ can be neglected, and the equation (263) takes the form

$$r_0 = \frac{k_2[E_0][S_0]}{[S_0]} = k_2[E_0] \tag{266}$$

Hence, the reaction will be zero-order w.r.t. substrate concentration and first order w.r.t the total enzyme concertation.

*ii) At low* $[S_0]$: When initial substrate concentration is low, $[S_0]$ can be neglected, and the equation (263) takes the form

$$r_0 = \frac{k_2}{K_m}[E_0][S_0] \tag{267}$$

Hence, the reaction will be the first-order w.r.t. substrate concentration and first order w.r.t the total enzyme concertation as well.

> ➤ *Calculation of Catalytic Efficiency*

The catalytic efficiency of enzyme-catalyzed reactions is simply the maximum overall reaction rate per unit of enzyme concentration. When initial substrate concentration is very high, $K_m$ can be neglected, and the equation (263) takes the form

$$r_{max} = \frac{k_2[E_0][S_0]}{[S_0]} = k_2[E_0] \tag{268}$$

$$k_2 = \frac{r_{max}}{[E_0]} \tag{269}$$

Hence, $k_2$ is simply equal to the catalytic efficiency of enzyme-catalyzed reactions.

> ➤ *Calculation of Michaelis-Menten Constant*

The Michaelis-Menten constant of enzyme-catalyzed reactions can simply be calculated once the maximum initial reaction rate is known. When initial substrate concentration is very high, $K_m$ can be neglected, and the equation (263) takes the form

$$r_{max} = \frac{k_2[E_0][S_0]}{[S_0]} = k_2[E_0] \tag{270}$$

After using the $k_2[E_0] = r_{max}$ in equation (263), we have

$$r_0 = \frac{r_{max}[S_0]}{K_m + [S_0]} \tag{271}$$

When the initial reaction rate is half of the maximum initial rate i.e. $r_0 = r_{max}/2$, we have

$$\frac{r_{max}}{2} = \frac{r_{max}[S_0]}{K_m + [S_0]} \tag{272}$$

$$K_m + [S_0] = 2[S_0] \tag{273}$$

$$K_m = [S_0] \tag{274}$$

Hence, the concentration at which the initial rate in half of the maximum initial rate, will be equal to $K_m$.

## ❖ Evaluation of Michaelis's Constant for Enzyme-Substrate Binding by Lineweaver-Burk Plot and Eadie-Hofstee Methods

The conventional approach to determine the Michaelis's constant ($K_m$) involve the plot of initial reaction rate vs initial substrate concentration. Then the Michaelis-Menten constant of enzyme-catalyzed reactions can simply be calculated once the maximum initial reaction rate is known. The typical Michaelis-Menten is used to fit the data is given below.

$$r_0 = \frac{r_{max}[S_0]}{K_m + [S_0]} \tag{275}$$

Where $r_{max} = k_2[E_0]$ is the maximum initial rate at $[E_0]$ enzyme concentration whereas $[S_0]$ is the initial substrate concentration. Finally, the concentration at which the initial rate in half of the maximum initial rate, will be equal to $K_m$.

This conventional method, however, suffers from serious limitations like the need for lots of data points until the initial rate becomes constant. Therefore, two other methods are quite popular to find Michaelis's constant without getting the actual plateau in the initial reaction rate. In this section, we will discuss two such methods for the evaluation of Michaelis's constant for enzyme-substrate binding.

### ➢ *The Lineweaver-Burk Plot to Evaluate Michaelis's Constant*

In order to understand the Lineweaver-Burk method for the determination of the maximum initial rate ($r_{max}$) and Michaelis's constant, we need to rearrange the general Michaelis-Menten equation first i.e. the reciprocal of equation (275) as given below.

$$\frac{1}{r_0} = \frac{K_m + [S_0]}{r_{max}[S_0]} \tag{276}$$

or

$$\frac{1}{r_0} = \frac{K_m}{r_{max}[S_0]} + \frac{[S_0]}{r_{max}[S_0]} \tag{277}$$

or

$$\frac{1}{r_0} = \frac{K_m}{r_{max}} \frac{1}{[S_0]} + \frac{1}{r_{max}} \tag{278}$$

Which is the equation of the straight line ($y = mx + c$). Therefore, if we plot the reciprocal of initial reaction rate vs the reciprocal of initial substrate concentration; the slope and intercepts will give $K_m/r_{max}$ and $1/r_{max}$, respectively.

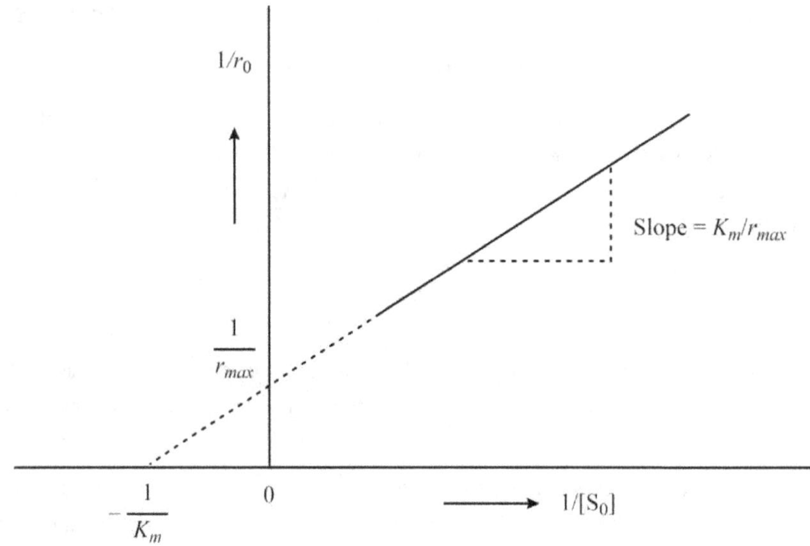



Figure 3. The Lineweaver-burk plot for enzyme-catalyzed reactions to evaluate the maximum initial rate $(r_{max})$ and Michaelis's constant.

It is also worthy to mention that if the intercept is further extrapolated, it will lead to the intercept on the $x$-axis that equals to $-1/K_m$.

> ### ➢ *The Eadie-Hofstee Plot to Evaluate Michaelis's Constant*

In order to understand the Eadie-Hofstee method for the determination of the maximum initial rate $(r_{max})$ and Michaelis's constant, we need to rearrange the general Michaelis-Menten equation first i.e. the reciprocal of equation (275) as given below.

$$\frac{1}{r_0} = \frac{K_m + [S_0]}{r_{max}[S_0]} \tag{279}$$

or

$$\frac{1}{r_0} = \frac{K_m}{r_{max}[S_0]} + \frac{[S_0]}{r_{max}[S_0]} \tag{280}$$

or

$$\frac{1}{r_0} = \frac{K_m}{r_{max}}\frac{1}{[S_0]} + \frac{1}{r_{max}} \tag{281}$$

Multiplying both sides by $r_0$, we get

$$\frac{r_0}{r_0} = \frac{K_m}{r_{max}} \frac{r_0}{[S_0]} + \frac{r_0}{r_{max}} \tag{282}$$

Rearranging for $r_0/[S_0]$, we get

$$\frac{r_0}{r_0} - \frac{r_0}{r_{max}} = \frac{K_m}{r_{max}} \frac{r_0}{[S_0]} \tag{283}$$

$$\frac{r_0}{[S_0]} = \frac{r_0 \, r_{max}}{r_0 \, K_m} - \frac{r_0 \, r_{max}}{r_{max} \, K_m} \tag{284}$$

$$\frac{r_0}{[S_0]} = \frac{r_{max}}{K_m} - \frac{r_0}{K_m} \tag{285}$$

$$\frac{r_0}{[S_0]} = -\frac{1}{K_m} r_0 + \frac{r_{max}}{K_m} \tag{286}$$

Which is the equation of the straight line ($y = mx + c$). Therefore, if we plot the ratio of initial reaction rate to initial substrate concentration vs the initial reaction rate; the slope and intercepts will give $-1/K_m$ and $r_{max}/K_m$, respectively.

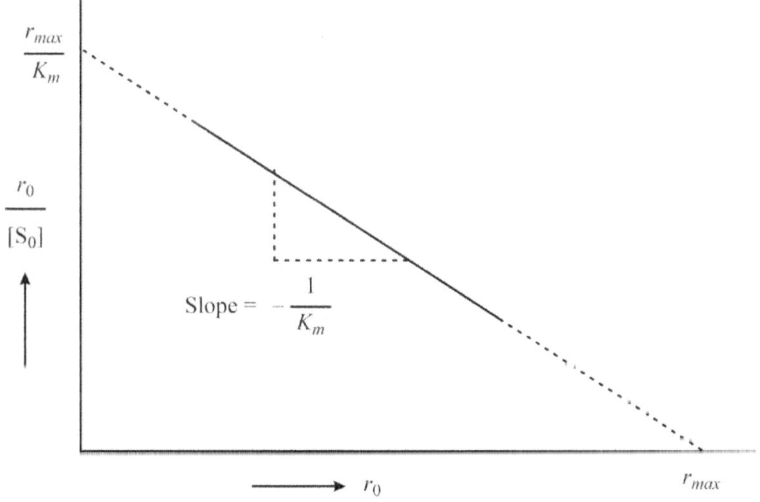


Figure 4. The Eadie-Hofstee plot for enzyme-catalyzed reactions to evaluate the maximum initial rate ($r_{max}$) and Michaelis's constant.

It is also worthy to mention that if the intercept is further extrapolated, it will lead to the intercept on the $x$-axis that equals to $r_{max}$.

## ❖ Competitive and Non-Competitive Inhibition

An enzyme inhibitor is a compound that binds to an enzyme and decreases its overall activity, and the phenomenon is typically known as "enzyme inhibition". Two of the most common enzyme inhibition processes will be discussed in this section.

### ➤ *Competitive Inhibition*

In the case of competitive enzyme inhibition, the binding of an inhibitor prevents the binding of the substrate and the enzyme. This type of behavior is actually achieved by blocking the binding site of the target molecule (the active site) by some means. The competitive enzyme inhibition can be classified into two types as discussed below.

**1. Fully competitive inhibition:** The fully competitive inhibition occurs when an enzyme ($E$) binds with the substrate ($S$) and inhibitor ($I$) separately, and it is only the enzyme-substrate complex ($ES$) that will convert into the product. This whole process can be described mathematically as

$$E + S \underset{k_{-1}}{\overset{k_1}{\rightleftharpoons}} ES \overset{k_2}{\longrightarrow} E + P \tag{287}$$

$$E + I \underset{k_{-3}}{\overset{k_3}{\rightleftharpoons}} EI \tag{288}$$

After applying the steady-state approximation on $ES$, we have

$$\frac{d[ES]}{dt} = k_1[E][S] - k_{-1}[ES] - k_2[ES] = 0 \tag{289}$$

$$k_1[E][S] = k_{-1}[ES] + k_2[ES] \tag{290}$$

$$[ES] = \frac{k_1[E][S]}{k_{-1} + k_2} \tag{291}$$

When $[S_0] \gg [E_0]$, we can assume $[S_0] \approx [S]$, and also

$$[E_0] = [E] + [ES] + [EI] \tag{292}$$

$$[E] = [E_0] - [ES] - [EI] \tag{293}$$

Now recalling the equilibrium constant for inhibition equilibria i.e.

$$K_3 = \frac{k_3}{k_{-3}} = \frac{[EI]}{[E][I]} \tag{294}$$

$$K_I = \frac{1}{K_3} = \frac{[E][I]}{[EI]} \tag{295}$$

Hence, we can say

$$[EI] = \frac{[E][I]}{K_I} \tag{296}$$

After using the value of $[EI]$ from equation (296) into equation (293), we have

$$[E] = [E_0] - [ES] - \frac{[E][I]}{K_I} \tag{297}$$

$$[E] = \frac{K_I[E_0] - K_I[ES] - [E][I]}{K_I} \tag{298}$$

$$K_I[E] = K_I[E_0] - K_I[ES] - [E][I] \tag{299}$$

$$K_I[E] + [E][I] = K_I[E_0] - K_I[ES] \tag{300}$$

$$[E] = \frac{K_I[E_0] - K_I[ES]}{K_I + [I]} \tag{301}$$

Using the above-derived result in equation (291), we have

$$[ES] = \frac{k_1[S]}{k_{-1} + k_2} \frac{K_I[E_0] - K_I[ES]}{K_I + [I]} = \frac{[S]}{K_m} \frac{K_I[E_0] - K_I[ES]}{K_I + [I]} \tag{302}$$

$$[ES] = \frac{K_I[E_0][S] - K_I[ES][S]}{K_m K_I + K_m[I]} \tag{303}$$

Now rearranging further for $[ES]$, we get

$$K_m K_I[ES] + K_m[I][ES] = K_I[E_0][S] - K_I[ES][S] \tag{304}$$

$$K_m K_I[ES] + K_m[I][ES] + K_I[ES][S] = K_I[E_0][S] \tag{305}$$

$$[ES] = \frac{K_I[E_0][S]}{K_m K_I + K_m[I] + K_I[S]} \tag{306}$$

Since the rate of formation of product is

$$r_0 = k_2[ES] \tag{307}$$

Using the value of $[ES]$ from equation (306) in equation (307), we get

$$r_0 = \frac{k_2 K_I[E_0][S]}{K_m K_I + K_m[I] + K_I[S]} \tag{308}$$

Since $k_2[E_0]$ is $r_{max}$, the equation (308) takes the form

$$r_0 = \frac{r_{max}\, K_I [S]}{K_m K_I + K_m [I] + K_I [S]} \tag{309}$$

Taking the reciprocal to get Lineweaver-Burk plot

$$\frac{1}{r_0} = \frac{K_m K_I + K_m [I] + K_I [S]}{r_{max}\, K_I [S]} \tag{310}$$

$$\frac{1}{r_0} = \frac{K_m K_I}{r_{max}\, K_I [S]} + \frac{K_m [I]}{r_{max}\, K_I [S]} + \frac{K_I [S]}{r_{max}\, K_I [S]} \tag{311}$$

$$\frac{1}{r_0} = \left( \frac{K_m}{r_{max}} + \frac{K_m [I]}{r_{max}\, K_I} \right) \frac{1}{[S]} + \frac{1}{r_{max}} \tag{312}$$

Rearranging further and using initial substrate concentration, we get

$$\frac{1}{r_0} = \frac{K_m}{r_{max}} \left( 1 + \frac{[I]}{K_I} \right) \frac{1}{[S_0]} + \frac{1}{r_{max}} \tag{313}$$

After comparing with equation without enzyme inhibition i.e.

$$\frac{1}{r_0} = \frac{K_m}{r_{max}} \frac{1}{[S_0]} + \frac{1}{r_{max}} \tag{314}$$

it is clear that the intercepts in both the equations are the same whereas the slope has been increased from $K_m/r_{max}$ to $(K_m/r_{max})(1 + [I]/K_I)$. This implies that enzyme inhibition changed the $K_m$ to $K_m(1 + [I]/K_I)$.

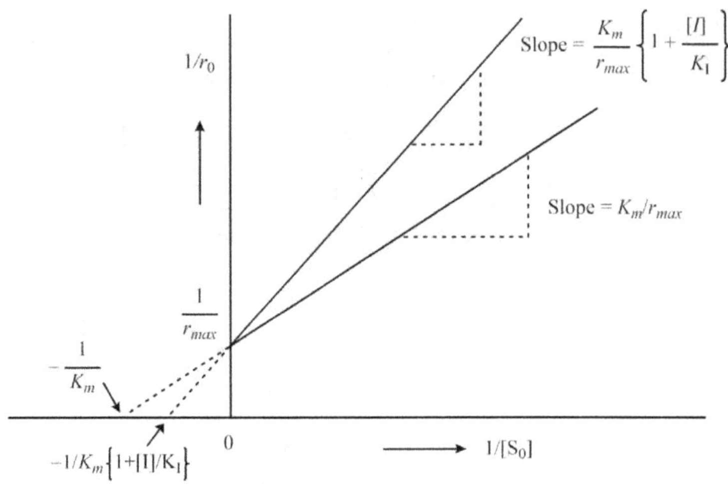


Figure 5. Lineweaver-Burk plot for enzyme-catalyzed reaction with fully competitive enzyme inhibition.

**2. Partially competitive inhibition:** The partially competitive inhibition occurs when the enzyme ($E$) binds with the substrate ($S$) and inhibitor ($I$) simultaneously, and complex $ES$ and $EI$ also combine with $I$ and $S$ to give $EIS$. Also, besides the enzyme-substrate complex ($ES$), the complex $EIS$ will also convert into the product with the same rate of reaction. This whole process can be described mathematically as

$$E + S \;\overset{K_1'}{\rightleftharpoons}\; ES \;\overset{k_2}{\longrightarrow}\; E + P \tag{315}$$

$$E + I \;\overset{K_2'}{\rightleftharpoons}\; EI \tag{316}$$

$$EI + S \;\overset{K_3'}{\rightleftharpoons}\; EIS \tag{317}$$

$$ES + I \;\overset{K_4'}{\rightleftharpoons}\; EIS \tag{318}$$

$$EIS \;\overset{k_2}{\longrightarrow}\; EI + P \tag{319}$$

According to classical Michaelis-Menten equation $K_m = (k_2 + k_{-1})/k_1$; however, if $k_2 \ll k_{-1}$, we have $K_m = k_{-1}/k_1$ or $K_m = 1/K_1'$. Now, set the following results

$$K_m = \frac{1}{K_1'} = \frac{[E][S]}{[ES]} \tag{320}$$

$$K_2 = \frac{1}{K_2'} = \frac{[E][I]}{[EI]} \tag{321}$$

$$K_3 = \frac{1}{K_3'} = \frac{[EI][S]}{[EIS]} \tag{322}$$

$$K_4 = \frac{1}{K_4'} = \frac{[ES][I]}{[EIS]} \tag{323}$$

Following enzyme conservation, we have

$$[E_0] = [E] + [ES] + [EI] + [EIS] \tag{324}$$

Using values of $[ES]$, $[EI]$ and $[EIS]$ from equation (320-322) into equation (324), we get

$$[E_0] = [E] + \frac{[E][S]}{K_m} + \frac{[E][I]}{K_2} + \frac{[E][I][S]}{K_2 K_3} \tag{325}$$

$$[E] = \frac{K_m [E_0]}{K_m(1 + [I]/K_2) + [S](1 + K_m[I]/K_2 K_3)} \tag{326}$$

The overall reaction rate of product formations should be

$$r_0 = k_2[ES] + k_2[EIS] \tag{327}$$

After using values of $[ES]$ and $[EIS]$ from equation (320, 322) into equation (327), we get

$$r_0 = k_2\frac{[E][S]}{K_m} + k_2\frac{[E][I][S]}{K_2K_3} \tag{328}$$

Substituting the value of $[E]$ from equation (326) in (328), and rearranging at initial substrate concentration

$$r_0 = \frac{k_2[E_0][S_0]}{\{K_m(1 + [I]/K_2)/(1 + K_m[I]/K_2K_3)\} + [S_0]} \tag{329}$$

Since $k_2[E_0]$ is $r_{max}$, equation (329) takes the following form after the reciprocal to get Lineweaver-Burk plot

$$\frac{1}{r_0} = \frac{\{K_m(1 + [I]/K_2)/(1 + K_m[I]/K_2K_3)\}}{r_{max}}\frac{1}{[S_0]} + \frac{1}{r_{max}} \tag{330}$$

After comparing with equation without enzyme inhibition i.e.

$$\frac{1}{r_0} = \frac{K_m}{r_{max}}\frac{1}{[S_0]} + \frac{1}{r_{max}} \tag{331}$$

it is clear that intercepts in both the equations are the same whereas the slope has been varied from $K_m/r_{max}$ to$\{K_m(1 + [I]/K_2)/(1 + K_m[I]/K_2K_3)\}/r_{max}$. This implies that enzyme inhibition changed the $K_m$ to $\{K_m(1 + [I]/K_2)/(1 + K_m[I]/K_2K_3)\}$.

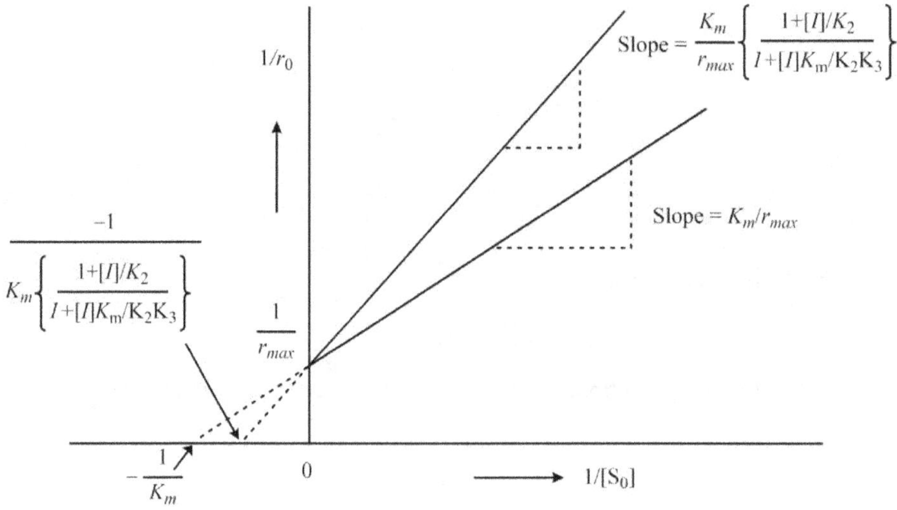


Figure 6. The Lineweaver-Burk plot for partially competitive enzyme inhibition.

> ➤ *Non-Competitive Inhibition*

Non-competitive inhibition may simply be defined as the enzyme inhibition where the inhibitor decreases the activity of the enzyme catalysis and binds equally well to the enzyme whether or not it has already bound the substrate. The non-competitive enzyme inhibition can be classified into two types as discussed below.

**1. Fully non-competitive inhibition:** The fully non-competitive inhibition occurs when an enzyme ($E$) binds with inhibitor ($I$), and complex $ES$ and $EI$ also combine with $I$ and $S$ to give $EIS$. However, it is only the enzyme-substrate complex ($ES$) that converts into the product. This whole process can be described mathematically as

$$E + S \; \underset{}{\overset{K_1'}{\rightleftharpoons}} \; ES \; \overset{k_2}{\longrightarrow} \; E + P \tag{332}$$

$$E + I \; \overset{K_2'}{\rightleftharpoons} \; EI \tag{333}$$

$$EI + S \; \overset{K_1'}{\rightleftharpoons} \; EIS \tag{334}$$

$$ES + I \; \overset{K_2'}{\rightleftharpoons} \; EIS \tag{335}$$

According to classical Michaelis-Menten equation $K_m = (k_2 + k_{-1})/k_1$; however, if $k_2 \ll k_{-1}$, we have $K_m = k_{-1}/k_1$ or $K_m = 1/K_1'$. Now, set the following results

$$K_m = \frac{1}{K_1'} = \frac{[E][S]}{[ES]} = \frac{[E][S]}{[EIS]} \tag{336}$$

$$K_I = \frac{1}{K_2'} = \frac{[E][I]}{[EI]} = \frac{[ES][I]}{[EIS]} \tag{337}$$

Following enzyme conservation, we have

$$[E_0] = [E] + [ES] + [EI] + [EIS] \tag{338}$$

$$[E] = [E_0] - [ES] - [EI] - [EIS] \tag{339}$$

Using values of $[ES]$, $[EI]$ and $[EIS]$ from equation (336-337) into equation (338), we get

$$[E_0] = [E] + \frac{[E][S]}{K_m} + \frac{[E][I]}{K_I} + \frac{[E][I][S]}{K_m K_I} \tag{340}$$

$$[E] = \frac{[E_0]}{(1 + [S]/K_m)(1 + [I]/K_I)} \tag{341}$$

The overall reaction rate of product formations should be

$$r_0 = k_2 [ES] \qquad (342)$$

After using values of $[ES]$ from equation (336) into equation (342), we get

$$r_0 = k_2 \frac{[E][S]}{K_m} \qquad (343)$$

Substituting the value of $[E]$ from equation (341) in (343), and rearranging at initial substrate concentration

$$r_0 = \frac{k_2 [E_0][S_0]}{(K_m + [S_0])(1 + [I]/K_I)} \qquad (344)$$

Since $k_2 [E_0]$ is $r_{max}$, the equation (344) takes the form

$$r_0 = \frac{r_{max}[S_0]}{(K_m + [S_0])(1 + [I]/K_I)} \qquad (345)$$

Taking the reciprocal to get Lineweaver-Burk plot

$$\frac{1}{r_0} = \frac{K_m}{r_{max}}\left(1 + \frac{[I]}{K_I}\right)\frac{1}{[S_0]} + \frac{1}{r_{max}}\left(1 + \frac{[I]}{K_I}\right) \qquad (346)$$

After comparing with equation without enzyme inhibition i.e.

$$\frac{1}{r_0} = \frac{K_m}{r_{max}}\frac{1}{[S_0]} + \frac{1}{r_{max}} \qquad (347)$$

it can be clearly seen that intercept is increased from $1/r_{max}$ to $(1/r_{max})(1 + [I]/K_I)$ whereas the $K_m$ has remained the same.

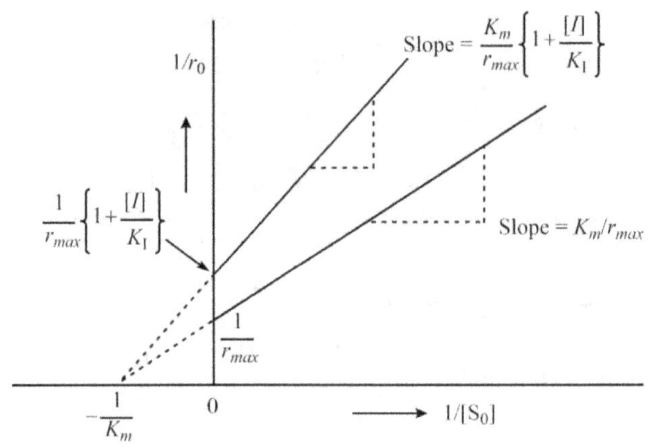


Figure 7. The Lineweaver-Burk plot for fully non-competitive enzyme inhibition.

**2. Partially non-competitive inhibition:** The partially non-competitive inhibition occurs when an enzyme ($E$) binds with inhibitor ($I$), and complex $ES$ and $EI$ also combine with $I$ and $S$ to give $EIS$. However, unlike fully non-competitive inhibition, besides the enzyme-substrate complex ($ES$), the complex will also convert into the product. This whole process can be described mathematically as

$$E + S \;\overset{K_1'}{\rightleftharpoons}\; ES \;\overset{k_2}{\longrightarrow}\; E + P \tag{348}$$

and

$$E + I \;\overset{K_2'}{\rightleftharpoons}\; EI \tag{349}$$

and

$$EI + S \;\overset{K_1'}{\rightleftharpoons}\; EIS \tag{350}$$

and

$$ES + I \;\overset{K_2'}{\rightleftharpoons}\; EIS \tag{351}$$

and

$$EIS \;\overset{k'}{\longrightarrow}\; EI + P \tag{352}$$

According to classical Michaelis-Menten equation $K_m = (k_2 + k_{-1})/k_1$; however, if $k_2 \ll k_{-1}$, we have $K_m = k_{-1}/k_1$ or $K_m = 1/K_1'$. Now, set the following results

$$K_m = \frac{1}{K_1'} = \frac{[E][S]}{[ES]} \tag{353}$$

and

$$K_I = \frac{1}{K_2'} = \frac{[E][I]}{[EI]} \tag{354}$$

and

$$K_m = \frac{1}{K_1'} = \frac{[EI][S]}{[EIS]} \tag{355}$$

and

$$K_I = \frac{1}{K_2'} = \frac{[ES][I]}{[EIS]} \tag{356}$$

Following enzyme conservation, we have

$$[E_0] = [E] + [ES] + [EI] + [EIS] \tag{357}$$

Using values of $[ES]$, $[EI]$ and $[EIS]$ from equation (353-356) into equation (357), we get the following expression.

$$[E_0] = [E] + \frac{[E][S]}{K_m} + \frac{[E][I]}{K_I} + \frac{[E][I][S]}{K_m K_I} \tag{358}$$

or

$$[E] = \frac{[E_0]}{(1 + ([S]/K_m)(1 + [I]/K_I)} \tag{359}$$

The overall reaction rate of product formations should be

$$r_0 = k_2[ES] + k'[EIS] \tag{360}$$

After using the values of $[ES]$ and $[EIS]$ from equation (353-356) into equation (360), we get the following result.

$$r_0 = k_2 \frac{[E][S]}{K_m} + k' \frac{[E][I][S]}{K_m K_I} \tag{361}$$

Now, substituting the value of $[E]$ from equation (359) in (361), and rearranging at the initial substrate concentration

$$r_0 = \frac{(k_2[E_0][S_0] + k'[E_0][S_0][I]/K_I)/(1 + [I]/K_I)}{K_m + [S_0]} \tag{362}$$

Using $k_2[E_0]$ is $r_{max}$ and then taking the reciprocal to get the Lineweaver-Burk plot, we get the following relation.

$$\frac{1}{r_0} = \frac{K_m(1 + [I]/K_I)}{(r_{max} + k'[E_0][I]/K_I)} \frac{1}{[S_0]} + \frac{(1 + [I]/K_I)}{(r_{max} + k'[E_0][I]/K_I)} \tag{363}$$

After comparing the result given above with the general equation without enzyme inhibition, i.e., we can conclude some important points.

$$\frac{1}{r_0} = \frac{K_m}{r_{max}} \frac{1}{[S_0]} + \frac{1}{r_{max}} \tag{364}$$

it can be clearly seen that intercept is increased from $1/r_{max}$ to $(1 + [I]/K_I)/(r_{max} + k'[E_0][I]/K_I)$ whereas the $K_m$ has remained the same.

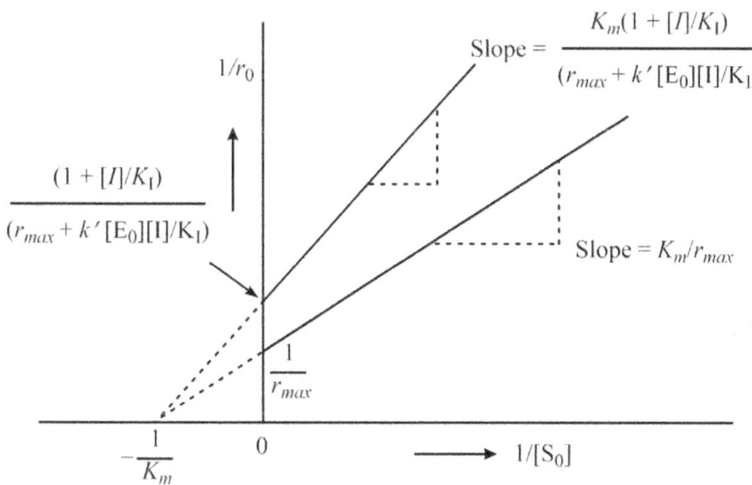


Figure 8. The Lineweaver-Burk plot for partially non-competitive enzyme inhibition.

## ❖ Problems

Q 1. What are chemical chain reactions? Discuss their general kinetics.

Q 2. Derive and discuss the rate law for the decomposition of ethane.

Q 3. Define photochemical reactions. How they are different from thermochemical reactions?

Q 4. What is the photochemical quantum yield of a reaction? Derive and discuss the same for photochemical combination $H_2$-$Br_2$ case.

Q 5. Derive and discuss the Rice-Herzfeld mechanism of decomposition of Acetaldehyde.

Q 6. Define the chain length.

Q 7. Calculate the expression for the chain length for the dehydrogenation of ethane.

Q 8. What are the branching chain reactions? How they lead to the situation of the explosion?

Q 9. Derive and discuss the conventional Michaelis-Menten equation for enzyme-catalyzed reactions.

Q 10. How would you treat Michaelis-Menten equations to get Lineweaver-Burk plot?

Q 11. Discuss the Eadie-Hofstee method for the evaluation of Michaelis-Menten constant in enzyme-catalyzed reactions.

Q 12. What is enzyme inhibition? Discuss the fully competitive enzyme inhibition in detail.

### ❖ Bibliography

[1] B. R. Puri, L. R. Sharma, M. S. Pathania, *Principles of Physical Chemistry*, Vishal Publications, Jalandhar, India, 2008.

[2] P. Atkins, J. Paula, *Physical Chemistry*, Oxford University Press, Oxford, UK, 2010.

[3] E. Steiner, *The Chemistry Maths Book*, Oxford University Press, Oxford, UK, 2008.

[4] S. A. Arrhenius, *Über die Dissociationswärme und den Einfluß der Temperatur auf den Dissociationsgrad der Elektrolyte*. Z. Phys. Chem. 4 (1889) 96–116.

[5] S. A. Arrhenius, *Über die Reaktionsgeschwindigkeit bei der Inversion von Rohrzucker durch Säuren Z. Phys. Chem.* 4 (1889) 226–248.

[6] K. J. Laidler, *Chemical Kinetics*, Harper & Row, New York, USA, 1987.

[7] K. J. Laidler, *The World of Physical Chemistry*, Oxford University Press, Oxford, UK, 1993.

[8] H. Eyring, *The Activated Complex in Chemical Reactions*, J. Chem. Phys. 3 (1935) 107-115.

[9] K. L. Kapoor, *A Textbook of Physical Chemistry Volume 5*, Macmillan Publishers, New Delhi, India, 2011.

# CHAPTER 8

# Electrochemistry – II: Ion Transport in Solutions

## ❖ Ionic Movement Under the Influence of an Electric Field

In order to imagine the conduction process in electrolytic solutions at the atomic level, we can follow two approaches which are somewhat different in their initial assumptions.

The first approach includes the visualization of ionic movements governed by the diffusion phenomenon first and then studying the perturbation of this ionic movement by an externally applied electric field. Since it is a well-known fact that the diffusion of ions is simply the movement of ions from a high-numbered region to a low-numbered region. In other words, we can say the ionic diffusion is the result of a concentration-gradient in which a particular type of ions travel from a high concentration region towards a low concentration region until a homogeneity in the concentration is reached. Now, although the net movement of ions stops after the loss of concentration gradient, the individual ionic movement still happens but with zero mean displacements. In other words, we can say that in a homogeneous ionic solution, the ions can move randomly in any direction resulting in a zero net diffusion.

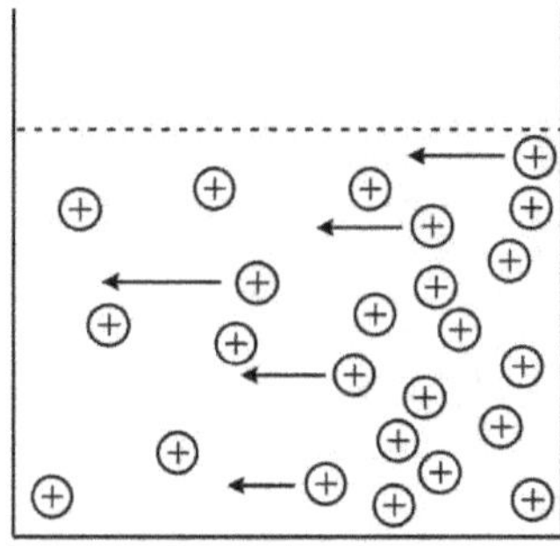

Diffusion of cations

Figure 1. The movement of positive ions from higher concentration to lower concentration.

Now since the ions are charged particles, the movements of these ions are strongly affected when an electric field is applied. From the laws electrostatic interactions, we can conclude that the cations will prefer to move towards the negative electrode whereas the anions will prefer to move towards the positive electrode. More specifically, the application of an electric field makes the ions to adopt a single direction in space, which is a direction along or opposite to the direction of the applied field. Therefore, the ions drift under the applied field and stop their random walk.

The second approach to study the phenomenon of electrolytic conduction at the atomic level includes the framing of the drift of only one ion under the externally applied field. The electric field would make the ion to accelerate as per Newton's second law. Now if the ion is in the vacuum, it would show an acceleration until it strikes with the respective electrode. However, it will not happen since a large number of other ions also present in the same electrolytic solution along with the solvent as well. Consequently, the ion is almost bound to collide with other ions or solvent particles in its journey. The ion will stop for some time and then will start to accelerate again. This stop-start phenomenon will impart a discontinuity in the speed and direction of this moving ion. It means that ionic movement is not very much smooth but actually a resistance is offered by the surrounding medium. Therefore, we can say that the application of an external electric field will make the ion move towards the oppositely charged electrode but in a stops-starts and zigzag fashion.

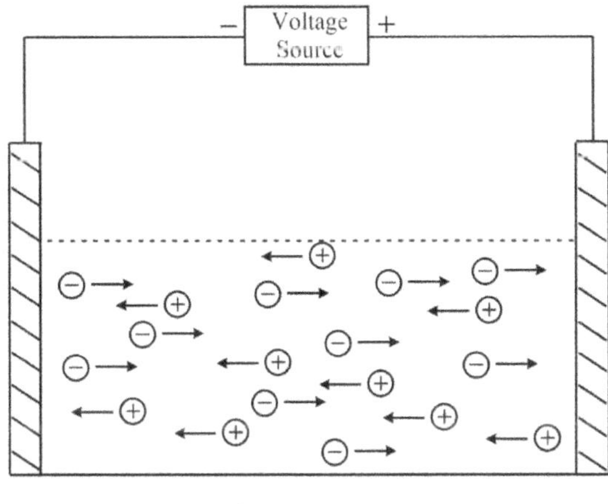


Figure 2. The general depiction movement of ions under the influence of the external electric field.

Since an ion starts moving towards the positively charged electrode only after the application of the electric field; the initial velocity before that can simply be neglected because it arises from random collisions, which can be in any random direction. However, after applying the electric field, the ion feels a force that makes it move in the same direction, i.e., the direction of the electrostatic force. In other words, the electric field will create an additional velocity component on the ion under consideration that drifts the ion to the oppositely charged electrode. Now, let $\vec{F}$ be the force vector that imparts a drift velocity $v_d$; and then using Newton's second law of motion states that this force divided by the particle's mass is simply equal to the acceleration. From the general expressions for acceleration, we have

$$a = \frac{dv}{dt} \tag{1}$$

And

$$a = \frac{\vec{F}}{m} \tag{2}$$

From equation (1) and equation (2), we have

$$\frac{\vec{F}}{m} = \frac{dv}{dt} \tag{3}$$

Now although the time between two collisions may vary significantly, we can use a mean time $\tau$ for simplicity which can be formulated as (if $N$ collisions take place in '$t$' time) given below.

$$\tau = \frac{t}{N} \tag{4}$$

Now because the drift velocity is imparted to the ion by the external force, its value must be equal to the product of meantime and the acceleration due to force, i.e.,

$$v_d = \frac{dv}{dt}\tau \tag{5}$$

Using the value of $dv/dt$ from equation (3) in equation (5), we get

$$v_d = \frac{\vec{F}}{m}\tau \tag{6}$$

It is obvious from the above relation that the meantime is related to the drift velocity showing that the jumps between collisions affect ionic movement. Besides, it is also clear that the drift velocity is directly proportional to the driving force of the applied electric field. The ionic flux can be formulated in terms of drift velocity as given below.

$$Flux = Ionic\ cencentration \times Drift\ velocity \tag{7}$$

Hence, since the drift velocity is directly proportional to the electric force simulating conduction, the flux must also be proportional to the magnitude of the electric field, i.e.,

$$Flux \propto Electric\ field \tag{8}$$

The nature of equation (6) also unveils the situation where the flux or the drift velocity no longer holds the direct proportionality with the applied electric field. For equation (6), it is very important to assume that during the collision, the velocity component imparted to the ion by applied electric field vanishes completely and the ion starts as a full-fresher each time. If this is not satisfied, these leftover velocity components would add up after every collision and the real velocity, in that case, would be much greater than the calculation given by equation (6). In other words, equation (6) will no longer be valid. Therefore, we can conclude that for a reasonable guess for drift velocity, the magnitude of the applied electric field must be very small.

## ❖ Mobility of Ions

It has already been discussed in the previous section that the ions in a homogeneous electrolytic solution move randomly with zero net displacements, and the situation changes when the external electric field is applied. The applied field imparts a directive velocity component to the ion under consideration and makes it move towards the oppositely charged electrode. This ion collides with other ions and drifts towards the oppositely charged electrode with a stop-start and zig-zag fashion. The drift velocity ($v_d$) of such ion is given by the following relation.

$$v_d = \frac{\tau}{m}\vec{F} \tag{9}$$

Where $\vec{F}$ is the force exerted upon the ion by applied field and $m$ is the mass of the ion. The symbol $\tau$ represents the mean lifetime between to collisions during the ionic drift.

It is obvious from the equation (9) that drift velocity is proportional to forces exerted by the electric field and $\tau/m$ is the constant of proportionality. The physical significance of the proportionality constant lies in the fact that it becomes equal to the drift velocity when the force is unity, and therefore, represents the "mobility nature" of the ion considered. In other words, we can say that the proportionality constant in equation (9) represents the absolute mobility ($\bar{u}_{abs}$) of the ion i.e.

$$\bar{u}_{abs} = \frac{\tau}{m} = \frac{v_d}{\vec{F}} \tag{10}$$

The units of absolute mobility are cm s$^{-1}$ dyne$^{-1}$. Now, since the functional electric force is equal to the electric force per unit charge; or the electric field ($X$) multiplied with the charge on the ion ($z_i e_0$) i.e.

$$\vec{F} = z_i e_0 X \tag{11}$$

After using the value of $\vec{F}$ from equation (11) in equation (10), we get

$$\bar{u}_{abs} = \frac{v_d}{z_i e_0 X} \tag{12}$$

$$v_d = \bar{u}_{abs}\, z_i e_0 X \tag{13}$$

When, $X = 1\ volt$, the above equation takes the form

$$(v_d)_{1\text{volt cm}^{-1}} = \bar{u}_{abs}\, z_i e_0 = \bar{u}_{conv} \tag{14}$$

Where $\bar{u}_{conv}$ represents the conventional mobility of the ion with units cm$^2$ Volt$^{-1}$ s$^{-1}$. Now although the expressions of both types of mobilities are quite similar, it is worthy to note that the "absolute mobility" has a broader domain of application because any force that governs the drift velocity can be used. On the other hand, the conventional mobility is pretty much limited to the electric force only.

## ❖ Ionic Drift Velocity and Its Relation with Current Density

In this section, we will try to explain how the ionic movement is quantitatively related to current density flowing through an electrolytic solution under the influence of an applied electric field. To do so, we need to understand the ionic drift velocity and then its relationship with current density.

### ➢ *Ionic Drift Velocity*

When an ion in the electrolytic solution is placed under the externally applied electric field, the electric field will make the ion to accelerate as per Newton's second law. Now if the ion is in the vacuum, it would show an acceleration until it strikes with the respective electrode. However, it will not happen since a large number of other ions also present in the same electrolytic solution along with the solvent as well. Consequently, the ion is almost bound to collide with other ions or solvent particles in its journey. The ion will stop for some time and then will start to accelerate again. This stop-start phenomenon will impart a discontinuity in the speed and direction of this moving ion. It means that ionic movement is not very much smooth but actually a resistance is offered by the surrounding medium. Therefore, we can say that the application of the external electric field will make the ion move towards the oppositely charged electrode but in a stops-starts and zigzag fashion. An ion starts moving towards the positively-charged electrode only after the application of the electric field. The initial velocity before that can simply be neglected because it arises from random collisions which can be in any random direction. However, after applying the electric field, the ion feels a force that makes the ion move in the same direction, i.e., the direction of the electrostatic force. Now let $F$ be the force vector that imparts drift velocity $v_d$, then form Newton's second law of motion states that this force divided by the particle's mass is simply equal to the acceleration.

$$\frac{\vec{F}}{m} = \frac{dv}{dt} \tag{15}$$

Now although the time between two collisions may vary significantly, we can use a mean time $\tau$ for simplicity which can be formulated as (if $N$ collisions take place in '$t$' time) given below.

$$\tau = \frac{t}{N} \tag{16}$$

Now because the drift velocity is imparted to the ion by the external force, its value must be equal to product meantime and the acceleration due to this force i.e.

$$v_d = \frac{dv}{dt}\tau \tag{17}$$

Using the value of $dv/dt$ from equation (15) in equation (17), we get

$$v_d = \frac{\vec{F}}{m}\tau \tag{18}$$

It is obvious that the drift velocity is directly proportional to the driving force of the applied electric field.

> ### ➤ *The Relationship between Ionic Drift Velocity and Current Density*

It is obvious from the equation (18) that drift velocity is proportional to forces exerted by the electric field and $\tau/m$ is the constant of proportionality. The physical significance of the proportionality constant lies in the fact that it becomes equal to the drift velocity when the force is unity, and therefore, represents the "mobility nature" of the ion considered. In other words, we can say that the proportionality constant in equation (18) represents the absolute mobility ($\bar{u}_{abs}$) of the ion, i.e.,

$$\bar{u}_{abs} = \frac{\tau}{m} = \frac{v_d}{\vec{F}} \tag{19}$$

The units of absolute mobility are cm s$^{-1}$ dyne$^{-1}$. Now, since the functional electric force is equal to the electric force per unit charge; or the electric field ($X$) multiplied with the charge on the ion ($z_i e_0$), i.e.,

$$\vec{F} = z_i e_0 X \tag{20}$$

After using the value of $\vec{F}$ from equation (20) in equation (19), we get

$$\bar{u}_{abs} = \frac{v_d}{z_i e_0 X} \tag{21}$$

$$v_d = \bar{u}_{abs}\, z_i e_0 X \tag{22}$$

When, $X = 1\ volt$, the above equation takes the form

$$(v_d)_{1\text{volt cm}^{-1}} = \bar{u}_{abs}\, z_i e_0 = \bar{u}_{conv} \tag{23}$$

Where $\bar{u}_{conv}$ represents the conventional mobility of the ion with units cm$^2$ Volt$^{-1}$ s$^{-1}$. Now although the expressions of both types of mobilities are quite similar, it is worthy to note that the "absolute mobility" has a broader domain of application because any force that governs the drift velocity can be used. On the other hand, the conventional mobility is pretty much limited to the electric force only.

Now consider a plane with unit area perpendicular to the direction of ionic movement under the influence of the externally applied field. Now although the cations and anions move in opposite directions, both types of ions will pass through this transit. If $v_+$ is the drift velocity of cation, then all the cations present within $v_+$ cm of this transit plane will pass through it. Let $J_+$ be the cationic flux that represents the total number of mole of ions passing through this area in every second. Therefore, the cationic flux must be equal to the multiplication of the corresponding volume (1 cm$^2 \times v_+$ cm) and the concentration of cations ($c_+$) in moles cm$^{-3}$. Mathematically, we can say that

$$j_i = c_i v_i \tag{24}$$

The current density ($J_+$) or the charge flowing through this transit plane per second due to this cationic flux can simply be obtained by multiplying the cationic flux by the charge carried by 1 mole of ions ($z_+ F$), i.e.,

$$J_+ = c_+ v_+ z_+ F \tag{25}$$

Similarly, If $v_-$ is the drift velocity of anion and $c_-$ the concentration of anions in moles cm$^{-3}$, the anionic current density may simply be written as

$$J_- = c_- v_- z_- F \tag{26}$$

Where $z_-$ is the charge number of the anion and $F$ is Faraday constant. In general, we can write for the $i^{th}$ species as

$$J_i = c_i v_i z_i F \tag{27}$$

Where $z_i$ and $c_i$ are the charge number and concentration in moles cm$^{-3}$ of the $i^{th}$ type of ion.

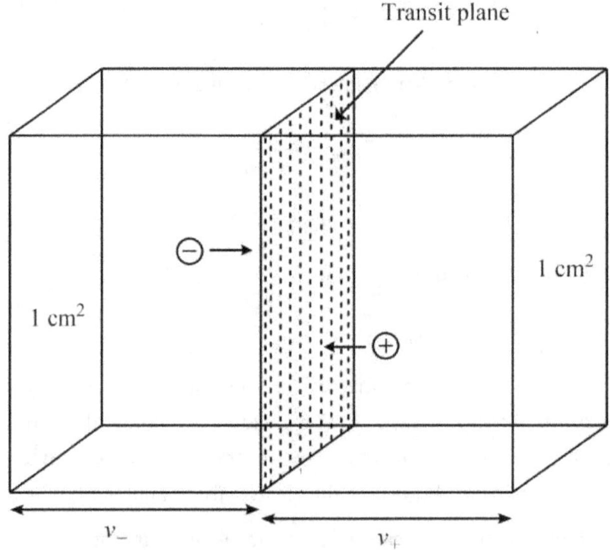


Figure 3. The general depiction ionic movement through transit plane under applied field.

Hence, the total current density from all the ionic species must be the summation of their individual current densities, i.e.,

$$J = \sum_i J_i = \sum_i c_i v_i z_i F \tag{28}$$

For univalent electrolytes like NaCl, we have $z_+ = z_- = z$ and $c_+ = c_- = c$; therefore, we can write

$$J = zcF(v_+ + v_-) \tag{29}$$

Now we can start to relate the ionic drift velocity with more generally measurable quantities in the phenomenon of electrolytic conductance like molar conductivity, equivalent conductivity or simply the conductance. To do so, use the fundamental expression for the ionic drift velocity form equation (23) in equation (28), i.e.,

$$J = \sum_i z_i F c_i (u_{conv})_i \, X \tag{30}$$

Now since the $J/X$ is equal to specific conductivity ($\sigma$), the above equation takes the form

$$\sigma = \frac{J}{X} = \sum_i z_i F c_i (u_{conv})_i \tag{31}$$

Therefore, for univalent electrolytes, we have

$$\sigma = \frac{J}{X} = zFc[(u_{conv})_+ + (u_{conv})_-] \tag{32}$$

It is clear from the above expression that the specific conductivity is directly proportional to the concentration of the electrolytic solution, which can be explained in terms of the fact that the number of ions per unit volume changes with dilution.

Now, in order to connect the ionic drift velocities with molar conductivity ($\Lambda_m$) for monovalent electrolytes, recalling the expression for molar conductivity here, i.e.,

$$\Lambda_m = \frac{\sigma}{c} \tag{33}$$

After putting the value of specific conductivity from equation (32), the above equation takes the form

$$\Lambda_m = \frac{zFc[(u_{conv})_+ + (u_{conv})_-]}{c} \tag{34}$$

$$\Lambda_m = zF[(u_{conv})_+ + (u_{conv})_-] \tag{35}$$

For equivalent conductivity ($\Lambda_{eq}$) of monovalent electrolytes, recall its correlation with molar conductivity i.e.

$$\Lambda_{eq} = \frac{\Lambda_m}{z} \tag{36}$$

Substituting the value of molar conductivity from equation (35) in equation (36), we get

$$\Lambda_{eq} = \frac{zF[(u_{conv})_+ + (u_{conv})_-]}{z} \tag{37}$$

$$\Lambda_{eq} = F[(u_{conv})_+ + (u_{conv})_-] \tag{38}$$

Hence, $\Lambda_{eq}$ will be independent of concentration only if the ionic mobility doesn't vary with concentration.

### ❖ Einstein Relation Between the Absolute Mobility and Diffusion Coefficient

Since it is a well-known fact that the diffusion of ions is simply the zig-zag walking of ions from a high-numbered region to a low-numbered region. In other words, we can say the ionic diffusion is the result of a concentration gradient in which a particular type of ions travel from a high concentration region towards a low concentration region until a homogeneity in concentration is reached. On the other hand, the conduction or the ionic migration is a result of the drift velocity component imparted to the ions by the electric force. However, it is important here to recall the fact that this velocity component does not stop the zig-zag walk of diffusion but actually gets superimposed on it. Albert Einstein understood this and formulated a relation between ionic mobility ($\bar{u}_{abs}$) and diffusion coefficient ($D$).

Now, since the conduction, as well as the diffusion, are irreversible processes, they cannot be treated by equilibrium statistical mechanics or by the equilibrium thermodynamics. However, the situation can be considered as a pseudo-equilibrium if the conduction and diffusion take place in the opposite direction but with same rates. To do so, consider an electrolytic solution of salt MX in which some of the cations are radioactive in nature. Now assume that $M^+$ ions are present in higher concentrations in one region and in lower concentration in some other region. In other words, the tracer ions are present with a concentration gradient. According to Fick's law of diffusion, the overall diffusion flux ($J_D$) must be

$$j_D = -D\frac{dc}{dx} \tag{39}$$

After applying the electric field, the tracer ions will feel the field and will start to move towards the opposite electrode. The drift velocity can be given as

$$v_d = \bar{u}_{abs}\vec{F} \tag{40}$$

The current density produced by this drift velocity is

$$J = z_+cFv_d \tag{41}$$

The conduction flux can be obtained by dividing the current density by charge carried by one mole of ions i.e.

$$j_c = \frac{z_+cFv_d}{z_+F} \tag{42}$$

or

$$j_c = cv_d \tag{43}$$

After using the expression of drift velocity from equation (40) in the above expression, we get

$$j_c = c\,\bar{u}_{abs}\vec{F} \tag{44}$$

The strength of the applied electric field is varied in such a way that the conduction flux and diffusion flux are equal and opposite. Mathematically, it should be like

$$j_c = -j_D \tag{45}$$

$$j_c + j_D = 0 \tag{46}$$

After using values of $j_D$ and $j_c$ from equations (39, 44) in the above expression, we get

$$c\,\bar{u}_{abs}\vec{F} - D\frac{dc}{dx} = 0 \tag{47}$$

or

$$\frac{dc}{dx} = \frac{c\bar{u}_{abs}\vec{F}}{D} \tag{48}$$

Since there is no net flow of ions, and therefore, this pseudo-equilibrium can be studied by Boltzmann law.

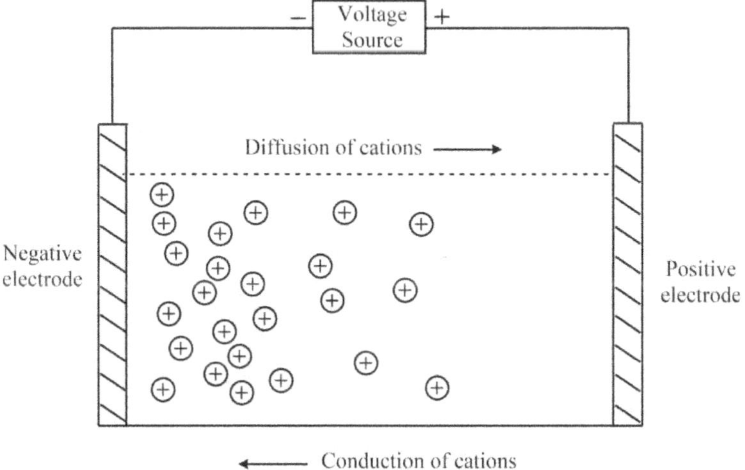


Figure 4. The pseudo-equilibrium when diffusion flux and conduction flux are equal and opposite.

Owing to the $x$ dependent variation, recall the ionic concentration at distance $x$, i.e.,

$$c = c_0\, e^{-U/kT} \tag{49}$$

Where $c_0$ is the ionic concentration in the zero potential region while $U$ is the potential energy of the ion under consideration in the externally applied electric field. Differentiating the above equation w.r.t. $x$, we have

$$\frac{dc}{dx} = -c_0\, e^{-U/kT}\frac{1}{kT}\frac{dU}{dx} \tag{50}$$

Replacing $c_0\, e^{-U/kT}$ by $c$ i.e. using equation (49), we get

$$\frac{dc}{dx} = -\frac{c}{kT}\frac{dU}{dx} \tag{51}$$

Since the force is $F = -dU/dx$, the above equation takes the form

$$\frac{dc}{dx} = \frac{c}{kT}F \tag{52}$$

From equation (48) and equation (52), we get

$$\frac{c\bar{u}_{abs}\vec{F}}{D} = \frac{c}{kT}\vec{F} \tag{53}$$

$$\frac{\bar{u}_{abs}}{D} = \frac{1}{kT} \tag{54}$$

or

$$D = \bar{u}_{abs}kT \tag{54}$$

Which is the famous Einstein relation between the absolute mobility and diffusion coefficient.

Furthermore, from the phenomenological treatment of the diffusion coefficient, it is also a quite well-known correlation that

$$D = BRT \tag{55}$$

Where $B$ represents the undetermined phenomenological coefficient and $R$ is the gas constant. Now, comparing equation (54) and equation (55), we have

$$\bar{u}_{abs}kT = BRT \tag{56}$$

or

$$B = \frac{\bar{u}_{abs}kT}{RT} = \frac{\bar{u}_{abs}k}{R} \tag{57}$$

Since $N = R/k$, the above equation can also be written as

$$B = \frac{\bar{u}_{abs}}{N} \tag{58}$$

It is obvious from the above equation that the phenomenological coefficient $B$ can simply be defined as the ratio of absolute mobility to the Avogadro number. Furthermore, The Einstein relation also connects the phenomena of diffusion with force arising from viscous drag and force of electric field on the ion during its drifting movement. Therefore, the formulation also forms the basis of Stokes-Einstein (viscosity and diffusion) and Nernst–Einstein relation (equivalent conductivity and diffusion).

## ❖ The Stokes-Einstein Relation

Albert Einstein realized that an ion moving in an electrolytic solution is somewhat analogous to a macroscopic sphere moving in a liquid medium. A macroscopic sphere can travel very fast if it is outside of water or any other liquid; however, its velocity will definitely be affected if it is put in some liquid. The reduced velocity of the sphere in liquid can be attributed to an opposing force exerted upon it by the diameter of the sphere ($d$), viscosity of the medium ($\eta$), density of the medium ($\rho$) and the speed of the ion itself ($v$). All these factors are correlated mathematically to give "Reynolds number ($R_e$)" as

$$R_e = vd\frac{\rho}{\eta} \tag{59}$$

If the hydrodynamics has a profile that makes the $R_e \ll 1$, Stokes proved that the dragging force ($F$) on the sphere can be formulated by the following relation.

$$F = 6\pi r\eta v \tag{60}$$

Which is the famous Stokes' law.

Now although the equation (60) is very useful in case of a macroscopic sphere moving in water or in any other liquid, the applicability of the same to microscopic ions requires some testing first. One condition that must be satisfied for the practicality of Stokes law to ion in solution is the very small value of Reynolds number i.e. $R_e \ll 1$. After using the ionic diameter and typical drift velocity of ion, it is found that Reynolds number is very small in comparison to unity, and therefore, suggests that Stokes law can be applied to ionic movement. However, equation (60) would show deviations from experimental results if the ions tagged are not completely spherical, and therefore, should be modified for such particles. It has been shown that for cylindrical particles, the factor $6\pi$ must be replaced by $4\pi$ to get reasonable results. It is also worthy to note that the Stokes law fails to explain the viscous drag on extremely small ions, justifying the need for some other advanced models. Furthermore, besides the viscous drag, the presence of other ions also creates collisions and stop-start zig-zag movement which makes it very difficult to apply Stokes' law.

Albert Einstein developed a modified approach to correlate the viscosity with the diffusion coefficient by suggesting that a driving force ($-d\mu/dx$) operates on the particles during diffusion which can be formulated as given below.

$$-\frac{d\mu}{dx} = 6\pi r\eta v_d \tag{61}$$

Where $v_d$ is the steady-state velocity of the ion under consideration. The right-hand side of the above equation means that the driving force we assumed must be opposed by an equal and opposite resistive force given by Stokes' law.

Besides, when a charged particle moves in a polar solvent, solvent dipoles surround it from oppositely charged ends; and this surrounding environment is destroyed and built up again and again due to movement,

and takes time for it. In this relaxation process, a relaxation force is in operation which can be considered as an additional frictional force on the tagged ion. Therefore, the dragging force expression is modified to

$$F = 6\pi\eta v r - 6\pi\eta v \frac{s}{\varepsilon} \tag{62}$$

Where $s = (4/9)(\tau/6\pi\eta)e_0^2/r^3$ whereas $\varepsilon$ represents the dielectric constant of the solvent. The correction factor sometimes can be very large but will be neglected for the simplicity of the derivation Stokes-Einstein law. Now, from the definition of absolute mobility, i.e.,

$$\bar{u}_{abs} = \frac{v_d}{\vec{F}} \tag{63}$$

The drift velocity can be divided either by the diffusional driving force or by the equal and opposite viscous force given by Stokes law. Therefore, using equation (61) in equation (63), we get

$$\bar{u}_{abs} = \frac{v_d}{-d\mu/dx} = \frac{v_d}{6\pi r\eta v_d} = \frac{1}{6\pi r\eta} \tag{64}$$

At this stage, recall the famous Einstein's relation between the absolute mobility and diffusion coefficient, i.e.,

$$D = \bar{u}_{abs}kT \tag{65}$$

Substituting the value of $\bar{u}_{abs}$ from equation (64) in the above expression, we have

$$D = \frac{kT}{6\pi r\eta} \tag{66}$$

Which is the famous Stokes-Einstein's relation between the viscosity and diffusion coefficient.

The Stokes-Einstein relation inspired the pioneering work of Perrin who studied the random walk of a colloidal particle using an ultramicroscope and found that the mean square distance ($< x^2 >$) covered in '$t$' time is correlated with the diffusion coefficient as given below.

$$D = \frac{< x^2 >}{2t} \tag{67}$$

From the knowledge of the weight of colloidal particles and corresponding density, the magnitude of radius $r$ can be obtained, which in turn can be employed (along with medium's viscosity) to find the value of Boltzmann constant from the rearranged form of equation (66)

$$k = \frac{6\pi r\eta D}{T} \tag{68}$$

Since $k = R/N_A$, the value of the Avogadro number can be obtained from $N_A = R/k$. Furthermore, the Stokes-Einstein law can also be used to find the value of conventional ionic mobility ($\bar{u}_{conv}$). To do so, recall the expression for conventional mobility i.e.

$$\bar{u}_{conv} = \bar{u}_{abs} z_i e_0 \tag{69}$$

Now using the value of $\bar{u}_{abs}$ from equation (64), we have

$$\bar{u}_{conv} = \frac{z_i e_0}{6\pi r \eta} \tag{70}$$

The mobility given is typically labeled as Stokes mobility. The physical significance of the above equation lies in the fact that it shows the correlation of conventional mobility with the charge on the ion, radius of the ion and viscosity of the solvent used. It should also be noted that the equation does not explain the concentration dependence of ion-ion interaction, and therefore, is an oversimplified approach.

## ❖ The Nernst-Einstein Equation

Just like the Stokes-Einstein's relation found the connection between the viscosity and the diffusion coefficient; another important equation, popularly called as Nernst-Einstein relation, correlated the diffusion coefficient with equivalent conductivity. In order to derive the Nernst-Einstein's equation, recall the expression for equivalent conductivity ($\Lambda_{eq}$) in terms of conventional mobilities for $z: z$ valent electrolyte, i.e.,

$$\Lambda_{eq} = F[(u_{conv})_+ + (u_{conv})_-] \tag{71}$$

Where $F$ is the Faraday constant; whereas $(u_{conv})_+$ and $(u_{conv})_-$ are the conventional mobilities of cation and anion, respectively. Now, since for $z: z$ electrolyte $z_+ = z_- = z$, the conventional mobilities are

$$(u_{conv})_+ = z_+ e_0 (\bar{u}_{abs})_+ = z e_0 (\bar{u}_{abs})_+ \tag{72}$$

$$(u_{conv})_- = z_- e_0 (\bar{u}_{abs})_+ = z e_0 (\bar{u}_{abs})_- \tag{73}$$

Using equation (72, 73) in equation (71), we get

$$\Lambda_{eq} = F[z e_0 (\bar{u}_{abs})_+ + z e_0 (\bar{u}_{abs})_-] \tag{74}$$

$$\Lambda_{eq} = z e_0 F[(\bar{u}_{abs})_+ + (\bar{u}_{abs})_-] \tag{75}$$

From Einstein's relation, we know that

$$(\bar{u}_{abs})_+ = \frac{D_+}{kT} \tag{76}$$

also

$$(\bar{u}_{abs})_- = \frac{D_-}{kT} \tag{77}$$

Using equation (76, 77) in equation (75), we get

$$\Lambda_{eq} = ze_0F\left[\frac{D_+}{kT} + \frac{D_-}{kT}\right] \tag{78}$$

$$\Lambda_{eq} = \frac{ze_0F}{kT}(D_+ + D_-) \tag{79}$$

Which is the popular Nernst-Einstein relation that allows us to find the value of equivalent conductivity just by knowing the diffusion coefficient of cation and anion only.

Another popular form of the Nernst-Einstein equation can be obtained by multiplying and dividing the right-hand side of equation (79) by Avogadro number as given below.

$$\Lambda_{eq} = \frac{ze_0FN_A}{kTN_A}(D_+ + D_-) \tag{80}$$

Since $e_0N_A = F$ and $kN_A = R$, the above equation takes the form

$$\Lambda_{eq} = \frac{zF^2}{RT}(D_+ + D_-) \tag{81}$$

It is also worthy to note that although nature is same, the equation (81) is more popular in electrochemical literature than equation (79).

### ❖ Walden's Rule

*The Walden's rule states that the product of the equivalent conductivity and the viscosity of the solvent for a specific electrolyte at a given temperature is constant.*

The Stokes-Einstein relation found the connection between the viscosity ($\eta$) and the diffusion coefficient ($D$); whereas the Nernst-Einstein relation correlates the equivalent conductivity ($\Lambda$) and diffusion coefficient. Therefore, a remarkable possibility is to eliminate the diffusion coefficient to correlate the viscosity of the solvent with the equivalent conductivity. In order to do so, recall the Stokes-Einstein relation first i.e.

$$D = \frac{kT}{6\pi r\eta} \tag{82}$$

Where $k$ is the Boltzmann constant and $\eta$ is the coefficient of viscosity. The symbol $r$ represents the radius of the ion and T is the temperature of the electrolytic solution. Now, recall the Nernst-Einstein relation i.e.

$$D = \frac{\Lambda_{eq}RT}{zF^2} \tag{83}$$

Where $z$ is the charge number and $F$ is the Faraday constant. From equation (82) and equation (83), we get

$$\frac{kT}{6\pi r\eta} = \frac{\Lambda_{eq}RT}{zF^2} \tag{84}$$

$$\Lambda_{eq} = \frac{kTzF^2}{6\pi r\eta RT} \tag{85}$$

Since $e_0 N_A = F$ and $kN_A = R$, the above equation can be written in the following form

$$\Lambda_{eq} = \frac{kTzFe_0 N_A}{6\pi r\eta TkN_A} = \frac{zFe_0}{6\pi r\eta} \tag{86}$$

$$\Lambda_{eq}\eta = \frac{zFe_0}{6\pi r} \tag{87}$$

Putting $zFe_0/6\pi = constant$, the above equation can also be written as

$$\Lambda_{eq}\eta = \frac{constant}{r} \tag{88}$$

Now, if the radius of the ion the solvated ion is same in solvents of different viscosities, the equation (88) is reduced to

$$\Lambda_{eq}\eta = constant = \frac{zFe_0}{6\pi r} \tag{89}$$

Which is the empirical Walden's rule. The experimental data for potassium iodide in various solvents is given for more clear picture is given below.

Table 1. The experimental data for KI in different solvents.

| Solvent used | Equivalent conductivity $(\Lambda_{eq})$ | Viscocity of the solvent $(\eta)$ | $\Lambda_{eq}\eta$ |
|---|---|---|---|
| Acetophenone | 39.8 | 0.01620 | 0.64476 |
| Ethanol | 50.9 | 0.01096 | 0.55786 |
| Pyridino | 71.4 | 0.00938 | 0.68401 |
| Methanol | 114.5 | 0.00545 | 0.62403 |
| Propanone | 185.4 | 0.00316 | 0.58586 |
| Acetonitrile | 198.3 | 0.00345 | 0.68414 |

It is obvious for the data listed in 'Table 1' that the product of equivalent conductivity and viscosity of the solvent is almost constant with slight deviation, which can be attributed to different solvated radii in different solvents.

## ❖ The Rate-Process Approach to Ionic Migration

In order to understand the rate-process approach to ionic migration, recall the fundamental relation between the ionic drift velocity ($v_d$) and current density ($J$) i.e.

$$J = zcFv_d \tag{90}$$

Where $z$ is the charge number and $c$ is the concentration of the ions. The symbol $F$ represents the Faraday constant. The drift velocity ($v_d$) is related to the macroscopic force

$$v_d = \frac{\tau}{m}\vec{F} \tag{91}$$

Where $m$ is the mass of the ion. The symbol $\tau$ represents the mean lifetime between to collisions during the ionic drift. The symbol $\vec{F}$ is the viscose or electric force that can be formulated as

$$\vec{F} = 6\pi r\eta v \qquad or \qquad \vec{F} = ze_0 X \tag{92}$$

Where $\eta$ is the coefficient of viscosity and $r$ is the radius of the cation. In the right-hand relation, $z$ is the charge number of the ion under consideration, $e_0$ is the electronic charge and $X$ is the applied electric field.

Furthermore, the drift velocity can also be assumed as the resultant velocity of the velocity of ions in the direction of the force field ($\vec{v}$) and the velocity of ions in the opposite direction ($\overleftarrow{v}$). Mathematically, we can say as given below.

$$v_d = \vec{v} - \overleftarrow{v} \tag{93}$$

Now, since the velocity is also the ratio of average jump distance ($l$) to the average time between two successive jumps ($\tau$), we can also write as

$$\vec{v} = \frac{l}{\tau} \tag{94}$$

Furthermore, the jump frequency i.e. number of jumps per unit of time ($k = 1/\tau$) is simply the reciprocal of the mean time between successive jumps. Therefore, the velocities can also be written as

$$\vec{v} = l\,\vec{k} \tag{95}$$

or

$$\overleftarrow{v} = l\,\overleftarrow{k} \tag{96}$$

Now in case of diffusion, the jumps of ions can be considered as a rate phenomenon for which the participating ion must possess a minimum amount of free energy to activation to do so. It was found that

$$\vec{k} = \frac{kT}{h}e^{-\Delta G^*/RT} \tag{97}$$

For the diffusion only (from right to left), the above equation can be relabelled with subscript $D$ i.e.

$$\overleftarrow{k}_D = \frac{kT}{h} e^{-\Delta G_D^*/RT} \tag{98}$$

However, when the electric field is applied, the diffusion of ions from right to left will be opposed. Therefore, the work-done ($w$) in moving the ion from the equilibrium state to the barrier-maximum will be equal to the product of the ionic charge ($z_+e_0$), and the potential difference between the activated state and equilibrium position. Now assume that this potential difference is simply a fraction $\beta$ of the overall active potential ($Xl$).

$$w = z_+e_0\beta Xl \tag{99}$$

Where $X$ is the electric field and $l$ is the distance between the two equilibrium states. For one mole of ions, the equation (99) takes the form

$$W = N_A z_+ e_0 \beta Xl \tag{100}$$

or

$$W = z_+ F \beta Xl \tag{101}$$

The overall depiction of the rate-process approach to the ionic migration phenomenon under the influence of an external electric field is shown below.

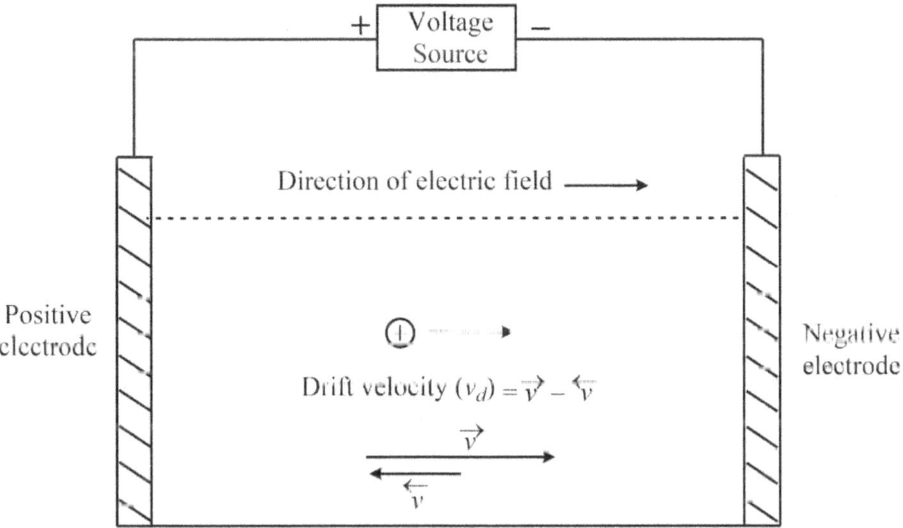


Figure 5. The general depiction of the rate-process approach to ionic migration under the influence of an external electric field.

The process of activation can be described by the diagram given below.

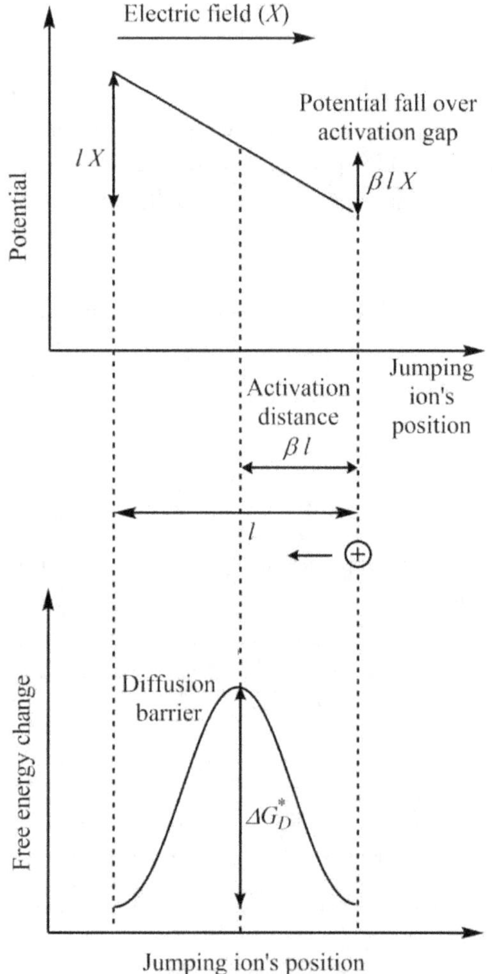


Figure 6. The general depiction ionic movement from the right to left facing the activation barrier under the influence of an external electric field.

The work done obtained in the abovementioned route will induce a free-energy change ($\Delta G_e^*$) that will contribute to the free energy of activation, i.e.,

$$\Delta G_{total}^* = \Delta G_D^* + \Delta G_e^*$$
(102)

or

$$\Delta G^*_{total} = \Delta G^*_D + z_+F\beta Xl \tag{103}$$

After using $\Delta G^*_{total}$ from equation (103), the equation (97) for "right to left" jumps in the presence of applied electric field take the form

$$\overleftarrow{k} = \frac{kT}{h} e^{-(\Delta G^*_D + z_+F\beta Xl)/RT} \tag{104}$$

or

$$\overleftarrow{k} = \frac{kT}{h} e^{-\Delta G^*_D/RT}\, e^{-z_+F\beta Xl/RT} \tag{105}$$

Recalling expression for $\overleftarrow{k}_D$ from equation (98), the equation (105) can also be written as

$$\overleftarrow{k} = k_D\, e^{-z_+F\beta Xl/RT} \tag{106}$$

Similarly, the jump frequency for the "left to right" movement may be found. However, since the cations are moving along the field (left to right) in case, their movement is favored. The fraction of the barrier these ions need to climb will be $(1 - \beta)$. Finally, it is very important to note that the electrical work of activation should be negative in "left to right" because field supports the ion. Therefore, the jump frequency for left to right movement should look like

$$\overrightarrow{k} = k_D\, e^{z_+F(1-\beta)Xl/RT} \tag{107}$$

Now if $\beta = 1/2$, then $1 - \beta$ will also be equal to $1/2$. This transforms equation (106, 107) as

$$\overleftarrow{k} = k_D e^{-pX} \tag{108}$$

and

$$\overrightarrow{k} = k_D\, e^{pX} \tag{109}$$

Where $p = z_+Fl/2RT$. From equation (108), it is very obvious that $\overleftarrow{k} < k_D$ whereas equation (109) implies that $\overrightarrow{k} > k_D$. Hence $\overrightarrow{k} > \overleftarrow{k}$.

Hence, we can conclude that the jumping frequency becomes anisotropic in the presence of externally applied filed. The jumping frequency in the applied field's direction is higher than what is against. Nevertheless, in the absence of field, the value of jump frequency is equal in all directions which is a characteristic feature of the random walk. When the field is applied, this isotropy is abolished and the walk is not completely random anymore. The ions start moving along the field creating a net current density. In other words, we can say that ionic drift due to the field is the multiplication of perturbation induced by the field and the random walk in the field's absence.

### ❖ The Rate-Process Equation for Equivalent Conductivity

The rate-process equation for equivalent conductivity can be derived by recalling the fundamental relation between the ionic drift velocity ($v_d$) and current density ($J$) for cation first i.e.

$$J = z_+ cF v_d \tag{110}$$

Where $z$ is the charge number and $c$ is the concentration of the ions. The symbol $F$ represents the Faraday constant. Also, the drift velocity can be assumed as the resultant velocity of the velocity of ions in the direction of the force field ($\vec{v}$) and the velocity of ions in the opposite direction ($\overleftarrow{v}$). Mathematically, we can say as given below.

$$v_d = \vec{v} - \overleftarrow{v} \tag{111}$$

Now, since the velocity is also the ratio of average jump distance ($l$) to the average time between two successive jumps ($\tau$), we can also write as

$$\vec{v} = \frac{l}{\tau} \tag{112}$$

Furthermore, the jump frequency i.e. number of jumps per unit of time ($k = 1/\tau$) is simply the reciprocal of the mean time between successive jumps. Therefore, the velocities can also be written as

$$\vec{v} = l\,\vec{k} \tag{113}$$

$$\overleftarrow{v} = l\,\overleftarrow{k} \tag{114}$$

From the rate-process approach to ionic migration, the values of $\vec{k}$ and $\overleftarrow{k}$ were found to be

$$\overleftarrow{k} = k_D\, e^{-pX} \tag{115}$$

and

$$\vec{k} = k_D e^{pX} \tag{116}$$

Where $k_D$ is the jumping frequency for diffusion and $X$ is simply the electric field. The expression for symbol $p = z_+ Fl/2RT$. Using values of $\vec{k}$ and $\overleftarrow{k}$ in equation (113, 114), we get

$$\vec{v} = l\, k_D e^{pX} \tag{117}$$

$$\overleftarrow{v} = l\, k_D\, e^{-pX} \tag{118}$$

Now substituting the value of $\vec{v}$ and $\overleftarrow{v}$ in equation (111), we have

$$v_d = l\, k_D\, e^{pX} - l\, k_D\, e^{-pX} \tag{119}$$

$$v_d = l\,k_D\,(e^{pX} - e^{-pX}) \tag{120}$$

Since $e^{pX} - e^{-pX} = 2\,Sinh\,pX$, the equation (120) can also be written as

$$v_d = 2l\,k_D\,2\,Sinh\,pX \tag{121}$$

Now using the value of drift velocity from equation (121) in equation (110), we get the current density as

$$J = z_+cF\,(2l\,k_D\,2\,Sinh\,pX) \tag{122}$$

Therefore, it is obvious from the above expression that the current density varies hyperbolically with the applied electric field.

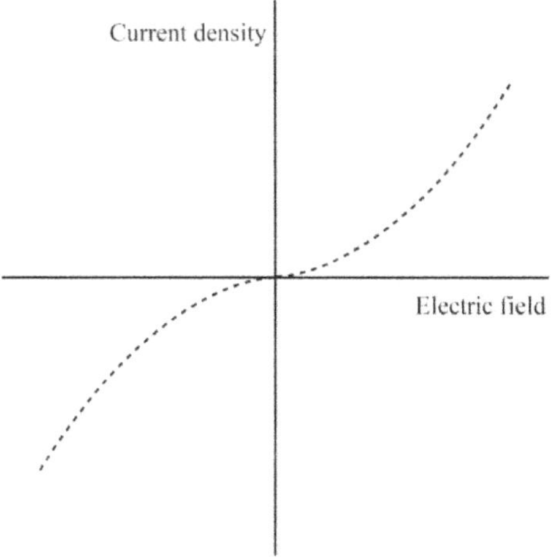


Figure 7. The variation of current density with the applied electric field.

Now since the $J/X$ is equal to specific conductivity ($\sigma$), the above equation takes the form

$$\sigma = \frac{J}{X} = z_+cF\,\frac{(2l\,k_D\,2\,Sinh\,pX)}{X} \tag{123}$$

Since molar conductivity is $\Lambda_m = \sigma/c$ and equivalent conductivity is $\Lambda_{eq} = \Lambda_m/z_+$; the equation (123) gives

$$\Lambda_{eq} = \frac{\Lambda_m}{z_+} = \frac{\sigma}{z_+c} = z_+cF\,\frac{(2l\,k_D\,2\,Sinh\,pX)}{z_+cX} \tag{124}$$

$$\Lambda_{eq} = F\,\frac{(2l\,k_D\,2\,Sinh\,pX)}{X} \tag{125}$$

Which is the equation for equivalent conductivity.

### ❖ Total Driving Force for Ionic Transport: Nernst-Planck Flux Equation

The rate perspective of the conduction process has already been discussed in the previous sections of this chapter. During the whole of the discussion, we assumed that the composition of the electrolyte was uniform throughout. However, the case will become somewhat different if we assume a concentration gradient w.r.t. tracer ions (cations in this case). Let the concentration of tracer cations is $(c_+)_x$ at a distance $x$ on the left of the barrier-maximum whereas the concentration $(c_+)_{x+l}$ on the right of the barrier maximum. Now, if we assume that $(c_+)_{x+l} > (c_+)_x$, we can say that

$$(c_+)_{x+l} = (c_+)_x + \frac{d(c_+)_x}{dx} \times l \tag{126}$$

Owing to the decreasing concentration gradient of tracer ions from right to left, we cannot use the simple expression for current density.

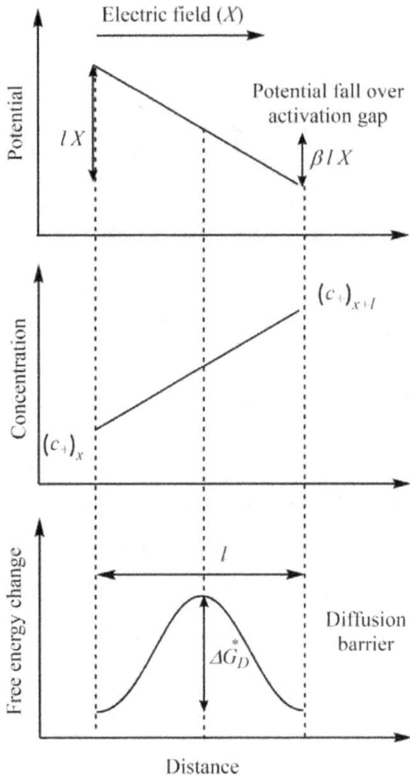


Figure 8. The general depiction of the ionic movement from right to left due to diffusion facing activation barrier under the influence of the external electric field.

Therefore, the fundamental relation between the ionic drift velocity ($v_d$) and current density ($J$) for cation must be recalled first i.e.

$$J = z_+ cF v_d \tag{127}$$

Where $z$ is the charge number and $c$ is the concentration of the ions. The symbol $F$ represents the Faraday constant. Also, the drift velocity can be assumed as the resultant velocity of the velocity of ions in the direction of the force field ($\vec{v}$) and the velocity of ions in the opposite direction ($\overleftarrow{v}$). Mathematically, we can say as given below.

$$v_d = \vec{v} - \overleftarrow{v} \tag{128}$$

Using the above result in equation (127), we get

$$J = z_+ cF(\vec{v} - \overleftarrow{v}) \tag{129}$$

$$J = z_+ cF\vec{v} - z_+ cF\overleftarrow{v}$$

Since concentration for left and on right are $(c_+)_x$ and $(c_+)_{x+l}$; the above equation can also be written as

$$J = z_+ (c_+)_x F\vec{v} - z_+ (c_+)_{x+l} F\overleftarrow{v} \tag{130}$$

Using the value of $(c_+)_{x+l}$ from equation (126) in equation (130), we get

$$J = z_+ (c_+)_x F\vec{v} - z_+ \left( (c_+)_x + \frac{d(c_+)_x}{dx} \times l \right) F\overleftarrow{v} \tag{131}$$

For simplicity, the label $(c_+)_x$ with $c_+$ i.e.

$$J = z_+ c_+ F\vec{v} - z_+ \left( c_+ + \frac{dc_+}{dx} l \right) F\overleftarrow{v} \tag{132}$$

Recalling the values of $\vec{v}$ and $\overleftarrow{v}$ i.e.

$$\vec{v} = l\, k_D e^{pX} \tag{133}$$

$$\overleftarrow{v} = l\, k_D\, e^{-pX} \tag{134}$$

Where $k_D$ is the jumping frequency for diffusion and $X$ is simply the electric field. The expression for symbol $p = z_+ Fl/2RT$. When the field strength is very low, $pX \ll 1$; and therefore, equations (133, 134) can be expended as given below.

$$\vec{v} = l\, k_D (1 + pX) \tag{135}$$

$$\overleftarrow{v} = l\, k_D\, (1 - pX) \tag{136}$$

Now, after rearranging equation (132) and then using equations (135, 136), we have

$$J = z_+c_+F\vec{v} - c_+z_+F\tilde{v} - \frac{dc_+}{dx}lz_+F\tilde{v} \tag{137}$$

$$J = z_+c_+Fl\,k_D(1 + pX) - c_+z_+Fl\,k_D\,(1 - pX) - \frac{dc_+}{dx}lz_+Fl\,k_D\,(1 - pX) \tag{138}$$

$$J = 2z_+c_+Fl\,k_DpX - z_+Fl^2\,k_D\,(1 - pX)\frac{dc_+}{dx} \tag{139}$$

Neglecting $pX$ in comparison to one for low-field approximation, we have

$$J = 2z_+c_+Fl\,k_DpX - z_+Fl^2\,k_D\,\frac{dc_+}{dx} \tag{140}$$

Since $p = z_+Fl/2RT$, the equation (140) can be transformed to

$$J = 2z_+c_+Fl\,k_D\,\frac{z_+Fl}{2RT}\,X - z_+Fl^2\,k_D\,\frac{dc_+}{dx} \tag{141}$$

$$J = z_+^2c_+F^2\,\frac{l^2k_D}{RT}\,X - z_+Fl^2\,k_D\,\frac{dc_+}{dx} \tag{142}$$

Since $l^2k_D = D_+$, the equation (142) takes the form

$$J = z_+^2c_+F^2\,\frac{D_+}{RT}\,X - z_+FD_+\,\frac{dc_+}{dx} \tag{143}$$

Now recalling the correlation between current density $(J_+)$ and flux $(j_+)$ of positive ions i.e.

$$j_+ = \frac{J_+}{z_+F} \tag{144}$$

Now because the current density given by equation (143) is only from cations, the corresponding flux can be obtained putting value of $J_+$ from equation (143) in equation (144), we get

$$j_+ = \frac{z_+^2c_+F^2D_+X}{z_+FRT} - \frac{z_+FD_+}{z_+F}\frac{dc_+}{dx} \tag{145}$$

$$j_+ = \frac{c_+D_+}{RT}z_+FX - D_+\frac{dc_+}{dx} \tag{146}$$

After multiplying and dividing the second term by $c_+RT$, we have

$$j_+ = \frac{c_+D_+}{RT}z_+FX - \frac{D_+c_+}{RT}\frac{RT}{c_+}\frac{dc_+}{dx} \tag{147}$$

$$j_+ = \frac{c_+D_+}{RT}z_+FX - \frac{D_+c_+}{RT}\frac{d(RT\,\ln c_+)}{dx} \tag{148}$$

$$j_+ = \frac{c_+D_+}{RT}z_+FX \;-\; \frac{D_+c_+}{RT}\frac{d(\mu_+^0 + RT\ln c_+)}{dx} \tag{149}$$

$$j_+ = \frac{c_+D_+}{RT}z_+FX \;-\; \frac{D_+c_+}{RT}\frac{d(\mu_+^0 + RT\ln c_+)}{dx} \tag{150}$$

Since $\mu_+^0 + RT\ln c_+ = \mu_+$, the above equation becomes

$$j_+ = \frac{c_+D_+}{RT}z_+FX \;-\; \frac{D_+c_+}{RT}\frac{d\mu_+}{dx} \tag{151}$$

It is a well-known fact in electrochemical theory that the electric field is simply equal to negative of the gradient of electrostatic potential i.e. $X = -d\psi/dx$. Therefore, the equation (151) takes the form

$$j_+ = \frac{c_+D_+}{RT}z_+F\left(-\frac{d\psi}{dx}\right) \;-\; \frac{D_+c_+}{RT}\frac{d\mu_+}{dx} \tag{152}$$

or

$$j_+ = -\frac{c_+D_+}{RT}\left(z_+F\frac{d\psi}{dx} + \frac{d\mu_+}{dx}\right) \tag{153}$$

Since the $-d\mu_+/dx$ and $-z_+F\,d\psi/dx$ are the driving forces for pure diffusion and pure conduction phenomena, respectively; the total driving force for ionic transport must be equal to the negative gradient of chemical potential and electrostatic potential. The sum of the two potentials is called as electrostatic-chemical potential $(\bar{\mu}_+)$, and is defined by

$$\bar{\mu}_+ = z_+F\psi + \mu_+ \tag{154}$$

Taking negative both sides and then differentiating w.r.t. $x$, we have

$$-\frac{d\bar{\mu}_+}{dx} = -\left(z_+F\frac{d\psi}{dx} + \frac{d\mu_+}{dx}\right) \tag{155}$$

Utilizing the above result in equation (153), we get

$$j_+ = -\frac{c_+D_+}{RT}\frac{d\bar{\mu}_+}{dx} \tag{156}$$

Since the Einstein relation is $D_+ = (\bar{u}_{abs})_+kT$, the above equation becomes

$$j_+ = -\frac{c_+(\bar{u}_{abs})_+kT}{RT}\frac{d\mu_+}{dx} \tag{157}$$

Moreover, the relationship between conventional $(\bar{u}_{conv})_+$ and absolute mobilities $(\bar{u}_{abs})_+$ is

$$(\bar{u}_{conv})_+ = (\bar{u}_{abs})_+\,z_+e_0 \tag{158}$$

Therefore, the use of equation (158) in equation (157) gives

$$j_+ = -\frac{c_+(\bar{u}_{conv})_+ k}{z_+ e_0 R}\frac{d\bar{\mu}_+}{dx} \tag{159}$$

Since $R = N_A k$, the above equation becomes

$$j_+ = -\frac{c_+(\bar{u}_{conv})_+}{z_+ e_0 N_A}\frac{d\bar{\mu}_+}{dx} \tag{160}$$

Putting $e_0 N_A = F$, we have

$$j_+ = -\frac{(\bar{u}_{conv})_+}{z_+ F}c_+\frac{d\bar{\mu}_+}{dx} \tag{161}$$

Which is the famous Nernst-Planck flux equation that relates the total driving force for the ionic transport with the overall flux.

## ❖ Ionic Drift and Diffusion Potential

In order to understand the link between ionic drift and diffusion potential, consider a solution of monovalent electrolyte with concentration $c$. Now assume that this solution is brought in contact with pure water and the boundary of contact is assumed to be $x = 0$. Owing to the concentration gradient, both cation as well anions will start moving into pure water immediately.

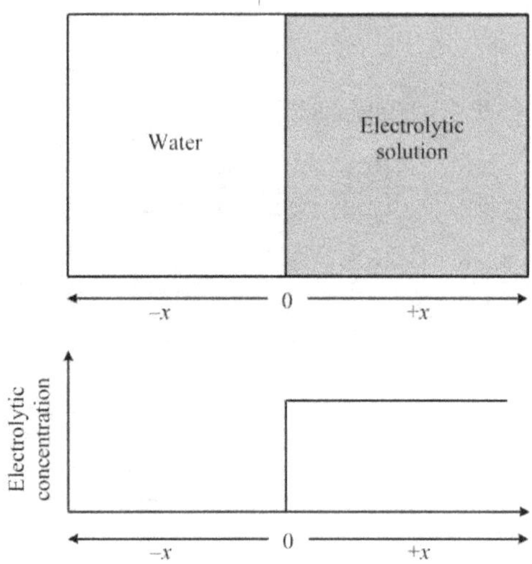


Figure 9. The general depiction electrolytic solution and concentration in contact with water in start.

Now owing to different absolute ionic mobilities of cations and anions (say $\bar{u}_+ > \bar{u}_-$), the Einstein relation can be written for cations and anions as given below.

$$D_+ = \bar{u}_+ kT \tag{162}$$

and

$$D_- = \bar{u}_- kT \tag{163}$$

Where $D_+$ and $D_-$ are the diffusion coefficients for cations and anions, respectively. The symbol $k$ is simply the Boltzmann constant and $T$ is the temperature. The symbol $\bar{u}_+$ and $\bar{u}_-$ represent the absolute ionic mobilities for cation and anion, respectively. Since we have assumed that $\bar{u}_+ > \bar{u}_-$, the following must be true

$$D_+ > D_- \tag{164}$$

This implies that the cations will move faster in comparison to anions, will lead the anions in their diffusion race. Now consider two unit-volume-elements at distance $-x_1$ and $-x_2$ in the water phase with $-x_2$ on more left than $-x_1$. Since the cations are moving faster than anions, the concentration ratio of the two ($c_+/c_-$) will be higher in volume element at $-x_2$ than in the volume element at $-x_1$ distance. In other words, the ratio $c_+/c_-$ will increase as we move from the boundary to the waterside of the system.

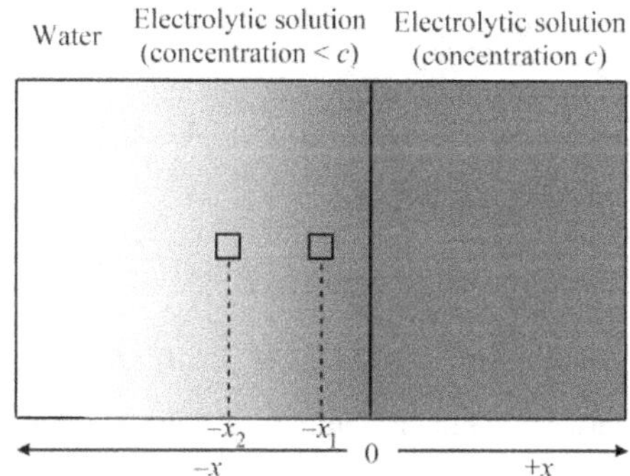


Figure 10. The development of diffusion potential due to different ionic mobilities.

Consequently, a situation will arise in which the positive and negative charges are separated with a negative layer on the left and positive layer on the right. All this will lead to the development of a potential difference that will oppose the faster movement of cations and will reinforce the slower movement of anions. This potential is generally called as the "diffusion potential" and tries to level the ionic mobilities of cations and anions; and hence, tries to maintain the electroneutrality in different parts of the solution. It should also be noted that diffusion potential is also called as the "liquid junction potential" in the case of concentration cells and "membrane potential" if the two solutions are separated by an uncharged membrane.

### ❖ The Onsager Phenomenological Equations

Since the ionic mobilities of cations are not independent but affect each other as we have studied in the previous section, the expression for the flux must also be modified. In order to understand the concept, recall the general form of Nernst-Planck equation for the ionic flux of $i$th species ($j_i$) i.e.

$$j_i = -\frac{\bar{u}_i}{z_i F} c_i \frac{d\bar{\mu}_i}{dx} \tag{165}$$

Where $\bar{u}_i$ is the conventional ionic mobility of $i$th species and $c_i$ represents the corresponding concentration. The symbol $z_i$ and $F$ are the charge number and Faraday constant, respectively. The symbol $d\bar{\mu}_i/dx$ is the total driving force for ionic transport. Correcting for the coupling between ionic mobilities arising from diffusion potential, we have

$$j_i = -\frac{\bar{u}_i}{z_i F} c_i \frac{d\bar{\mu}_i}{dx} + coupling\ correction \tag{166}$$

These mutually interactive flows can be treated by the methods of near-equilibrium thermodynamics in a phenomenological or macroscopic framework. This procedure is simple and can be covered in the postulates given below.

**Statement 1:** When the system is in near-equilibrium and ionic mobilities are independent of each other, the fluxes can be treated as proportional to the driving forces (Nernst-Planck flux equation). Mathematically, we can say about the ionic flux of $1^{st}$ species ($j_1$) that

$$j_1 = L_{11}\,\vec{F}_1 \tag{167}$$

Where $L_{11} = \bar{u}_1 c_1/z_1 F$ is phenomenological constant and $\vec{F}_1 = d\bar{\mu}_1/dx$ is the corresponding driving force.

**Statement 2:** If the coupling between ionic mobilities is considered, the fluxes should be treated as proportional to the driving forces (Nernst-Planck flux equation) plus the contribution from the coupling. Mathematically, we can say about the ionic flux of $1^{st}$ species ($j_1$) that

$$j_1 = L_{11}\,\vec{F}_1 + coupling\ correction \tag{168}$$

Now if there are many types of species then above equation takes the form

$$j_1 = L_{11}\,\vec{F}_1 + Flux\ of\ 1\ due\ to\ driving\ force\ of\ 2nd\ species \tag{169}$$
$$+ Flux\ of\ 1\ due\ to\ driving\ force\ of\ 3rd\ species + \cdots$$

**Statement 3:** The proportionality of fluxes is also valid for the contributions of the forces from other ionic species. Hence, equation (169) can also be written as

$$j_1 = L_{11}\,\vec{F_1} + \left[L_{12}\,\vec{F_2} + L_{13}\,\vec{F_3} + L_{14}\,\vec{F_4} \ldots \ldots L_{1n}\,\vec{F_n}\right] \tag{170}$$

Where $L_{12}, L_{13}\ L_{14}\ \ldots\ldots\ L_{1n}$ are the phenomenological constants for the interactions on flux from the flux of other ionic species. The symbol $\vec{F_1}, \vec{F_2}, \vec{F_3}\ \ldots\ldots \vec{F_n}$ are the corresponding driving forces.

**Statement 4:** If a monovalent electrolyte like NaCl is dissolved into water, the fluxes of cation $(j_+)$, anion $(j_-)$ and water $(j_0)$ can be written as

$$j_+ = L_{++}\,\vec{F_+} + L_{+-}\,\vec{F_-} + L_{+0}\,\vec{F_0} \tag{171}$$

$$j_- = L_{--}\,\vec{F_-} + L_{-+}\,\vec{F_+} + L_{-0}\,\vec{F_0} \tag{172}$$

$$j_0 = L_{00}\,\vec{F_0} + L_{0+}\,\vec{F_+} + L_{0-}\,\vec{F_-} \tag{173}$$

The equations (171-173) are typically known as the Onsager phenomenological equations.

**Statement 5:** According to Onsager's reciprocity relation, all symmetrical coefficients are equal. Mathematically, we can say that

$$L_{ij} = L_{ji} \tag{174}$$

Which is an experimentally proved result.

## ❖ The Basic Equation for the Diffusion

The basic equation for the diffusion potential can be obtained by using the Onsager phenomenological equations very easily. To do so, imagine an electrolytic solution with $M^{z+}$ type cations, $A^{z-}$ type anions and water as the solvent with $j_+, j_-$ and $j_0$ fluxes, respectively. Now if the solvent is assumed to be non-moving, its flux can simply be put equal to zero i.e. $j_0 = 0$. Such a solution will only have two types of ionic fluxes which can be formulated as given below.

$$j_+ = L_{++}\,\vec{F_+} + L_{+-}\,\vec{F_-} \tag{175}$$

$$j_- = L_{--}\,\vec{F_-} + L_{-+}\,\vec{F_+} \tag{176}$$

Where $L_{++}$ and $L_{--}$ are coefficients for mutually independent flow, whereas $L_{+-}$ and $L_{-+}$ are for the mutually coupled flow.

Now, under steady state approximation, the magnitude of positive charge flowing through the unit volume must be equal to the amount of the negative charge flowing through the same element but in the opposite direction. Mathematically, we can say that

$$z_+ F j_+ = -(z_- F j_-) \tag{177}$$

Where $F$ is Faraday constant. The minus sign is for the mutually opposite directions.

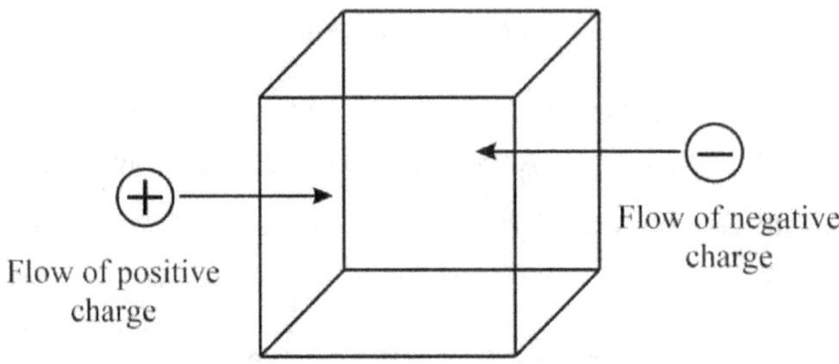


Figure 11. The flow of charge under steady-state approximation.

Rearranging equation (177) and then using values of $j_+$ and $j_-$ form equation (175-176), we get

$$z_+F\left(L_{++}\,\vec{F}_+ + L_{+-}\,\vec{F}_-\right) + z_-F\left(L_{--}\,\vec{F}_- + L_{-+}\,\vec{F}_+\right) = 0 \tag{178}$$

Putting $z_+F = q_+$ and $z_-F = q_-$ for simplicity, we have

$$q_+\left(L_{++}\,\vec{F}_+ + L_{+-}\,\vec{F}_-\right) + q_-\left(L_{--}\,\vec{F}_- + L_{-+}\,\vec{F}_+\right) = 0 \tag{179}$$

$$q_+L_{++}\,\vec{F}_+ + q_+L_{+-}\,\vec{F}_- + q_-L_{--}\,\vec{F}_- + q_-L_{-+}\,\vec{F}_+ = 0 \tag{180}$$

$$\vec{F}_+\,(q_+L_{++} + q_-L_{-+}) + \vec{F}_-(q_+L_{+-} + q_-L_{--}) = 0 \tag{181}$$

Using short symbols $p_+ = q_+L_{++} + q_-L_{-+}$ and $p_- = q_+L_{+-} + q_-L_{--}$, we have

$$p_+\vec{F}_+ + p_-\vec{F}_- = 0 \tag{182}$$

Now recalling the expressions for driving forces i.e.

$$\vec{F}_+ = q_+\frac{d\psi}{dx} + \frac{d\mu_+}{dx} \tag{183}$$

$$\vec{F}_- = q_-\frac{d\psi}{dx} + \frac{d\mu_-}{dx} \tag{184}$$

Where $d\mu_+/dx$ and $q_+d\psi/dx$ are the driving forces for pure diffusion and pure conduction phenomena for the cations; whereas, $d\mu_-/dx$ and $q_-d\psi/dx$ are the driving forces for pure diffusion and pure conduction phenomena for the anions. Using equation (183, 184) in equation (182), we get

$$p_+q_+\frac{d\psi}{dx} + p_+\frac{d\mu_+}{dx} + p_-q_-\frac{d\psi}{dx} + p_-\frac{d\mu_-}{dx} = 0 \tag{185}$$

or

$$-p_+q_+\frac{d\psi}{dx} - p_-q_-\frac{d\psi}{dx} = p_+\frac{d\mu_+}{dx} + p_-\frac{d\mu_-}{dx} \tag{186}$$

$$-\frac{d\psi}{dx}\left(p_+q_+ + p_-q_-\right) = p_+\frac{d\mu_+}{dx} + p_-\frac{d\mu_-}{dx} \tag{187}$$

$$-\frac{d\psi}{dx} = \frac{p_+}{p_+q_+ + p_-q_-}\frac{d\mu_+}{dx} + \frac{p_-}{p_+q_+ + p_-q_-}\frac{d\mu_-}{dx} \tag{188}$$

Since we know that

$$\frac{p_+}{p_+q_+ + p_-q_-} = \frac{t_+}{z_+F} \tag{189}$$

or

$$\frac{p_-}{p_+q_+ + p_-q_-} = \frac{t_-}{z_-F} \tag{190}$$

Where $t_+$ and $t_-$ are transport numbers of cation and anion, respectively. Using equation (189, 190) in equation (188), we get

$$-\frac{d\psi}{dx} = \frac{t_+}{z_+F}\frac{d\mu_+}{dx} + \frac{t_-}{z_-F}\frac{d\mu_-}{dx} \tag{191}$$

Or in general, we can conclude as

$$-\frac{d\psi}{dx} = \sum \frac{t_i}{z_iF}\frac{d\mu_i}{dx} \tag{192}$$

The minus sign of the electric field is because it is in the opposite direction to the chemical potential gradients of ionic diffusion. Furthermore, equation (192) can also be written as

$$\frac{d\psi}{dx} = \frac{1}{F}\sum \frac{t_i}{z_i}\frac{d\mu_i}{dx} \tag{193}$$

or

$$-d\psi = \frac{1}{F}\sum \frac{t_i}{z_i}d\mu_i \tag{193}$$

In terms of activity ($a_i$), the above equation can be written as

$$-d\psi = \frac{RT}{F}\sum \frac{t_i}{z_i}d\ln a_i \tag{194}$$

Which is the basic equation of diffusion potential.

## ❖ Planck-Henderson Equation for the Diffusion Potential

The basic equation for diffusion potential is applicable only if the potential difference ($d\psi$) is considered over a very small distance ($dx$). However, the problem of obtaining an overall potential difference ($\Delta\psi = \psi^0 - \psi^l$) that develops from $x = 0$ to $x = l$ was still there.

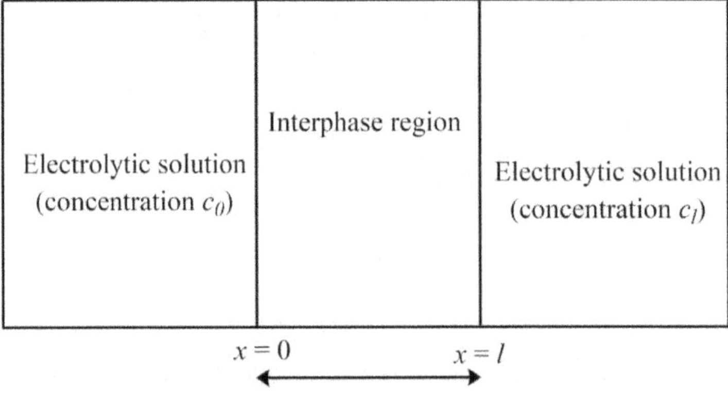


Figure 12. The overall potential difference across the complete interphase domain between electrolytes with concentration $c_0$ and $c_l$.

This was overcome by Planck-Henderson equation which can be obtained by recalling the basic equation for diffusion first i.e.

$$-d\psi = \frac{1}{F}\sum \frac{t_i}{z_i}d\mu_i \tag{195}$$

Where $t_i$ and $z_i$ are the charge number of $i$th species whereas $F$ represents the Faraday constant. Integrating equation (195), we get

$$-\Delta\psi = \psi^0 - \psi^l = \frac{1}{F}\sum_i \int_{x=0}^{x=l} \frac{t_i}{z_i}\frac{d\mu_i}{dx}dx \tag{196}$$

or

$$-\Delta\psi = \frac{RT}{F}\sum_i \int_{x=0}^{x=l} \frac{t_i}{z_i}\frac{d\ln a_i}{dx}dx \tag{197}$$

or

$$-\Delta\psi = \frac{RT}{F}\sum_i \int_{x=0}^{x=l} \frac{t_i}{z_i}\frac{1}{f_i c_i}\frac{d\,(f_i c_i)}{dx}\,dx \tag{198}$$

At this stage, the things we need to evaluate the equation (198) are the concentration of all species in the interphase region, the variation of activity coefficient and transport number with concentration. For simplicity, the activity coefficients can be taken as unity and transport numbers as constant. In addition to these assumptions, the variation of concentration of $i$th species with distance is considered as linear i.e.

$$c_i(x) = k_i x + c_i(0) \tag{199}$$

For constant $k_i$, differentiate above equation i.e.

$$\frac{dc_i}{dx} = k_i = \frac{c_i(l) - c_i(0)}{l} \tag{200}$$

Now using equation (199, 200) in equation (198), we get

$$-\Delta\psi = \frac{RT}{F}\sum_i \int_{x=0}^{x=l} \frac{t_i}{z_i}\frac{k_1}{c_i(0) + k_1 x}\,dx \tag{201}$$

or

$$-\Delta\psi = \frac{RT}{F}\sum_i \frac{t_i}{z_i} \int_{x=0}^{x=l} \frac{d[k_1 x + c_i(0)]}{k_1 x + c_i(0)} \tag{202}$$

or

$$-\Delta\psi = \frac{RT}{F}\sum_i \frac{t_i}{z_i} \left\{\ln\left[k_1 x + c_i(0)\right]\right\}_{x=0}^{x=l} \tag{203}$$

or

$$\Delta\psi = \frac{RT}{F}\sum_i \frac{t_i}{z_i}\ln\frac{c_i(l)}{c_i(0)} \tag{204}$$

Which is the general form of the Planck-Henderson equation for diffusion potential. Using $c_+ = c_+ = c$ and $z_1 = z_- = z$ for $z\!:\!z$ electrolyte, we have

$$-\Delta\psi = \frac{RT}{zF}(t_+ - t_-)\ln\frac{c_i(l)}{c_i(0)} \tag{205}$$

Furthermore, putting $t_+ + t_- = 1$, the equation (205) takes the form

$$-\Delta\psi = \frac{RT}{zF}(2t_+ - 1)\ln\frac{c_i(l)}{c_i(0)} \qquad (206)$$

Which is the another form of Planck-Henderson equation for simple systems.

## ❖ Problems

Q 1. Discuss the ionic movement under the influence of an electric field.

Q 2. What are absolute and conventional ionic mobilities? How they are related?

Q 3. Derive and discuss the relationship between ionic drift velocity and current density.

Q 4. Define the diffusion coefficient. How it is related to the absolute mobility.

Q 5. Derive Stokes-Einstein relation.

Q 6. What is the Nernst-Einstein relation? What is its significance?

Q 7. State and explain Walden's rule.

Q 8. Explain the rate process approach to ionic migration in detail.

Q 9. Derive the relationship between ionic drift and diffusion potential.

Q 10. What are Onsager phenomenological equations? How they can be used to derive the basic equation of diffusion potential?

Q 11. Write down the Planck-Henderson equation for monovalent electrolytes.

### ❖ Bibliography


[1] E. Steiner, *The Chemistry Maths Book*, Oxford University Press, Oxford, UK, 2008.

[2] M. R. Wright, *An Introduction to Aqueous Electrolyte Solutions*, John Wiley & Sons Ltd, Sussex, UK, 2007.

[3] P. Debye, E. Hückel, *The Theory of Electrolytes. I. Lowering of Freezing Point and Related Phenomena*, Physikalische Zeitschrift., 24 (1923) 185-206.

[4] V. S. Bagotsky, *Fundamentals of Electrochemistry*, John Wiley & Sons, New Jersey, USA, 2006.

[5] J. Bockris, A. Reddy, *Modern Electrochemistry – Volume 1: Ionics*, Kluwer Academic Publishers, New York, USA 2002.

[6] B. R. Puri, L. R. Sharma, M. S. Pathania, *Principles of Physical Chemistry*, Vishal Publications, Jalandhar, India, 2008.

[7] P. Atkins, J. Paula, *Physical Chemistry*, Oxford University Press, Oxford, UK, 2010.

[8] A. Fick, *On liquid diffusion*, Annalen der Physik und Chemie., 94 (1855) 59.

# INDEX

DALAL
INSTITUTE

DALAL
INSTITUTE

DALAL
INSTITUTE

DALAL
INSTITUTE

DALAL
INSTITUTE